Telecommunications Network Management into the 21st Century

Telecommunications Network Management into the 21st Century

Techniques, Standards, Technologies, and Applications

Edited by

Salah Aidarous
NEC America
and
IEEE Communications Society
Committee on Network Operations
and Management

Thomas Plevyak
Bell Atlantic
and
IEEE Communications Society,
Communications Systems
Engineering Committee

IEE

The Institution of Electrical Engineers

IEEE PRESS
The Institute of Electrical and Electronics Engineers, Inc., New York

This book is a copublication of the
IEEE Press and IEE.

Distribution for IEEE Press is North and South America.
ISBN 0-7803-1013-6.   IEEE Order No.: PC0362-4

Distribution elsewhere is IEE.
ISBN 85296-814-0

Printed in the United States of America

10 9 8 7 6 5 4

Library of Congress Cataloging-in-Publication Data

Telecommunications network management into the 21 century:
  techniques, standards, technologies, and applications / edited by
  Salah Aidarous, Thomas Plevyak
    p. cm.
  Includes bibliographical references and index.
  ISBN 0-7803-1013-6
  1. Telecommunications systems—Management. 2. Computer networks—
  Management. I. Aidarous, Salah. II. Plevyak, Thomas
  TK5102.5.T3955 1994
  621.382—dc20   93-41040
                                                        CIP

The Editors and Authors
Dedicate This Book to Their Families:
The Cornerstone of Successful Societies

Contents

Contents

Introduction

This book presents an orchestrated collection of original chapters written expressly for the book by leading authors in the critically important field of network management and its related disciplines. It is a reference document for individuals at all levels of skill in the worldwide information industry.

The genesis for *Telecommunications Network Management* into the 21st Century was a workshop held at Globecom'91 in Phoenix, Arizona, on December 2, 1991. The workshop, titled "Network Management Techniques and Standards," brought together most of the authors appearing in this book. It was a successful, highly interactive workshop that underscored the global interest in network management capabilities and the growing necessity for timely understanding and application to new technologies, products, networks, and services. Several additional topics and authors were identified among NOMS'92 participants (Memphis, Tennessee, April 7–10, 1992).

It is often said, and we heartily agree, that future information industry capabilities will not be fully realized without timely application of forward-looking network management techniques. Gone are the days when technology and product developers as well as network service providers can consider network management as an afterthought. The growing complexity of information networks demands new approaches to operations management. Fast-paced technology developments, multivendor implementations, competition, open architectures, and sophisticated business user requirements are among the driving forces. These forces have created the need for a high level of end-to-end integrated automation of routine operations, administration maintenance, and the network elements they manage; and employee and customer access to operations systems in a universal manner. To accomplish these multifaceted goals, it becomes necessary to restructure fundamental approaches to network operations and management, literally from the ground up.

In the current setting, computer communications hardware and software are provided by a growing number of vendors. Proprietary interfaces between devices manufactured by a given vendor are prevalent. This has lead to "closed systems" with specific operations and maintenance capabilities. They are critically useful to date, but not able to attain the goals discussed above. The increasing development of "open systems," based on industry-specified or standard interfaces, permits communication between different vendors' equipment and thus sets the stage to meet the above goals in an open systems interconnection (OSI) paradigm. OSI-based object-oriented architectures, interfaces, protocols, and information models become the fundamental approach needed to advance information networks into the 21st century.

This book begins with a chapter on the principles of network management, which is followed by tutorial chapters on network management problems and technologies and the telecommunications management network (TMN). It goes on to address a multiplicity of subject areas in research and practice, from domestic and international standards to modeling frameworks and architecture. The book covers a lot of ground and will give the reader or reference user an excellent understanding of the field. But like most complex subject areas, it cannot be considered a full treatise.

While topics generally relate to telecommunications network management, the concepts are applicable to data network management. For further information on data network management, the reader is referred to the book by William Stallings, entitled *Network Management*, and published by IEEE Computer Society Press.

We are confident this book will prove interesting, informative, and effective.

Salah Aidarous   Thomas Plevyak

Acknowledgments

The editors wish to thank the following reviewers for their time and energy in refining the manuscript.

Dr. Robert Aveyard
US West Advanced Technologies

Dr. Maged Beshai
Bell-Northern Research

Dr. Walter Bugga
AT&T Bell Laboratories

Dr. Roberta Cohen
AT&T Bell Laboratories

Mr. Donald Coombs
Bell-Northern Laboratories

Dr. Gordon Fitzpatrick
Bell-Northern Research

Dr. Shri Goyal
GTE Laboratories

Dr. Alan Johnston
AT&T Bell Laboratories

Prof. Kim Kappel
Georgia Institute of Technology

Dr. Fred Kaudel
Bell-Northern Research

Dr. Kenneth Lutz
Bellcore

Dr. Robert Mallon
AT&T Bell Laboratories

Mr. Don Proudfoot
Bell-Northern Research

Mr. Robert Weilmayer
GTE Laboratories

Dr. Stephen Weinstein
Bellcore

Dr. Doug Zuckerman
AT&T Bell Laboratories

List of Acronyms

AC	Access Concentrator
ACA	Automated Cable Analysis
ACSE	Association Control Service Element
ADM	Add-drop Multiplexers
AIN	Advanced Intelligent Network
ANM	Autonomous Network Manager
AP	Application Program
APS	Automatic Protection Switch
ASC	Available Spare Capacity
ASN.1	Abstract Syntax Notation 1
ASR	Automatic Set Relocation
ATM	Asynchronous Transfer Mode
ATMS	Assumption-based Truth Maintenance System
B-DCN	Backbone Data Communications Network
B-ISDN	Broadband ISDN
BF	Basic Functions
BML	Business Management Layer
BO EOC	Bit-Oriented Embedded Operations Channel
CCSN	Common Channel Signaling Network
CDF	Cumulative Distribution Function
CDT	Connection-Dropping Threshold

CLASS	CUSTOM LOCAL AREA SIGNALING SYSTEM
CMIP	COMMON MANAGEMENT INFORMATION PROTOCOL
CMISE	COMMON MANAGEMENT INFORMATION SERVICE ELEMENT
CNC	CONCENTRATOR
CO	CENTRAL OFFICE
COT	CENTRAL OFFICE TERMINAL
CPE	CUSTOMER PREMISE EQUIPMENT
CRC	CYCLICAL REDUNDANCY CHECKING
CRUD	CREATE READ UPDATE DELETE
CSMA	CARRIER SENSE MULTIPLE ACCESS
CSMA/CD	CARRIER SENSE MULTIPLE ACCESS WITH COLLISION DETECTION
CV	CODE VIOLATIONS
DCC	DATA COMMUNICATIONS CHANNEL
DCF	DATA COMMUNICATION FUNCTION
DCN	DATA COMMUNICATIONS NETWORK
DCS	DIGITAL CROSS-CONNECT SYSTEM
DDD	DIRECT DISTANCE DIALING
DDL	DATA DEFINITION LANGUAGE
DES	DISCRETE EVENT SIMULATION
DFS	DEPTH-FIRST SEARCH
DLC	DIGITAL LOOP CARRIER
DML	DATA MANIPULATION LANGUAGE
DN	DISTINGUISHING NAME
DPP	DISTRIBUTED PREPLAN
DQDB	DISTRIBUTED QUEUE DUAL BUS
DR	DISTRIBUTED RESTORATION
DRA	DISTRIBUTED RESTORATION ALGORITHMS
DSI, DS3	DIGITAL SIGNAL 1, DIGITAL SIGNAL 3
DTF	DIGITAL TRANSMISSION FACILITY
EADAS	ENGINEERING AND ADMINISTRATION DATA ACQUISITION SYSTEM
ECC	EMBEDDED COMMUNICATIONS CHANNEL
EDP	ELECTRONIC DATA PROCESSING
EF	ENHANCED FUNCTIONS
EL	ELEMENT LAYER
EM	ELEMENT MANAGER
EML	ELEMENT MANAGEMENT LAYER (SEE NML AND SML)
EOC	EMBEDDED OPERATIONS CHANNEL
EQN	EXTENDED QUEUING NETWORK
ES	ERRORED SECOND
ETSI	EUROPEAN TELECOMMUNICATIONS STANDARDS INSTITUTE
FDDI	FIBER DISTRIBUTED DATA INTERFACE
FFM	FIRST FAILURE TO MATCH
FIFO	FIRST-IN, FIRST-OUT

FITL	FIBER IN THE LOOP
FMO	FUTURE MODE OF OPERATION
FSM	FINITE STATE MACHINE
FTAM	FILE TRANSFER ACCESS AND MANAGEMENT
GDMO	GUIDELINES FOR THE DEFINITION OF MANAGED OBJECTS
GNE	GATEWAY NETWORK ELEMENT
GOS	GRADE OF SERVICE
ICE	INTEGRATED CAPACITY EXPANSION
IDLC	INTEGRATED DIGITAL LOOP CARRIER
IEC	INTEREXCHANGE CARRIER
IETF	INTERNET ENGINEERING TASK FORCE
IN	INTELLIGENT NETWORK
INA	INFORMATION NETWORKING ARCHITECTURE
INE	INTELLIGENT NETWORK ELEMENT
INM	INTEGRATED NETWORK MONITORING
INMS	INTEGRATING NETWORK MANAGEMENT SYSTEM
INS	INTEGRATED NETWORK SERVICING
IP	INTERNET PROTOCOL
IPF	INFORMATION PROCESSING FUNCTION
IPX	INTERNET PACKET EXCHANGE
IRV	INITIAL REPEAT VALUE
ISDN	INTEGRATED SERVICES DIGITAL NETWORK
ISDN NT1S	ISDN NETWORK TERMINATION 1S
ISO	INTERNATIONAL STANDARD ORGANIZATION
ITF	INFORMATION TRANSPORT FUNCTION
ITP	INTEGRATED TECHNOLOGY PLANNING
ITU-T	INTERNATIONAL TELECOMMUNICATIONS UNION-TELECOMMUNICATIONS
KB	KNOWLEDGE-BASED
KSP	K-SUCCESSIVELY-SHORTEST LINK-DISJOINT PATH
LAPB	LINK ACCESS PROCEDURES BALANCED
LAPD	LINK ACCESS PROTOCOL ON THE D-CHANNEL
LATA	LOCAL ACCESS AND TRANSPORT AREA
LDS	LOCAL DIGITAL SWITCH
LEN	LINE EQUIPMENT NUMBER
LHS	LEFT-HAND SIDE
LLA	LOGICAL LAYERED ARCHITECTURE
LLC	LOGICAL LINK CONTROL
LME	LAYER MANAGEMENT ENTITY
LP	LINEAR PROGRAMMING
LSSGR	LATA SWITCHING SYSTEMS GENERIC REQUIREMENTS

M-A	MANAGER-AGENT
MAC	MEDIA ACCESS CONTROL
MAN	METROPOLITAN AREA NETWORK
MAP	MANUFACTURING AUTOMATION PROTOCOL
MBD	MANAGEMENT BY DELEGATION
MCMF	MULTICOMMODITY MAXIMUM FLOW
MD	MEDIATION DEVICE
MDF	MAIN DISTRIBUTION FRAME
MF	MEDIATION FUNCTION
MIB	MANAGEMENT INFORMATION BASE
MIT	MANAGEMENT INFORMATION TREE
MO	MANAGED OBJECT
MO-EOC	MESSAGE-ORIENTED EMBEDDED OPERATIONS CHANNEL
MPPF	MULTIPOINT PROTOCOL POLLING FUNCTION
NACK	NEGATIVE ACKNOWLEDGMENT
NCC	NETWORK CONTROL CENTER
NE	NETWORK ELEMENT
NEF	NETWORK ELEMENT FUNCTION
NEL	NETWORK ELEMENT LAYER
NID	NODE ID
NM	NETWORK MANAGEMENT
NML	NETWORK MANAGEMENT LAYER (SEE EML AND SML)
NMS	NETWORK MANAGEMENT SYSTEM
NOC	NETWORK OPERATION CENTER
NPA	NETWORK PLANNING AREA
NPV	NET PRESENT VALUE
NRR	NETWORK RESTORATION RATIO
OAM	OPERATIONS, ADMINISTRATION, AND MAINTENANCE
OAM&P	OPERATIONS, ADMINISTRATION, MAINTENANCE & PROVISIONING
OC	OPERATIONS CHANNEL
OIM	OPERATIONS INTERFACE MODULE
OIM CF	OIM COMMON FUNCTIONS
OIM DF	OIM DEDICATED FUNCTIONS
OO	OBJECT-ORIENTED
OS	OPERATIONS SYSTEM
OSF	OPERATIONS SYSTEMS FUNCTION
OSI	OPEN SYSTEM INTERCONNECTION
OSI FaM	OSI FAULT MANAGER
OTCs	OPERATING TELEPHONE COMPANIES
PACK	POSITIVE ACKNOWLEDGMENT
PCS	PERSONAL COMMUNICATIONS SERVICE
PDU	PROTOCOL DATA UNIT
PLE	PATH LENGTH EFFICIENCY

PM	PERFORMANCE MONITORING
PMO	PRESENT MODE OF OPERATION
PNE	PATH NUMBER EFFICIENCY
POTS	PLAIN OLD TELEPHONE SERVICE
PPSN	PUBLIC PACKET-SWITCHED NETWORK
PS	PROTOCOL STACK
PSN	PACKET-SWITCHED NETWORK
PSTN	PUBLIC SWITCHED TELEPHONE NETWORK
PWAC	PRESENT WORTH OF ANNUAL CHARGES
PWCE	PRESENT WORTH OF CAPITAL EXPENDITURES

QA	Q ADAPTOR
QAF	Q ADAPTOR FUNCTION
QOS	QUALITY OF SERVICE

RA	RETURN ALARM
RDN	RELATIVE DISTINGUISHING NAME
RDT	REMOTE DIGITAL TERMINALS
RHS	RIGHT-HAND SIDE
RKB	RING KNOWLEDGE BASE
RL	REPEAT LIMIT
ROSE	REMOTE OPERATIONS SERVICE ELEMENT
RS	RECEIVE SIGNATURE

SAP	SERVICE ACCESS POINT
SC	SHARED CHANNEL
SCE	SERVICE CREATION ENVIRONMENT
SCP	SERVICE CONTROL POINT
SEAS	SIGNALING ENGINEERING AND ADMINISTRATION SYSTEM
SEF&I	SOFTWARE, ENGINEERED, FURNISHED & INSTALLED
SES	SEVERELY ERRORED SECONDS
SG	STUDY GROUP (OF THE ITU-T)
SHN	SELF-HEALING NETWORK
SIP	SMDS INTERFACE PROTOCOL
SM	STATISTICAL MULTIPLEXING
SMDS	SWITCHED MULTIMEGABIT DATA SERVICE
SMI	STRUCTURE OF MANAGED INFORMATION
SMK	SHARED MANAGEMENT KNOWLEDGE
SML	SERVICE MANAGEMENT LAYER (SEE EML AND NML)
SMS	SERVICE MANAGEMENT SYSTEM
SNI	SUBSCRIBER NETWORK INTERFACE
SNML	SUBNETWORK MANAGEMENT LAYER (SEE NEML)
SNMP	SIMPLE NETWORK MANAGEMENT PROTOCOL
SONET	SYNCHRONOUS OPTICAL NETWORK
SP	SIGNALING POINTS
SPC	STORED PROGRAM CONTROL
SQL	STRUCTURED QUERY LANGUAGE

SSP	SERVICE SWITCHING POINT
STDM	SYNCHRONOUS TIME DIVISION MULTIPLEXING
SYNTRAN	SYNCHRONOUS DS3 TRANSMISSION
TAS	TERMINAL ACTIVATION SERVICE
TCP	TRANSMISSION CONTROL PROTOCOL
TDB	TEMPORAL DATABASE
TINA	TELECOMMUNICATIONS INFORMATION NETWORKING ARCHITECTURE
TMN	TELECOMMUNICATION MANAGEMENT NETWORK
TMS	TRUTH MAINTENANCE SYSTEM
TN	TELECOMMUNICATIONS NETWORK
TSA	TIME SLOT ASSIGNMENT
TSI	TIME SLOT INTERCHANGE
U-SHR	UNIDIRECTIONAL SELF-HEALING NETWORK
UAS	UNAVAILABLE SECOND
UDP	USER DATAGRAM PROTOCOL

CHAPTER 1

Principles of Network Management

Salah Aidarous
NEC America
Irving, TX 75038

Thomas Plevyak
Bell Atlantic
Arlington, VA 22201

1.1 INTRODUCTION

Telecommunications networks have become essential to the day-to-day activities of the enterprise and individuals. Many corporations, agencies, universities, and other institutions now rely on voice, data (e.g., facsimile transmission, electronic funds transfer), and video (e.g., video teleconferencing) services to ensure their growth and survival. This trend will accelerate as personal communications services (PCS), LAN-to-LAN interconnectivity, image file transfer, and other innovative services are developed and are standardized throughout the operation of the enterprise.

In parallel with rapid advances and increased reliance on telecommunications services, network technologies continue to evolve. For example, transport technologies, such as the synchronous optical network (SONET), will support asynchronous transfer mode (ATM) and frame relay to deliver broadband services at both constant and variable bitrates. Innovative access technologies are emerging to accommodate customer-premises equipment access to higher bandwidth services and an expanding range of mobility services (e.g., PCS), and to provide seamless access to fiber optic and satellite networks. In addition, the advanced switching technologies required for ATM technology and switched multimegabit digital service (SMDS) are now being deployed.

1

Network management is one of the most important yet confusing topics in telecommunications today. It includes operations, administration, maintenance, and provisioning (OAM&P) functions required to provide, monitor, interpret, and control the network and the services it carries. These OAM&P functions provide operating telephone companies (OTCs) and their corporate customers and end-users with efficient means to manage their resources and services to achieve objectives. There have been different approaches and strategies taken by OTCs, equipment vendors, and users to manage their networks and equipment. Management solutions are often specific to each vendor's networking product environment.

Traditionally, the public network was designed to handle voice and data services using both analog and digital technologies. Network management methods were introduced according to each technology and service. The outcome was multiple overlays of circuit-switched, packet-switched, and slow-switched connectivity nodes. Private networks, on the other hand, were built to provide enterprise information networking using PBXs, mainframes, terminals, concentrators, and bridges. The public network was used to provide the wide area backbone network. From an OAM&P perspective, interoperability between these networks has been a major challenge for the telecommunications and computing industries.

Figure 1-1 shows a typical enterprise network that includes both the private corporate data network (usually managed by the corporate telecommunications group) and the public part of the corporate network which is usually managed by the OTC and the interexchange carrier (IEC). PBXs may also be owned and managed by the OTC. The network may carry different services that require different management methods and may cross jurisdictional boundaries involving different management organizations [1].

As the pace of technological development quickens, new products are brought to market even faster and support of several generations of equipment and software is required. The current network environment is complex, diverse, competitive, and characterized by different service subnetworks, multiple overlays, and multiple media. These factors have increased the cost of network management (e.g., operations costs are exceeding capital costs) making it the primary concern of many corporations, OTCs, equipment suppliers, and standards organizations.

This chapter addresses overall principles of network management. In that sense, it is an overview, not intended as introductory material to other chapters. Instead, Chapters 2 and 3 are tutorial and introductory. This book focuses on telecommunications OAM&P and network management. Data network management is treated where relationships overlap, but this area is not a central focus.

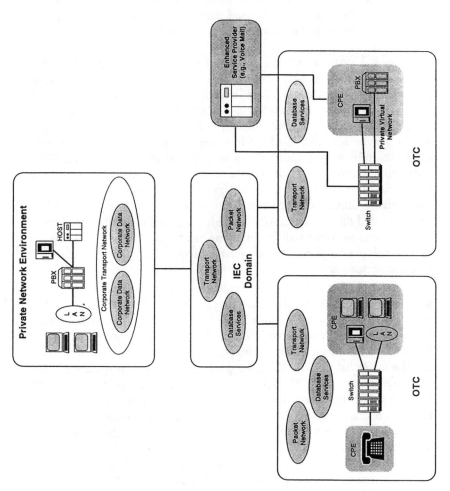

Figure 1-1 Typical Network Configuration

1.2 DRIVERS FOR NETWORK MANAGEMENT

In today's dynamic telecommunications environment, changes are occurring on many fronts. Services and network technologies are advancing rapidly, competition among service providers is intensifying, and customer demand for network access and customized services is increasing. A fundamental means of achieving these desirable changes is through an evolution of existing network management architectures [2].

Services Evolution: Network end-users are employing a spectrum of increasingly "complex" services that range from low-speed, bursty services to high-speed, continuous services and to high-bandwidth pay-per-view video. Conversely, large- and medium-business customers desire simpler, cheaper, yet even higher bandwidth services to link LANs for video conferencing and to enable effective information networking and distributed computing. They also want the consolidation of data, voice, image, and video traffic on their enterprise networks. New services are being developed at an ever-quickening pace. For example, virtual private network services are provided as an alternative to dedicated facilities. Personal communications services will provide the subscriber accessibility to his or her telephone services as well as reachability through a single telephone number.

Technology Evolution: Network technology is undergoing consolidation. For example, integrated circuit, frame, and packet switches, capable of carrying all services, is achievable with current technology. Coupling narrowband ISDN (basic and primary rate) and broadband ISDN provides a consolidated set of access technologies. On the other hand, SONET transport systems will provide a consistent digital core network. This is augmented on the private network side by integrated PBX technologies, LANs that carry all services, and use of high bandwidth services to consolidate wide area traffic. Network technology evolution is exploiting recent advances in distributed systems technology [open software foundation (OSF), distributed computing environment/distributed management environment (DCE/DME), telecommunications management network (TMN) [3-4], telecommunications information networking architecture/information networking architecture (TINA/INA)[5]], internet technology [extended internet protocol (IP), simple network management protocol (SNMP)], database driven services systems [advanced intelligent network (AIN)], and radio access systems.

Customer Requirements: Business customers are pushing for bandwidth- and service-on-demand with electronic interfaces to the network for requesting services or changes, reporting troubles, billing, and making payments. They want provisioning times in the order of minutes and services that do not fail. Residential customers and corporate network end-users want to set up basic or enhanced services such as call management or custom local area signaling system (CLASS), when and where they want them, through a simple, one-step process.

Competitiveness: The competitive landscape is changing for network/service providers. Business pressures are forcing many service providers to find ways to reduce operations costs, use resources more efficiently, streamline the implementation of new services and technologies, and identify new revenue-generating opportunities. Private network operators wish to use their networks as strategic elements in their own business areas but are being forced to reduce overhead wherever possible. These pressures will increase in the future.

Considering the rapid deployment of new services and technologies, escalating competitive pressures, and the broadening demands of customers, service providers face an immediate and pressing need to streamline, simplify, and automate network management operations.

1.3 TRADITIONAL APPROACHES TO NETWORK MANAGEMENT

Today's telecommunications networks (see Fig. 1-2) are characterized by a tight coupling of specific services to specific network resources, typically deployed in a series of multiple overlays; multiple OAM&P networks and operation systems for each of these service and resource overlays; and organizational structures made up of separate groups performing similar functions. This duplication of overlay structures was related to the operational characteristics of older technologies. In addition, specific vendor elements had their own proprietary approaches to OAM&P and network management created multiple administrative domains with poor interoperability between them. The total was the sum of all these independent, resource consuming partial solutions that have contributed to a network management environment that is inefficient, complex, and expensive to administer.

Traditional network management practices deal with a wide array of procedures, processes, and tools for configuration, fault detection, performance monitoring, security, accounting, and other management functions and are based on a "master–slave" relationship between management or operations systems[1] (OSs) and network elements (NEs). Network ele-

---

[1]The terms operations systems (OSs) and network management systems (NMSs) are used interchangeably in this book.

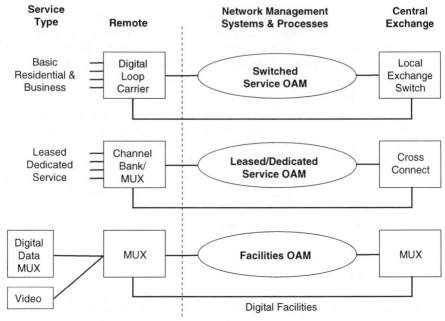

Figure 1-2 Current Service-Resource Relationship

ments typically have had only basic operations functionality with little ability to control activities or make decisions beyond the scope of call processing and information transmission. Accordingly, operations systems perform the bulk of the OAM&P work—processing raw data provided by individual network elements, making decisions, and instructing each individual network element to undertake specific actions.

This master–slave relationship contributes to operating inefficiencies in a number of ways. For example, there is little sharing of logical resources, such as data, because network elements and operations systems have been designed independently. In addition, each vendor's equipment has unique configuration and fault management interfaces as well as specific performance requirements. Network management systems must characterize each network element and vendor's interfaces on an individual basis, adding considerable time and complexity in introducing new services or technologies.

Other factors have compounded this complexity. For example, network management systems were generally constructed to optimize the work of individual service provider organizations or work groups at a particular point in time for a particular suite of technology. This type of development was undertaken independently by each organization and little attention was paid to system level interworking. Many copies of data, each tied to specific systems or job functions and to specific equipment vintages or implementations, were incorporated throughout the network, creating a major data synchronization problem. As a result, it has become increasingly difficult for the service provider, as a whole, to evolve services, network technologies, and network management processes in a cost-effective, timely, competitive manner in response to rapid changes in the telecommunications business.

1.4 REQUIREMENTS FOR AN EVOLUTIONARY NETWORK MANAGEMENT ARCHITECTURE

In developing an evolutionary network management architecture that will overcome the inefficiency, costliness, and complexity of the existing environment, it is essential to address key service, technical, and business aspects.

Service aspects include:

- enabling rapid new service deployment within both the network and network management system environments and
- promoting faster service activation.

Management or operations systems must be flexible and have a distributed, modular architecture that allows service providers to adapt to future customer needs. These needs may include, for example, rapid service deployment and activation, enhanced billing, and end-user online feature access. New software and features should ensure that customer services can be added in minutes rather than days or weeks.

Technology aspects include:

- the challenge of efficiently managing and distributing data throughout the network and
- elimination of physical overlay networks currently required for service/resource deployment and associated management systems.

Data management represents a major cost item for service providers due to the volume, redundancy, and difficulty in ensuring accuracy throughout a network/service provider's operation. Therefore, evolutionary architecture should allow for distribution of data throughout all layers of the network management environment and provide for intelligent network elements that can process data and pass information to network management systems on a peer-to-peer basis. Manual administration and alignment of redundant databases should be eliminated.

Given the sophistication and rapid growth of services, a more flexible operations environment must be established (i.e., multiple, single-function overlay networks must be eliminated). A distributed operations environment that correctly uses the capabilities of all components will remove the current interoperability bottleneck resulting from the proliferation of overlay networks.

An important step in creating this flexibility is to eliminate discrete overlay networks, by introducing network technology capable of providing generic resource capacity. This capacity will be logically assigned to a broad range of service types (i.e., the network will be provisioned in bulk and the services will be logically assigned to the network resources). Furthermore, incorporating intelligent network elements in the evolving operations architecture and repartitioning operations functions between network elements and network management systems will add momentum to the process of decoupling network management from service- and vendor-specific implementations. Successful achievement of this objective will be dependent upon utilization of standard open interfaces (covered in more detail in Chapters 2 and 3).

Business aspects include:

• reducing operations costs,
• enhancing the flexibility of the OAM&P environment, and
• providing services in a competitive, timely manner.

Cost reduction can be addressed on a number of fronts. One means is through simplifying the network, i.e., replacing service and technology-specific resources with generic resources capable of carrying a wide range of services. For example, replacing plesiosynchronous transport with SONET technology will reduce the need to manage multiplexors. In the access domain, software-controlled service-adaptive access technologies (e.g., those that enable service characteristics to be electronically downloaded to the network element) will simplify the network further and reduce the frequency and complexity of personnel dispatches. Another means of reducing cost is by integrating and simplifying operations processes and functions. Cost/benefit can also be achieved through elimination of redundant databases and amalgamation of processes and work forces so that these align with the network/service provider's business objectives. In addition to streamlining functions and costs, another benefit is an improvement in service responsiveness and quality. This could be achieved by providing near real-time service provisioning, automatic service restoral in the event of network disruption, and just-in-time resource provisioning.

An important means of enhancing OAM&P flexibility is to incorporate more intelligence into network elements. This redistribution of management functionality will enable network management systems to maintain a high-level, end-to-end view of services and resources (as opposed to the current scenario in which management systems must understand the implementation details of each individual network element's technology).

For network and service providers, this flexibility will mean that today's organizationally based operations structures and systems will move to a functionally based structure that spans a variety of services and technologies. One operations group could manage network surveillance, for instance, across all access, transport, and switching domains rather than having different surveillance groups for each domain. This distribution of functionality and data closer to the point of origin and application will pave the way for a simplification of operations systems and enable centralization of the operations workforce. Currently, many service providers are redesigning their business processes, an activity known as *process reengineering,* to achieve major gains in costs and services.

An evolutionary architecture should accomplish the overall objective of providing manageable flexibility points between the network management system, services, technologies, and service provider organizational structures. Experience has shown that services, technologies, and organizational structures traditionally evolve at independent rates; a modular operations architecture will facilitate each evolutionary step without necessitating a complete system redesign—a redesign that typically impacts each area highlighted in Fig. 1-3.

Future architectures are intended to guide the evolution of the existing network management environment to an advanced infrastructure that will enable service providers to achieve their business objectives and satisfy their long-term requirements for systems and organizational change. As illustrated in Fig. 1-4, future architectures orchestrate interactions

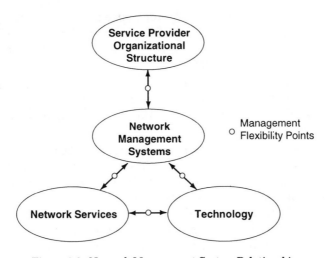

Figure 1-3   Network Management System Relationships

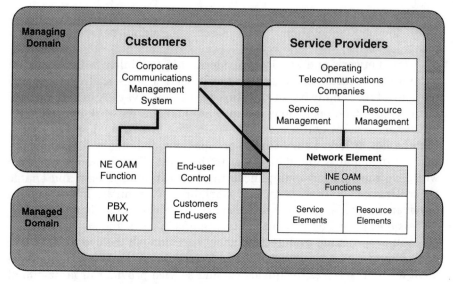

—— Management Relationship (Logical)

Figure 1-4 Network Management Domains

between the operations environment (comprising network management systems and intelligent network elements) in the managing domain and the base network elements in the managed domain. As well, the network management environment must also accommodate management requirements of corporate customers and individual end-users.

The OAM&P managing domain includes:

1. *Service provider's network management systems* that ensure a combined end-to-end view of the network for both service and resource management and reflect the primary business objectives of the service provider;

2. *Intelligent network elements* that have control applications capable of translating standard high-level messages from the network management applications to the vendor-specific technology implementations;

3. *Corporate communications management systems* for private networks. These systems provide corporate customers with functionality similar to network/service provider management systems (service and resource management, inventory management, directories, cost allocation, and traffic reports). Corporations currently have limited control over the services received from the public network provider. However, network/service provider service revenues could be increased by providing more

public network services to the corporate users if the services could be managed by the users in a secure and responsive manner; and

4. *Extended control functionality* that permits management of services and features by the *end-users,* either in conjunction with, or independent of, a corporate communications management system. These control functions could also enable residential or small business customers to modify their service profile directly from their terminals. The network element management function would update the service provider's network management system autonomously; increasing the customer involvement will reduce costs and improve responsiveness.

Both managing and managed domains are concerned with network service and resources. A key consideration is that the development of the interface between network management systems and the intelligent network element (INE) functions in a manner that does not stifle change or cause unnecessary disruption. For example, managed services typically span multiple versions of a single vendor's technology, not to mention technology from multiple vendors. Interfaces should change only if service capabilities evolve, not simply because a vendor's implementation changes.

In implementing the basic OAM&P architecture, the following requirements will be satisfied:

- communications will be established between network management systems and intelligent network elements;

- services will be managed as independently as possible from the resources on which they are implemented; and

- functional applications software will be incorporated within INEs to permit the mapping of generic high-level standard messages onto vendor-specific implementations.

Communications between network management systems and network elements will be standardized across vendor-specific implementations through a high-level open system interconnection (OSI) reference model (see Chapter 5), compliant with the TMN established by the International Telecommunications Union—Telecommunications (ITU-T—formerly CCITT) [4]. The reference model describes a communications architecture into which standard protocols, containing a clear description of data and data structures, can be placed or defined. Moreover, the model addresses the syntax and transfer of information and attempts to standardize the modeling and semantics of management information.

In addition to implementing a common communication framework that will ensure operations system and network element cooperation across OSI interfaces, the emerging management environment will incorporate functional applications that reside within the intelligent network elements. These functional applications will play a variety of roles from managing data to provisioning services to sectionalizing problems within specific network elements or transmission facilities.

For instance, each INE can be configured to steward its own data and to provide complementary OAM&P functionality to the network management systems. By optimizing data distribution and ensuring that the intelligent network element is able to autonomously update the network management systems on an as-required basis, operations performance is enhanced and redundant data is eliminated [6].

The architecture also provides for sharing of common supporting functions between OAM&P applications and the network management functional areas (i.e., OAM&P data management or security management). In addition, sharing of common functions applies to the complementary implementation in the network elements. This approach promotes consistency in implementation and minimizes development costs.

Prerequisites for data management include:

- ensuring accessibility of OAM&P data to multiple management applications;
- collecting and maintaining data from multiple sources;
- flexibility in aligning and synchronizing data; and
- management of data consistency between the network element data and the redundant or local copies, where necessary.

Additional requirements to be addressed in the emerging management systems include providing partitioned and secure data storage, OAM&P applications data views, OAM&P data formatting, and data alignment mechanisms.

Security management deals with protecting OAM&P applications, functions, and data against intentional or accidental abuse, unauthorized access, and communications loss. Security management can be implemented at different levels, including network provider and customer groups, network management systems, OAM&P applications, or functions and objects (see Chapters 4 and 5). User profiles, specifying access privileges and capabilities, will be based on hierarchical levels of authorization that reflect each group's administrative structure.

As customers place more stringent demands on the reliability and performance of the network, combined with the service provider's need to achieve more with the same number of or fewer people, the requirement for greater flexibility in assigning levels of access security increase. If the entire customer control issue can be considered to be an extension of the operations environment, partitioning of network management access—whether to a management system or directly to a network element—will simply be a matter of restricting access to their own service profiles and customer groups.

1.5 NETWORK MANAGEMENT FUNCTIONS

Architecture and control systems are key for evolving today's limited network management systems to meet tomorrow's objectives. Customers, as shown in Fig. 1-5, will have access to service providers' network management systems and applications. By repositioning network databases to take advantage of intelligent network elements, providing high-level standard interfaces, implementing standard protocols and messages, and sharing OAM&P functionality across operations systems and intelligent network elements, the evolving network will enable network/service providers to rapidly deploy new services, implement innovative technologies, reduce costs, enhance competitiveness, and meet the ever-increasing demands of customers.

This vision of an intelligent network will ultimately be realized in the telecommunications management network (TMN)[3], a management communications concept that defines the relationship between basic network functional building blocks (operations systems, data communications networks, and network elements) in terms of standard interfaces. The TMN also introduces the concept of subnetwork control that will play a pivotal role in evolving today's limited network management systems to meet future business objectives. A subnetwork is an aggregation of a group of NEs tied together by a common criteria (e.g., function, technology, supplier), and is viewed by the management application as a single entity. A device that implements subnetwork OAM&P functionality, known as element manager (EM), is an instrumental building block that will simplify the interworking between operations systems and network elements [7]. From an architectural perspective, EM provides the flexible management point between network management systems and the vendor implementation of technology. It uses the TMN framework for communications management with its generic information models and standard interfaces.

Behind the vision of an intelligent network lie a number of key tasks, including functional partitioning, high-level object-oriented messaging, autonomous updating or notifying, and functional applications, all of which must be performed effectively.

From a network management perspective, standards bodies address

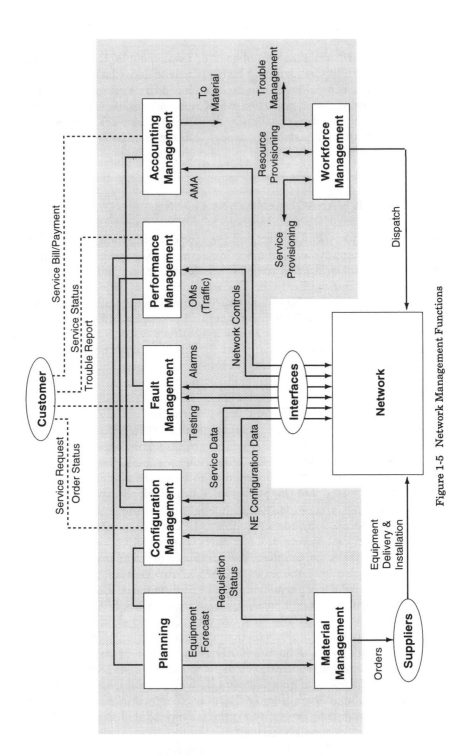

Figure 1-5 Network Management Functions

14

five functional areas, each of which represents a set of activities performed by operations personnel or customers. In many cases, both the network management system and the intelligent network element are involved in completing the functional task. Operations are separated from technologies by defining generic OSI-based OAM&P functions that are common to multiple technologies and services:

• *Configuration management* includes resource provisioning (timely deployment of resources to satisfy the expected service demand) and service provisioning (assigning services and features to end-users). It identifies, exercises control over, collects data from, and provides data to the network for the purpose of preparing for, initializing, starting, and providing for the operation and termination of services. Configuration management deals with logical, service, or custom networks such as the toll network, local public switched telephone network (PSTN), and private networks.

• *Fault management* includes trouble management, which looks after corrective actions for service, fault recovery, and proactive maintenance, which provides capabilities for self-healing. Trouble management correlates alarms to services and resources, initiates tests, performs diagnostics to isolate faults to a replaceable component, triggers service restoral, and performs activities necessary to repair the diagnosed fault. Proactive maintenance responds to near-fault conditions that degrade system reliability and may eventually result in an impact on services. It performs routine maintenance activities on a scheduled basis and initiates tests to detect or correct problems before service troubles are reported.

• *Performance management* addresses processes that ensure the most efficient utilization of network resources and their ability to meet user service-level objectives. It evaluates and reports on the behavior of network resources and at the same time ensures the peak performance and delivery of each voice, data, or video service.

• *Accounting management* processes and manipulates service and resource utilization records and generates customer billing reports for all services rendered. It establishes charges and identifies costs for the use of services and resources in the network.

• *Security management* controls access to and protects both the network and the network management systems against intentional or accidental abuse, unauthorized access, and communication loss. Flexibility should be built into security mechanisms to accommodate ranges of control and inquiry privileges that result from the variety of access modes by operations systems, service provider groups, and customers who need to be administratively independent.

Configuration management, fault management, and performance management are discussed in detail in Chapters 8, 9, and 10. Security management is treated in Chapter 4.

There are also several important network management functions that are not currently being addressed by standards or other forums, even though they are part of the conceptual framework:

• *Planning* encompasses the set of processes that permit the timely installation of resources to specify, develop, and deploy services in the network in response to service provider forecasts and end-user requirements.

• *Workforce management* plans and controls the activities of operations personnel. It deals with all workloads, personnel, and tools used in the management of the network. This includes repair (fault management), installation and cable locating (service provisioning), cable splicing and switch installation (resource provisioning), and field and central office technicians.

• *Material management* is concerned with procurement, control, and storage of equipment used in the installation and repair of the network. Material acquisition includes search, selection, and commitment of supplies and equipment from certified vendors. Material control monitors and updates inventory to ensure availability of material when and where required. Material distribution includes the handling of equipment from vendors and operations personnel, and the appropriate and timely delivery to the final destination.

Functional partitioning involves grouping functions into building blocks whose implementation can be moved across traditional boundaries in the physical architecture. Partitioning is essential to achieve effective, automated information flow-through on a complete system scale. This contrasts with today's approach in which attention is directed to isolated pockets of mechanization. Partitioning is also required in system development so that manageable portions of the operations architecture can be identified and allocated to specific projects.

Information models provide an abstraction of the telecommunications resources to be managed in the form of generic or technology-independent managed objects. To provide common solutions for switching and transport OAM&P, ITU-T has also generated an initial generic network information model in Recommendation M.3100 [4]. A key benefit is that this information model enables autonomous update and notification between the managed domains so that operating companies can provide corporate and, potentially, residential end-users, with immediate, independent access to services.

1.6 TRANSITION

The challenge facing network/service providers is how to manage the change in a continuously evolving network. Target architectures are being defined to support information networking services that will span multiple networks using equipment from several suppliers. Transition to these target architectures requires orchestrated automation, re-engineering of business processes, and introduction of new technologies. Transition strategies and alternatives need to be evaluated using prototyping, modeling, and simulation tools (see Chapters 6 and 7).

Realizing such hybrid environments is a large and complex undertaking with many challenges. Major advances in distributed data management, system platforms, interfaces, and security are needed in order to realize these challenges.

1.7 ORGANIZATION OF THE BOOK

Chapters 2 through 7 address new concepts and tools for network management. Chapter 2 describes network management problems, the different paradigms for network management, and provides a critical assessment for these directions. Chapter 3 defines the telecommunications management network (TMN) principles, associated implementation architectures, and applications. Chapter 4 provides an overview of OSI management activities within domestic and international standards forums. Chapter 5 provides a detailed overview of the object-oriented paradigm and its application to network management. Chapter 6 identifies the role of modeling and simulation in network management. Chapter 7 describes knowledge-based systems and applications to network management.

Chapters 8 through 11 deal with specific network management applications and their evolution as a result of these new concepts and paradigms. Chapter 8 describes configuration management and some associated network planning aspects. Chapter 9 provides a functional description of fault management and the associated interface specifications. Chapter 10 covers performance management and quality of service. Chapter 11 describes fast restoration techniques for high-speed networks.

REFERENCES

[1]. S. E. Aidarous, D. A. Proudfoot, and X. N. Dam, "Service Management in Intelligent Networks," IEEE Network Magazine, vol. 4, no. 1, January 1990.

[2]. D. A. Proudfoot, S. E. Aidarous, and M. Kelly, "Network Management in an Evolving Network," ITU - Europa Telecom, Budapest, October 1992.

[3]. CCITT, "Principles for a Telecommunications Management Network," M.3010, 1992.

[4]. CCITT, "Generic Network Information Model," M.3100, 1992.

[5]. Bellcore SR-NWT-002268, "Cycle 1 Initial Specifications for INA," issue 1, June 1992.

[6]. Bellcore TA-STS-000915, "The Bellcore OSCA Architecture," issue 3, March 1992.

[7]. Bellcore TA-TSV-001294, "Framework Generic Requirements for Element Management Layer (EML) Functionality and Architecture," issue 1, December 1992.

CHAPTER 2

A Critical Survey of Network Management Protocol Standards

Professor Yechiam Yemini
Distributed Computing & Communications (DCC) Lab
621 Schapiro Engineering Center
Columbia University
New York, NY 10027

2.1 INTRODUCTION

This chapter provides an introduction to network management from the point of view of management protocol standards, the Internet simple network management protocol (SNMP) and the open system interconnection common management information protocol (OSI CMIP). The goal of this chapter is to provide readers with sufficient detailed understanding of the problems, choices, and limitations of current management technologies and standards and to pursue more detailed references and work in the field. This chapter provides first an introduction to practical and technical problems of management (Section 2.1), reviews the central components of the SNMP (Section 2.2) and the OSI (Section 2.3) standards, illustrates their usage, and concludes with a critical comparative assessment (Section 2.4).

2.1.1 The Problem

The past decade saw a dramatic shift in the structure and role of computer communications systems within enterprises. From small and isolated data-processing islands, networked computing systems have grown into massively complex enterprise-wide systems. From a limited back end support role, these systems have become central in the conduct of critical business functions. The network, the computer, and the very enterprise have become increasingly indistinguishable.

These changes give rise to substantial exposure to new costs and risks associated with the operations of enterprise systems. As these systems grow and involve increasingly complex components and applications provided by multiple vendors, the likelihood of faults and inefficiencies increases. The costs of faults and inefficiencies can be enormous. For example, failures of a bank network can paralyze its operations, delays of security trades through brokerage system bottlenecks can cost in dollars and in customers, failures in Space Shuttle communications (e.g., 90-minute outage on May 4, 1993) could risk its mission, failures in a telecommunication network paralyzed New York Airports for a few hours, and loss of hospital lab reports can prevent timely diagnosis and care and result in malpractice suits.

Enterprises are also exposed to a new stressful dichotomy in the economies of scale and costs associated with operations. The cost of connectivity/computing per user has been driven down rapidly by a commodity market and fast technology changes. This stimulates rapid growth and changes of enterprise systems. Growth in scale and complexity of these systems leads to a nonlinear increase in operations support and expertise required to maintain enterprise systems. This drives up the costs of operations dramatically to consume an increasing share (65% by some estimates) of the MIS budgets of enterprises. As enterprises continue to downsize their data processing systems into massive distributed systems, a growing need arises to control operations cost by using management software to simplify and automate them.

Similar changes in complexity have emerged within telecommunication networks. Network providers have capitalized on the opportunities offered by enormous bandwidth availability, afforded by optical media, the separation of the physical layer from its control under signaling system 7, and the emergence of new networks and services such as cellular communications. Increasingly complex services and network structures have been deployed rapidly. Complexity, scale, and a shift of problems locus to layers above the physical media have exposed these networks too, to a discontinuity in operations management exposure and needs. A recent congressional study (March 1993) expressed great concern over the rapid escalation in network outages.

The goal of network management technologies is to reduce the risks and costs exposures associated with the operations of enterprise systems. The most critical management needs include handling of failures, performance inefficiences, configuration changes, security compromises, and uncontrollable costs. The demand for manageability has grown dramatically in recent years. Unfortunately, the technologies and standards needed to accomplish manageable systems are still in embryonic stages of development.

2.1.2 Platform-Centered Management and its Standards

Management systems are responsible for monitoring, interpreting, and controlling a network's operations. Figure 2-1 depicts a typical management system. Devices are equipped by their vendors with agent software to monitor and collect operational data (e.g., error statistics) into local databases, and/or detect exceptional events (e.g., error rates exceed threshold). The management platform queries device operational data or receives event notifications via management protocols. The management

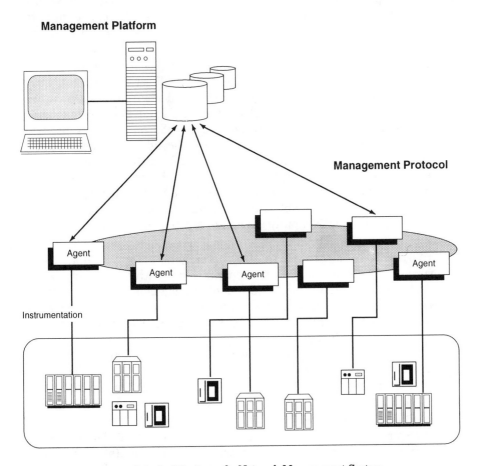

Figure 2-1 Architecture of a Network Management System

platform supports tools to display the data graphically, assist in its inter-pretation, and control the network operations (e.g., configuration control). Readers should consult [1] and [2] and subsequent chapters for additional details of platform functions.

The management paradigm reflected in Fig. 2-1 is platform centered. Management applications are centralized in platforms, separated from the managed data and control functions instrumented in the devices. Plat-form-centered management traces its roots to teleprocessing-mainframe network environments of the 1970s where devices (e.g., modems and con-centrators) were simple, lacked resources to run management software, were manageable through simple status and configuration variables, and, where network organizations could devote the human resources needed, develop the expertise to handle operations management. Device vendors provided proprietary network management systems to users. These ele-ment-management systems included specialized agent software, private management protocols, and (typically simple) monitoring platform workstations.

The 1980s saw discontinuities in enterprise systems that have redefined networks and their management needs. First, device complexity as well as the scale of complexity of networked environments have in-creased dramatically. A typical multiprotocol LAN hub can involve scores of different components, each involving hundreds of variables used to describe their operations and configuration. Second, the locus of opera-tions problems and complexity, too, has shifted from the physical layer to higher protocol stack layers (e.g., media-layer bridges, network layer routers, or even application layer elements such as directory servers). Third, the single-vendor stable environments of teleprocessing networks have been replaced with rapidly changing, multivendor network environ-ments.

With this rapid growth in size, heterogeneity, and complexity of enterprise systems, users often found themselves with a large number of such independent and incompatible vendor management systems. Instead of providing solutions, these systems often become part of the problem. It became necessary to unify management in a manner that would permit users to monitor, interpret, and control multivendor networks. A short term unification has been pursued through the development of integrated management systems (also known as managers of managers). The plat-forms of these systems collect and unify information from a variety of management systems via proxy agents that access private vendor agents, or element managers (consult [3] to [6] for examples).

Longer term unification of network management has been pursued by intense standardization efforts, starting in the late 1980s. Platform-centered management requires several standards to accomplish multiven-dor interoperability. First, the management protocols must be unified under a standard for accessing different vendor devices. Second, the structure of the agent's management databases, manipulated by the protocol, must be standardized. Such standards have been the subject of the Internet SNMP and the OSI management models. These standards enable platforms to query management information of heterogeneous multivendor devices in a uniform manner.

Merely standardizing the syntactic framework for manager-agent information transfer is, however, insufficient to accomplish integrated management. Management protocol standards enable a platform to collect data from multivendor devices. The explosive complexity of this operational data far transcends the abilities of platform operators to handle, however. The task of processing management data must be transferred to applications software.

The first problem to be solved is that of platform heterogeneity. Platform-centered management expects management applications software to interpret and control operations to execute at platforms. Device vendors wishing to introduce management applications for their products must provide them for different end-user platforms. This leads to intractable complexities in creating management applications software as an application needs be replicated for each platform. For example, a device vendor wishing to offer six applications over five platforms may need to develop and maintain a total of 30 product versions. To help reduce this problem, a few organizations (e.g., the OSF, XOPEN) have been pursuing development of management platform standards.

Even if management platform environments are standardized, management applications developers face the complexities of semantic heterogeneity of managed data. Management protocol standards such as SNMP and CMIP merely standardize the syntactic organization and access of managed data, not its meaning. An application managing a class of multivendor devices requires a uniform semantic model of their managed information. In the absence of such a unified semantic model, each device requires its own management applications software. Consider, for example, the problem of developing an application for an orderly shutdown of a section of a network or configuring the parameters of its devices in a coherent manner. The application will have to set appropriate managed variables to accomplish this. Since the specific variables and their meaning vary from device to device, an application developer must accommodate every possible combination of vendor choices. This, of course, leads to intractable software development problems. To resolve this semantic-heterogeneity barrier it is necessary for the very meaning of managed information, not just its syntax, to be standardized. A number of IEEE and International Telecommunications Union–Telecommunication (ITU-T–formerly CCITT) protocol committees have thus been developing standards for managed information associated with protocol entities [e.g., Ethernet, fiber distributed data interface (FDDI)].

Managed information is not limited to protocol interfaces, however, and involves parameters associated with the operations of the device (e.g., resource status). Various Internet committees have accordingly pursued definitions of managed information standards for certain classes of devices (e.g., LAN hubs, WAN interfaces). The need to standardize managed information often conflicts with the need of vendors to develop unique device operational features, however. This conflict leads to long term, unresolved difficulties. For example, consider a vendor wishing to offer certain unique device features that need to be managed (e.g., control of a LAN switch). The vendor must design platform applications to control these features and fit within the range of user platform environments. Users will need to integrate various management applications of multiple vendors into coherent management platforms. Obstacles in the development of management software thus produce difficulties in the introduction of innovative unique new devices and features.

Finally, integrated management requires that the network and the computer system be uniformly managed. While enterprise systems and device architectures rapidly remove the boundaries between the network and the computer, management technologies and systems are lagging in developing such vertical integration of management functions. Faults in enterprise systems, as illustrated in the next subsection, cannot differentiate the boundaries between the network and the computer. Management standards must concern themselves with development of both horizontal (multivendor) integration of management functions as well as vertical integration.

In summary, platform-centered management of heterogeneous networks requires the following standardization to accomplish integrated network management:

i.   Standards to unify manager-agent information access across multivendor devices. These include standardization of managed information database structure at the agent as well as the management protocol to access it.

ii.   Standards to unify the meaning of managed information across multivendor systems.

iii. Standards to unify platform-processing environments across multivendor platforms.

These standards, furthermore, must support vertical integration of network and system management. The Internet SNMP and OSI CMIP address the first category standards while complementary efforts are pursued by other organizations to address the second and third category of standards. This chapter focuses only on the first category of standards—those governing managed information organization and access protocols.

2.1.3 Why Is Management Difficult?

We use an example of a network "storm" to illustrate some of the complexities of management problems and the applications of SNMP and CMIP to their solutions. Network storms, involving rapid escalation of cascading failures, are not uncommon. Consider a T1 communication link used to multiplex a large number of connections [e.g., X.25 virtual circuits or transmission control protocol (TCP)] to a server/host. Suppose a long burst of noise disrupts the link, garbling T1 slots and causing a loss of packets (Fig. 2-2 (i)). Logical link level protocols, at layers above, will invoke automatic retransmission, resulting in a burst of retransmission tasks at the interface processor queue (Fig. 2-2 (ii)). The interface processor will see its queue loaded with hundreds of tasks, leading to its thrashing. Higher-layer transport entities will time-out, causing a burst of corrective activities, e.g., reset connections, executed by host CPUs (Fig. 2-2 (iii)), leading to their thrashing. In general, protocol stack mechanisms handle lower-layer problems through higher-layer corrective actions. These mechanisms can, as in the example, escalate the very problems that they are designed to solve.

How can such complex network fault behaviors be monitored, detected, and handled? Suppose that relevant operational variables, such as T1 bit-error rates and the size of the interface processor queue, can be observed as depicted in Fig. 2-3. The formation of the storm could be detected from the correlation among the sudden growth in error rates, followed by a delayed growth in queue size.

What management information should be used to capture these behaviors? SNMP uses a simple model for the structure of managed informa-

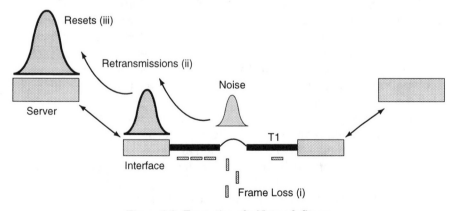

Figure 2-2 Formation of a Network Storm

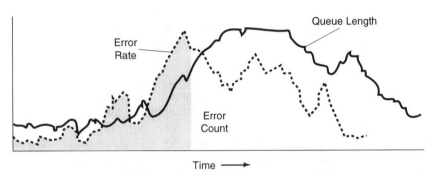

Figure 2-3 Temporal Behaviors Correlation

tion (SMI) [7], involving six application-defined data types and three generic ones. Temporal behaviors are described in terms of counters and gauges. An error counter represents the cumulative errors (integral) since device booting (the shaded area under the error rate curve). A gauge can model the queue length. The values of these managed variables can be recorded in an agent's management information base (MIB), where they can be polled by a platform. An error counter, however, is not very useful for detecting rapid changes in error rates to identify a storm. Changes in error rates are reflected in the second derivative of an error counter. A platform must sample the counter frequently to estimate its second derivative, leading to unrealistic polling rates. Recent SNMP extensions (RMON [8]) provide improvements to handle the difficulty.

OSI management uses an object-oriented model of managed information [9], [10]. The behaviors of interest—noise, errors, and queue length—are different forms of a time series. A generic managed object (MO) class may be defined to describe a general time series. This MO may include data attributes of the time series and operations (methods, actions) to compute functions of the time series (e.g., derivatives). This MO may also provide generic events notifications (e.g., whenever some function of time series exceeds threshold). The generic time series MO class may be specialized to define MO subclasses to model the bit-error rate of the T1 link and queue length of the interface processor. A management platform can create instances of these MOs, within device agents' databases. The device agent can monitor the respective network behaviors and record the respective values in these MO instances. The platform can, furthermore, enroll with the agent to receive notifications of events describing rapid changes of error rates and excessive processor queue.

To identify a storm it is necessary to detect not only error-rate and interface queue threshold events, but to identify a correlation among them. Unfortunately, the observation processes used for event detection may result in decorrelation of behaviors. Threshold excesses must be sustained over a sufficient window to result in event notification and avoid spurious alerts. Implementation may use different sampling rates and detection windows for the error-rate and for the queue length. As a result, either one of the events may be detected while the other may not, or the events may be detected in inverted temporal order. This decorrelating effect of observations can lead to erroneous interpretation. Often, hundreds or thousands of alerts may be generated by a fault. These events must be correlated to detect the source of the problem. The results of such analysis may be very sensitive to the choices of managed information provided by devices and the implementation details of monitoring processes.

To make things worse, the devices involved in the storm formation are typically manufactured by different vendors. The managed data and its meaning may vary greatly among devices, rendering interpretation difficult. Moreover, networks are often operated by multiple organizations, each responsible for a different domain. Faults escalating across such domain boundaries may be particularly difficult to detect and handle. In the example, the T1 links may be managed by a telephone company while the layers above are operated by a user organization. The user may be unable to observe the behavior of the T1 layer and the telephone company may be unable to observe the behaviors of higher layers.

To conclude, note some of the typical problems identified through the example. First, capturing the range of complex network behaviors may not be feasible within a narrow model of managed information. In the example, the counters and gauges of SNMP were too restrictive to support events detection and correlation. An information model of unbounded scope, however, may lead to complexity exceeding the computational capabilities of agents. Second, interpretation of network behaviors depends on a semantic model of its operations. Merely collecting vast amounts of management information, independent of the ability to interpret it, does not improve manageability. Third, the meaning of managed information may depend on subtle implementation decisions of its monitoring. Fourth, interpretation depends on the ability to correlate behaviors recorded by different managed objects. Relationships among managed information are central to interpretation.

Syntactic structures used to record and access managed information can play an important role in improving manageability. Encapsulation of managed information in MO class structures can help simplify the conceptual organization of, and access to, managed information. Such syntactic structures are not a panacea, however. The central challenges confronting network management are to develop effective and systematic answers to the questions: what should be monitored, how should it be interpreted, and how should this analysis be used to control the network behavior? These answers must be sufficiently general for arbitrary heterogeneous networks and remain valid under rapid network changes and within multidomain networks. A syntactic framework for managed information access can enable successful answers, but does not in itself provide them.

2.2 THE SNMP MANAGEMENT PARADIGM

This section provides an introduction to SNMP, its applications, its limitations, and recent extensions.

2.2.1 Overview

The overall organization of the simple network management protocol (SNMP) is depicted in Fig. 2-4. SNMP includes two major components: (1) an architecture for organizing managed information in management information bases at agents [7], [11] and (2) a query protocol to access these databases [12]. We review these briefly and then follow with a more detailed description.

The first SNMP component depicted in Fig. 2-4 is the MIB tree, located in the expanded agent's box. The MIB is organized as a directory tree of managed data located at the leaves, depicted as shaded cells. The internal nodes of the tree are used to group managed variables by related categories. For example, all information associated with a given protocol entity (e.g., Internet Protocol [IP], TCP, User Datagram Protocol [UDP]) is recorded on a given subtree. The tree structure provides a unique path identifier (index) for each leaf. This identifier is formed by concatenating the numerical labels of its path from the root. The managed information stored at the leaves can be accessed by passing the respective identifier to the agent.

The second SNMP component is depicted by the manager-agent protocol primitives on the left. They provide a query mechanism for platform programs (managers) to retrieve or modify MIB data. GET retrieves data, SET changes the data and GET-NEXT provides a means to traverse the MIB tree and retrieve data in order. The TRAP command provides a means for asynchronous notifications of significant events from agents to managers.

How does a manager communicate with agents? The GET/SET/GET-NEXT/TRAP information access primitives are packaged in SNMP protocol data units (PDUs). These PDUs must be packaged and transported between the manager and agents. In principle, the decision of which transport layer facility should be used to support these communications is independent of the PDUs themselves. SNMP managers and agents could use any datagram transport environment. Indeed, while the primary transport mechanism used to support SNMP manager-agent exchange is the UDP/IP stack, implementations of SNMP have been defined

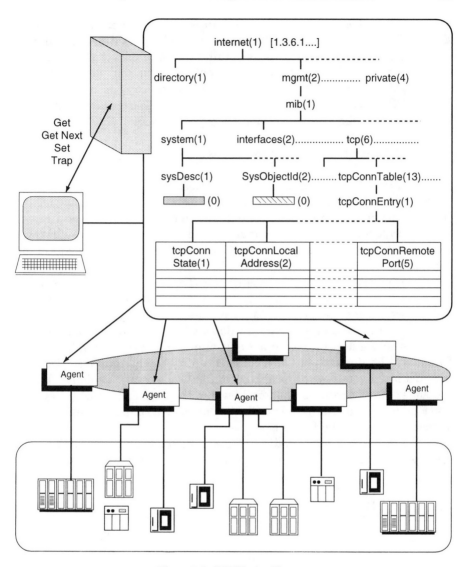

Figure 2-4 SNMP at a Glance

for practically all common transports from Internet Packet Exchange (IPX) and Appletalk to OSI.

How does a PDU carry managed information between manager and agents? A manager program can generate a GET/GET-NEXT/SET request to access the agent's MIB. These requests are packaged in PDUs, sent by the manager. The request PDUs carry an identifier of the respective request and a list of variable-binding pairs. A variable-binding pair serves as a container of managed information. A pair includes an identifier of a managed variable (MIB tree leaf) and a respective value. The pair {ifIn-Error.3, 357}, for example, identifies the third cell in the ifInError column within the interface table, and its counter value of 357. A GET request carries a variable-binding pair with an empty value, and the agent uses the identifier to retrieve the requested value and returns it via a GET-RESPONSE. In the sections following we expand this overview to describe the structure and operations of SNMP in greater detail.

2.2.2 Management Communications

How should a manager communicate with agents? From the point of view of communications, the manager and agents are applications, utilizing an application protocol. The management protocol requires a transport mechanism to support manager-agent interactions. SNMP typically uses transport services provided by UDP/IP transport. This protocol environment is depicted in Fig. 2-5. A manager application must identify the agent in order to communicate with it. The agent application is identified by its IP address and a UPD port assigned a well-known number 161, reserved for SNMP agents. A manager application packages an SNMP request inside a UDP/IP envelope, which contains its source port identifier and a destination IP address and port 161 identifier. The UDP frame will be dispatched via a local IP entity to the managed system, where it is delivered by the local UDP entity to the agent. Similarly, TRAP messages must identify managers to whom they are addressed. They use an IP address and a well-known SNMP manager UDP port identifier 162. This is depicted in the figure below.

Can a management protocol between manager and agent utilize arbitrary transport? The choice of transport is somewhat orthogonal to the

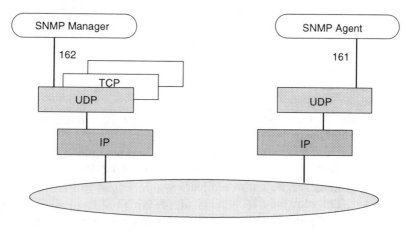

Figure 2-5 SNMP Communications Environment

management protocol itself. SNMP merely requires datagram transport (which is potentially unreliable) to transfer its PDUs between manager and agents. This permits mapping of SNMP to many protocol stacks. A datagram transport model reduces the complexity of transport mapping. Notice, however, that a number of transport-specific choices are involved. Different transport layers may use a variety of addressing techniques. Different transport layers may impose size restrictions on PDUs. Transport mapping requires handling these addressing and size considerations as well as a few other transport-specific parameters.

SNMPv.2 design [13] used the experience gained by SNMP to sharpen and simplify the mappings to different transports. The management protocol has been orthogonally separated from the transport environment, encouraging its use over practically any protocol stack.

How can manager-agent communications be secured against failures? Management communications are most critical at times when a network is least reliable. How can managers communicate reliably with agents? The use of UDP, by SNMP, means potentially unreliable transport. SNMP leaves the function of recovery from loss to the manager application. GET, GET-NEXT, and SET are all confirmed by a respective GET-RESPONSE frame. A platform can detect loss of a request when a response does not return. It may proceed to repeat the request or pursue other actions. TRAP messages, however, are generated by the agent and are not confirmed. If a TRAP is lost, the agent applications would not be aware (nor, of course, would the manager). Since TRAP messages often signal information of great significance, requiring manager attention, securing their reliable delivery is important.

How can delivery of TRAP messages be secured against loss? The agent can be designed to repeat the TRAP notification. One can assign a MIB variable to set the number of repetitions required. This variable could be configured by a manager SET command. Alternatively, the agent may repeat the TRAP until a designated MIB variable has been SET by a manager to turn the flood off. Notice that both methods provide only partial solutions to the problem. In the first case the number of repetitions may not be sufficient to ensure reliable delivery. In the second case a failure in the network can result in a massive storm of TRAPs (limited only by the speed at which agents have been programmed to generate them). This may cause escalation of the network stress that caused the TRAPs in the first place. In both cases, when TRAPs are to be delivered to multiple managers, complex patterns of loss and inconsistencies among managers may arise. Leaving the decisions on design of TRAPs recovery to agents' implementations furthermore increases the complexity of managing multivendor agents.

Accordingly, SNMPv.2 [13] has pursued an improved mechanism to handle event notifications. First, the TRAP primitive has been eliminated and replaced by an unsolicited GET-RESPONSE frame, generated by an agent and directed to the "trap-manager" (e.g., at UDP port 162). This reflects a view that event notifications can be unified as responses to virtual requests by event managers. By eliminating a primitive, the protocol has been simplified. A special TRAP MIB has been added to unify event handling, subscription by managers to receive events, and repetitions to improve their reliable delivery.

What is the impact of manager-agent transport on manageability? The use of the managed network to support management communications (in-band management) raises interesting issues. Note that the choice of in-band vs. out-of-band management is entirely orthogonal to the choice of management protocol. In-band management may lead to loss of an agent's access (depending on the source of trouble) precisely at a time when the agent needs management attention. The problem can be somewhat alleviated if the access path to agents is somewhat protected from the entities that they manage.

A more subtle impact on manageability arises in the context of transport addressing. Consider Fig. 2-5: the SNMP agent is uniquely identified by an IP address and a UDP port number. This means that only a single agent is accessible at a given IP address. This agent, furthermore, can maintain only a single MIB. Therefore, within a single IP address, only a single MIB instance may exist. This unique binding of an MIB to an IP address can limit the complexity of data that an agent can offer. Consider a situation where a system requires multiple MIBs to manage its different components. To be accessed via a single agent, these different MIBs will have to be unified under a single static MIB tree. Situations may arise where such unification cannot be accomplished. In such cases each MIB may require its own SNMP/UDP/IP stack, leading to great complexity in the organization of management (e.g., correlating information from multiple MIBs of a given system) and its access (e.g., via multiple IP addresses).

Alternatively, a single agent in a system can act as an extensible proxy for subagents encapsulating different MIBs associated with a given subsystem. This approach is supported by SNMPv.2 extensions to handle manager communications. These extensions permit an agent to act as a manager of local subagents, providing access to a collection of subagents.

2.2.3 The MIB Structure

MIBs provide a hierarchical database organization model of managed information [7], [11]. Managed information is recorded in managed variables and stored at the leaves of a static tree. SNMP utilizes the International Organization for Standardization (ISO) registration tree as a managed information directory. The ISO registration tree, depicted in

Fig. 2-6, is used to index definitions of various standards. Tree nodes are labeled with a name (global identifier) and a number (relative identifier). A node is uniquely identified by concatenating the numbers on its root path. For example, the *internet*-labeled subtree is identified by the path 1.3.6.1. This subtree has been allocated to the Internet organization to record its standards. The Internet tree supports 3 subtrees associated with management: *management, experimental,* and *private.* These subtrees are used to record, respectively, various Internet-standard MIBs (e.g., MIB-II), MIB standards under considerations (e.g., RMON), and private vendors MIBs of different products.

One can best view the tree structure as a means to accomplish two goals. First, it provides unique identification of managed information items. For example, Fig. 2-7 shows that the path 1.3.6.1.2.1.1.1 leads to the sysDesc cell containing a system-description information. Second, it provides grouping of related managed information items under a single subtree. For example (Fig. 2-7), all managed information associated with a given Internet system is recorded under the system subtree 1.3.6.1.2.1.1.

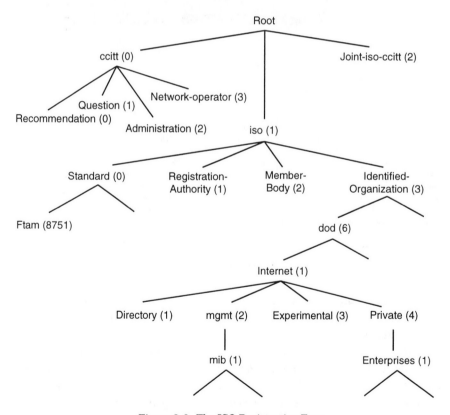

Figure 2-6 The ISO Registration Tree

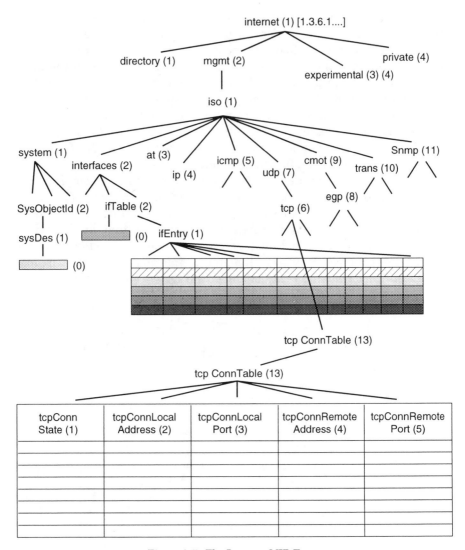

Figure 2-7 The Internet MIB Tree

Notice that this directory structure of managed information is static. The location of information on the MIB tree is determined at the time when the MIB is designed.

Figure 2-7 illustrates the organization of the Internet MIB-II tree [11]. This MIB aims to provide generic managed variables to handle an Internet stack (IP, UDP, TCP, ICMP). MIB-II is broken into 11 functional subtrees. Each of these subtrees represents a grouping of variables associated with a managed entity (e.g., a protocol entity such as IP or TCP).

These trees are further broken into subtrees. At the bottom of the tree, leaves are used to index managed variables of a given type. Some leaves (e.g., sysDesc, which contains a system description) index a single instance of a managed variable and require a single storage cell. Other leaves (e.g., tcpConnState, describing a TCP connection state) may index multiple instances. These different instances are organized as columns of cells. These columns form tables whose rows represent different instances of a given entity (e.g., a TCP connection or an interface).

Indexing access to nontabular (singular) instances is a simple problem. The cell associated with such a leaf is viewed as a child of the leaf, and is labeled 0. For example, the managed variable containing the system's object ID (cell located under sysObjectId) is identified via 1.3.6.1.2.1.1.2.0; that is, the path to sysObjectId concatenated with 0. Tabular data, however, requires somewhat more complex indexing since it is necessary to identify each row in a table (an instance of an entity) uniquely.

Instances of a given managed entity (e.g., interface) may vary from system to system and from time to time. Tables are used to support such dynamic changes; new entries may be added or deleted by the agent. Therefore, it is not possible to index table rows statically. Instead, table entries are uniquely identified by key columns. The contents of uniquely identifying (key) columns is used as an index. One possibility, used in the interface table, is to include a special index column to serve as a key. The value stored in this column provides unique identification of rows. For example, suppose the third row (instance) in the interface table has an ifInIndex value 12, then ifInError.12 identifies the third element in the ifInError column.

Another possibility, used in the TCP connections table, is to utilize a few columns as a key. The four columns: <tcpConnLocalAddress, tcpConnLocalPort, tcpConnRemoteAddress, tcpConnRemotePort> provide a key to the TCP connections table. This value is, again, concatenated after the column identifier to uniquely index a respective instance value. For example, suppose the third row of the TCP connection table includes the following key columns values: <128.10.15.12, 173, 128.32.50.01, 130>. The state of the respective TCP connection is uniquely identified by concatenating the column identifier tcpConnState [1.3.6.1.2.1.6.13.1.1] with the values of the four indexing columns 128.10.15.12.173.128.32.50.01.-130. Obviously, this indexing technique can result in inefficient identification and access of managed variables. Recent follow-up proposals advanced by SNMPv.2 [13] improve the indexing technique of tabular data to simplify access.

In summary, SNMP MIBs are organized as hierarchical databases with managed data stored at tree leaves. Subtrees are used to represent logical containment, while managed variables stored at the leaves are used to represent instances of the respective entities. The structure of the database (tree) is defined statically by the MIB designers. The only execution time extendability and changes are in the values stored in the database and in the creation/deletion of instances (rows) within tables.

2.2.4 MIB Access

SNMP can be viewed as a query language on the MIB tree. The primitives used by a manager application to retrieve data from the MIB are GET and GET-NEXT. Both result in a GET-RESPONSE frame carrying the data back in the form of variable-binding pairs. Both may be used to retrieve multiple managed variables. GET must specify the set of managed variables directly via their path identifier. This is useful for retrieval of nontabular data, where the path is statically known. For example, to retrieve the system description, a GET request must include the identifier of its cell 1.3.6.1.2.1.1.1.0.

GET-NEXT is used to traverse tabular data. Successive data is retrieved by traversing the MIB tree utilizing an intrinsic lexicographic order. The lexicographic order arranges a parent node ahead of its children, and the children from left to right. Table columns are ordered left to right and rows are ordered top down. This order, also known as *preorder*, is depicted in Fig. 2-8.

GET-NEXT applied to the items marked 10, 14 in Fig. 2-8 will retrieve items 11 and 15. For example, consider the following section of the interface table, depicted on the next page.

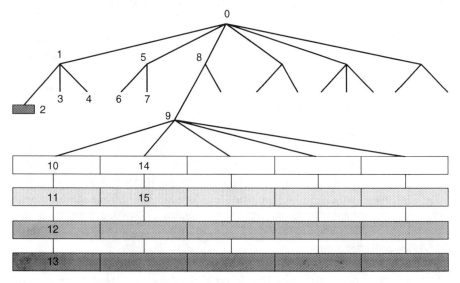

Figure 2-8   The Lexicographic Order (Preorder) over the MIB Tree

ifIndex	ifInOctets	ifInUcastPkts	ifInDiscards	IfInErrors	ifInUnknownProtos
1	354212		537	325	
2	6543232		822	93	
3	12428		73	11	

Successive applications of GET-NEXT will retrieve the following variable-binding pairs:

> IfInOctets, IfInDiscards, IfInErrors=>
> =>ifInOctets.1=354212, ifInDiscards.1=537, ifInErrors.1=325
> =>ifInOctets.2=6543232, ifInDiscards.2=822, IfInErrors.2=93
> =>ifInOctets.3=12428, ifInDiscards.3=73, ifInErrors.3=11
> =>ifInUcastPkts.1=135,ifInErrors.1=325, ifInUnknownProtos.1=2

GET and GET-NEXT provide means to retrieve MIB data. Control of device behavior is accomplished via the SET primitive. SET is often used to invoke the agent's actions as side effects of MIB changes. For example, by SET-ting a device administrative status to texting, a diagnostic testing routine can be invoked. This means that, in particular, agents must actively monitor MIB changes and invoke respective actions, unlike passive database systems where updates result in a mere recording of data.

A deficiency of SNMP's GET-NEXT retrieval is the need to poll a row at a time. When a table is large, this can significantly slow its traversal. Often, entire tables must be polled and scanned at the platform. This has been corrected by SNMPv.2, which replaced the GET-NEXT command with a GET-BULK command. GET-BULK will retrieve a number of specified successive rows that fit within a single UDP frame. This may be viewed as a generalization of GET-NEXT that improves retrieval time for tabular data.

2.2.5 The MIB Contents

The structure of managed information (SMI), defined in [7], provides a simple model of managed data. This model is defined in terms of ASN.1, a language to describe syntax of data structures (readers are advised to consult [14], or general textbooks on networks for an introduction to ASN.1). The SMI introduces six application data types to describe managed data: counter, gauge, time-ticks, network address, ip-address and opaque. Counters are used to represent cumulative sampling of time series, gauges represent samples of time series, time-ticks measure relative time, and opaque data types are used to describe arbitrary octet strings. Additionally, primitive universal types of ASN.1 including integer, octet string, and object identifier can be used in specifying managed data.

Limiting the data types used by an SMI, and restricting data item sizes in MIB definitions, greatly reduces the complexity of data encoding/decoding and storage organization. Simplicity and resource control are of great significance in a resource-constrained agent environment and thus play a central role in SNMP's design.

The SMI also includes a special macro extension of ASN.1 the OB-JECT-TYPE. This macro serves as the main tool for defining the managed objects at the MIB leaves. The OBJECT-TYPE macro provides facilities to define a managed variable and to associate with it a data type, an access mode (read, write, read/write), a status (required, optional), and a static MIB-tree location (path identifier). The definition of the OBJECT-TYPE macro (taken from [7]) and sample definitions of managed variables (taken from [11]) are described in the table below. The first part of the MIB definitions constructs the path identifiers for the internal nodes of the MIB trees. It defines these path identifiers as data of object-identifier type. The identifier of a node is obtained by concatenating a number to the identifier of its father. Once the internal nodes are defined, leaf nodes are constructed using the OBJECT-TYPE macro. These leaf nodes define the data-type (syntax) of the managed variables that they store, the access control, status, and object-identifier path to access the managed variable.

A few notes are useful to explain these definitions and their usage.

1. Object-identifiers define the location of internal nodes (e.g., "system," "interfaces") or leaves (e.g., sysDescr, ifInErrors) on the MIB tree. The path identifier is created by concatenating the parent path with a label of the node (e.g., sysDescr = {system 1}).

2. Tables are constructed as sequences of row entries. Row entries define the columns of the table. The interface table, for example, is constructed from columns devoted to different interface parameters (e.g., ifSpeed, ifInErrors). These different column parameters are registered as leaves under the ifEntry subtree describing the table.

3. The definitions of MIB structure provide a purely syntactic structure. The intended meaning of the different variables is specified in English. Two MIB implementations may interpret the meaning of a given variable differently. Semantic conformance may be impossible to validate.

4. The formal definitions of the MIB may be used by MIB compilers (see [15]) to construct MIBs and their access structures. The compiler uses the definitions to create a database structure to store the MIB. This greatly simplifies MIB development.

5. Implementation of the MIB is straightforward. Nontabular data may be stored in fixed linear data structures. Tabular data may need to extend or contract as new table rows are created/deleted.

A sorted linked list structure or a tree (with table records stored in leaves) can adequately represent such dynamic data.

It is useful to view the MIB structure in terms of traditional database systems [16]. A database system is defined by its data definition language (DDL), describing the database structure, and a data manipulation language (DML). The SMI model and ASN.1 extensions can be viewed as a DDL to build MIBs. An MIB compiler is analogous to a DDL compiler, used to generate database structures from abstract specifications. The protocol access primitives can be viewed as a query language (DML). From the perspective of traditional databases, SNMP can be viewed as a simple hierarchical database system whose essence is defined by the DDL (SMI) and DML (protocol primitives).

We conclude this section by providing a brief description of the Internet MIB (MIB-II). The table below summarizes the role of different subtrees [11]:

System Group: overall description of the managed system
Interface Group: static & dynamic operational interface data
Address Translation Group: map of IP addresses to physical net address
IP Group: IP operational statistics, address cache, routing table
ICMP Group: protocol statistics
TCP Group: TCP parameters, statistics, connection table
UDP Group: UDP statistics, UDP listener table
EGP Group: routing protocol statistics, EGP neighbors table
CMOT: place holder
Transmission: place holder
SNMP Group: SNMP protocol statistics

The system group provides an overall description of managed system attributes.

sysDescr: a text description of the managed system
sysObjectId: a vendor identification of the system in terms of private MIB subtree
sysUpTime: time in 1/100 sec since initialization of network management of system
sysContact: name & access information of responsible person
sysLocation: where it is

The interfaces group provides a table description of managed interfaces including static attributes as well as dynamic operational statistics

ifNumber: total # of managed interfaces

ifTable: summarizes operational statistics of interfaces. Its columns are described in the table below.

ifIndex: # assigned to interface
ifDescr: describes the if (name of manufacturer, product name, version..)

```
-- RFC 1155  defines SMI --
........
OBJECT-TYPE MACRO ::=
BEGIN
    TYPE NOTATION    ::=    "SYNTAX" type  (TYPE ObjectSyntax)
                           "ACCESS" Access
                           "STATUS" Status
    VALUE NOTATION ::= value (VALUE ObjectName)

    Access ::= "read-only" | "read-write" | "write-only" | "not-accessible"
    Status ::= "mandatory" | "optional" | "obsolete"
END
--This Macro provides a template to define a managed variable and associate with it
--a "syntax," a read/write "access," a standard "status," and a MIB-tree registration
--"value." The ObjectSyntax consists of 6 application defined types of the SMI as
--well as ASN.1 universal types: Integer, Octet String, and Object Identifier.
........

--RFC 1213 defines MIB-II --

........
mib-2 OBJECT-IDENTIFIER  ::= {mgmt 1} — Registration value concatenates 1 to
mgmt
system OBJECT-IDENTIFIER ::= {mib-2, 1} — Register system as 1st subtree of mgmt
interfaces OBJECT-IDENTIFIER ::= {mib-2 2}
at OBJECT-IDENTIFIER ::= {mib-2 3}
.....
-- Now define & register managed variables at the leaves

sysDescr OBJECT-TYPE
    SYNTAX DisplayString (SIZE (0 . . 255)) -- This managed variable is a text string
    ACCESS read-only -- It can be read only
    STATUS  mandatory -- It describes the system, so it must be included
    ::={system 1} -- It is registered as a first leaf on the system subtree
.....
sysUpTime  OBJECT-TYPE
    SYNTAX TimeTicks
    ACCESS read-only
    STATUS  mandatory
    ::={system 3} -- Register as the 3rd leaf of system subtree
.....
ifTable OBJECT-TYPE
    SYNTAX SEQUENCE OF IfEntry -- A table is a sequence of entries; entries
                                  -- describe the structure of a row
    ACCESS read-only
    STATUS mandatory
    ::= {interfaces 2} -- Register on interface subtree
.....
ifSpeed OBJECT-TYPE
    SYNTAX Gauge -- Describes instantaneous interface bandwidth in bits/sec
    ACCESS read-only
    STATUS  mandatory
    ::={ifEntry 5} -- Register on interface subtree devoted to interface table

IfInErrors OBJECT-TYPE
    SYNTAX Counter — Describes cumulative interface input error packets
    ACCESS read-only
    STATUS  mandatory
    ::={ifEntry 14} — Register on interface subtree devoted to interface table
```

ifType: e.g., ethernet-csmacd(6),iso88023-csmacd(7),..iso88026-man(10),fddi(15)..
ifMtu: maximal datagram that interface can process
ifPhysAddress: interface address at layer below ip
ifAdminStatus: desired state up/down/testing
ifOperStatus: actual state up/down/testing
ifLastChange: sysUpTime when current operational state entered
ifInOctet: total number of octets received
ifInUcastPkts: # unicast pkts delivered to higher layer protocols
ifInNUcastPkts: # non-unicast pkts delivered upward
IfInDiscards: # of pkts discarded (e.g., due to lack of buffer space)
ifInErrors: # in-bound pkts with errors
ifInUnknownProtos: # discarded because unknown protocol
ifOutOctets: # octets transmitted
....
ifOutQLen: length of outbound pkt queue
....

2.2.6   Example: How to Design an SNMP MIB

A useful exercise to integrate and clarify the discussions in previous sections is to design an MIB. Within our limited scope, only crude design elements may be illustrated. Readers are advised to refer to a more detailed MIB design [15]. The example selected is that of an MIB for a trivial telephone PBX.

Step 1: Develop the MIB Tree Structure Top-Down to Represent Logical Organization of the System. The MIB tree of a PBX can be developed by following the architecture of its components. A section of a hierarchical system decomposition is depicted in Fig. 2-9. The respective MIB tree, depicted in Fig. 2-10, derives directly from this system composition.

Step 2: Identify Similar Components (Instances) to Be Described by Tables. Examples of instances of similar managed elements include interface cards, extensions ports, and trunk ports. Each of these elements re-

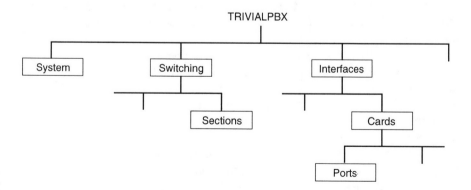

Figure 2-9   Hierarchical Composition of a Trivial PBX

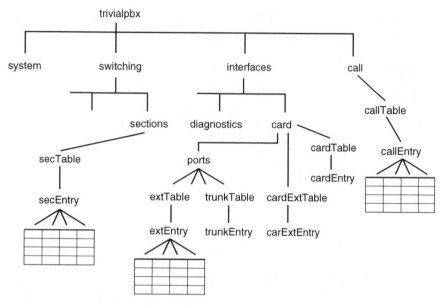

Figure 2-10   MIB Tree Skeleton for the Trivial PBX

quires an MIB table to describe different instances. Tables are built by
locating their attributes at column header leaves under a respective table
entry node (e.g., the extEntry).

At this stage some nontrivial choices in data modeling are encountered. Consider the problem of designing tables to handle operational
data of extension ports and of calls using them. The table for calls, executed by an extension, cannot be recorded under the respective extension
table row for the extension. MIB data can reside only under static leaves,
whereas extensions can change from one device to another. Therefore,
both tables must be separately placed at the leaves. The extensions table
will contain data concerning extensions, while the calls tables will contain
calls data. To retain the one to many relations among extensions and calls,
the calls table can include an attribute to identify the respective extension.

How good is this design? Consider the problem of retrieving calls
data meeting a criterion of interest (e.g., long distance calls of a given
extension). The entire calls table will need to be polled sequentially via
GET-NEXT since it is not possible to retrieve only calls meeting the
criterion. Both selection and processing must be performed at the
management platform. This may require as many GET-NEXT polls as
there are rows in the table. Additionally, correlated operational data may
be spread throughout the MIB. For example, the ports table may include
port error statistics while the calls table may include related call errors
statistics. To recognize faults, data from multiple tables may need to be

correlated. Entire tables may need to be retrieved and processed at the platform. This may render handling of real-time problems difficult.

Evidently, the design of MIBs to represent data correlated via different relationships is a central element in supporting effective access and interpretation [17]. In particular, data requiring real-time access (e.g., for fault management) must be organized to support fast access via a minimal number of GET commands. The problems of representing complex relationships within the framework of the SNMP MIB is yet to be explored. Readers are advised to consult traditional database literature [16] concerning related database design issues.

Step 3: Identify Managed Attributes to Be Located at Leaves (Tabular and Nontabular Data). Once the overall structure of the MIB tree is designed, the appropriate data attributes needed for each managed entity must be designed. Figure 2-11 depicts examples of such attributes. Physical entities such as cards and extension ports must include attributes to support configuration and fault management. Fault detection can be handled by monitoring respective fault indicators (e.g., error counters). Diagnosis as well as configuration control may require invocation of diagnostic software via side effects of SET requests. This, again, may lead to nontrivial MIB design problems. Suppose, for example, that a diagnostic procedure to test a given port must be invoked. The identity of the port must be passed as a parameter to the respective diagnostic procedure. A SET command may need to pass to an agent both the identity of the diagnostic procedure as well as the port identity. The value returned by

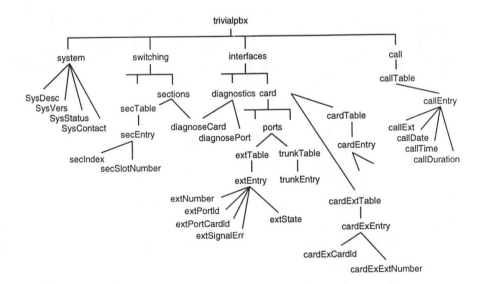

Figure 2-11 The Trivial PBX MIB Design

the procedure must be recorded in another MIB variable, to be accessed via a GET. The reader should contemplate the value of such a complex path to accomplish a remote procedure call.

Configuration monitoring often requires knowledge of configuration relations among managed objects, e.g., which switching section an interface card is attached to. Examples of typical configuration relations are "continued-in," "connected-to," "uses." Extracting such relations from managed data may be difficult or even impossible. For example, the Internet MIB does not represent a model of connectivity relationship. As a result, extracting the network topology requires a discovery process to analyze IP addresses. Explicit representation of configuration relations is of great significance in correlating configuration data.

Step 4: Establish ASN.1 Specifications of the Private Tree and Compile into the Agent. Once the MIB is designed, it is necessary to define it formally using the SMI. The definitions proceed top-down the tree. Internal nodes need only be registered on the tree, by assigning appropriate Object-identifier string. In the example of Fig. 2-11, registration of the internal nodes follows the pattern

```
. . . . . . . . . . . . . . . . .
card OBJECT IDENTIFIER::= {interfaces 2}
ports OBJECT IDENTIFIER::= {card 1}
extTable OBJECT IDENTIFIER::= {ports 3}
extEntry OBJECT IDENTIFIER::= {extTable 1}
. . . . . . . . . . . . . . . . .
```

Managed entities at the leaves are constructed via the OBJECT-TYPE macro, as in Section 2.2.3.

The variable, diagnosePort is used to invoke a diagnostic routine as a side effect of SET. The extNumber describes a column in the extension table. Similar definitions are used to complete the MIB description. At that time an MIB compiler can be invoked to create the appropriate MIB implementation.

```
. . . . . . . . . . . . . . . . . . .
diagnosePort OBJECT-TYPE
             SYNTAX INTEGER (0..65535) – – assign an integer value to cell
             ACCESS write-only
             Status mandatory
             ::={diagnostics 2}   – – register on MIB tree under diagnostics
. . . . . . . . . . . . . . . . .
extNumber OBJECT-TYPE
             SYNTAX INTEGER (0..9999) – – extension number is an integer
             ACCESS read-only
             Status mandatory    – – must be included
             ::={extEntry 1}     – – register on MIB tree as first column in extension table
. . . . . . . . . . . . . . . . .
```

2.2.7 Critical Assessment and Recent Directions

SNMP design sought to minimize the complexity of the management information base, the information access protocol, and the agent's implementation. Simplicity is of great significance in handling real-time management tasks within resource-constrained device environments. Design simplicity also eases implementation, interoperability, conformance, and market acceptance. SNMP has accomplished remarkable success along all these dimensions. Simplicity, however, results in some limitations. We seek in this section to assess some of these limitations. We also provide a brief discussion of recent developments of SNMP that resolve some of these limitations.

The Managed Information Model. The managed information model of SNMP leads to a few limitations on manageability. Some of these became apparent in previous sections.

The MIB Rigidity Problem. Managed data is defined at MIB design time. The data required by operators or applications software may not be directly available in the MIB. For example, an application may require information about changes in error rates, whereas the MIB only includes a cumulative error counter. The computations of the required information must be conducted at the manager. This means that large amounts of raw data, often unrealistic, must be polled. Applications often need to control the rate at which certain behaviors are sampled, rather than depend on rigid decisions of MIB designers. The design of the remote monitoring [8] MIB proposal accomplished an important step toward providing some flexibility in controlling monitoring processes. Monitoring procedures to collect operational data are flexibly parametrized. The parameters that control these procedures are organized in a control MIB. An application can SET entries in the control MIB tables to define which data are collected and in what manner. The data themselves are stored in a respective data MIB, where GET requests can access it. Notice, however, that the monitoring procedures themselves have been rigidly determined at MIB design time.

Management Actions Must Be Accomplished Through Side-Effects of SET Requests. The agent must trap SET accesses and invoke the respective management procedures. Simple actions that require no parameters and return no values (e.g., rebooting a system) can be usefully accomplished. The SET mechanism, however, is inadequate for operations that need to pass parameters. Consider, for example, a port diagnostic routine that must obtain some parameters upon invocation and return a value upon completion. Upon invocation, the routine will require the input parameters. They must be passed by a SET command prior to invocation.

When the routine is completed, its return values could be retrieved via a GET request. The SET of input variables, the SET of invocation variables, and the GET of results must be synchronized among manager and agent. Such remote synchronization can be difficult, time consuming, and, in multimanager environments, hazardous. Multiple ongoing invocations by different managers may interfere with each other, requiring concurrency control mechanisms via SET/GET access. Simulating remote procedure calls via side-effects of a SET can, therefore, result in both significant complexity and hazards.

The MIB Structure Does Not Include Provisions to Represent Complex Relations Among Managed Entities. In the trivial PBX example, the containment relation among ports and cards requires a separate table. If one wishes to retrieve data associated with ports of a given card, the tables associated with ports, cards, and cardsports will need to be retrieved and joined at the manager. Since tables are retrieved a row at a time, even a simple attempt to correlate data among different related managed entities may result in a significant amount of polling. From the perspective of classical databases theory, SNMP's query mechanism does not facilitate correlation of entities via relationships. Such correlation must be accomplished at a management platform and, thus, requires that all data to be correlated is polled first. Managed data interpretation often requires significant processing of data correlation. Correlating data at the platform may lead to situations where unrealistic amounts are to be polled.

The Information Access Model

SNMP Does Not Support Aggregated Retrieval Facilities. Retrieval of an entire table, for example, must proceed through a row-by-row GET-NEXT polling. Management information processing typically requires access to all entries of a given table. Large tables require a large number of polling activities. Support of unlimited bulk retrieval, however, may increase the complexity of the protocol. A single query may result in a very large transfer activity. This data may exceed the capacity of a single reply frame. A protocol will need to include provisions to support-linked replies. Proposed extensions of SNMP by SNMPv.2 [13] have thus pursued a limited support of bulk retrieval whereby the size of the bulk is limited by the size of a single frame retrieval.

SNMP Does Not Provide Means to Select (Filter) the Data Retrieved. Management involves primarily the handling of exceptional conditions. A manager is often interested only in data reflecting exceptional behaviors. A retrieval mechanism should ideally permit managers to select the data of interest, rather than force them to retrieve large amounts of useless data. For example, an SNMP manager of the PBX example may only be interested in extensions whose health parameters fall under threshold. The manager is forced to retrieve the entire extension table and scan it at the platform. If the table is large (e.g., 10,000 extensions) this leads to great inefficiency in management. Data should be maximally compressed at the source rather than dragged through the network to managers. Extending SNMP to support selective retrieval means somewhat greater complexity in agents functions and in protocol interfaces.

To conclude, one should view these and other limitations of SNMP in relation to its advantages and contributions. Nor should the limitations be viewed as critical reasons to seek replacements, but, rather as areas to be addressed by complementary technologies and mechanisms. While not a panacea, the SNMP effort has accomplished significant progress toward improved manageability of heterogeneous networks. Its simplicity permitted vast sections of industry to adopt and introduce it in products and establish it as a broad market standard for manager-agent interactions.

2.3 THE OSI MANAGEMENT PARADIGM

2.3.1 Overview

The OSI management model [18] is depicted in Fig. 2-12.

In Fig. 2-12, a managing platform on the left uses the common management information protocol (CMIP) to access managed information, provided by an agent residing in a LAN hub on the right. The agent maintains a management information tree (MIT) database. The MIT models a hub using managed objects (MOs) to represent LANs, interfaces, and ports. A platform can use CMIP to create, delete, retrieve, or change MOs in the MIT, invoke actions, or receive event notifications.

The MIT contains instances of managed objects (MOs) organized on a hierarchical database tree. The MIT is, however, very different from the SNMP MIB. First, managed information is stored in both internal and leaf nodes. Second, the structure of the MIT can change dynamically (unlike the MIB) as new nodes are added or deleted. Third, instances of the same MO are stored in individual nodes of the MIT rather than combined into tables as in SNMP. The MIT organizes these instances in a containment hierarchy. For example, the MIT in Fig. 2-12 represents hub information. Multiple instances of an Ethernet MO are located under the root. Each Ethernet may contain multiple instances of interface MO, and each of these may include multiple instances of a port MO.

The MIT provides, in similarity to the SNMP MIB, a unique identification of managed objects. An MO instance includes attributes that

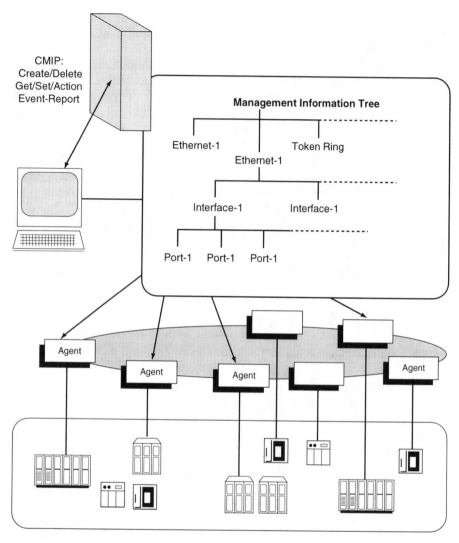

Figure 2-12   An Overall Architecture of an OSI Management System

serve as its relative distinguishing name (RDN). The RDN attributes
uniquely identify the instance among the siblings of its MIT parent. In the
hub example of Fig. 2-12, a port-ID number may be used as the RDN to
identify ports of a given interface MO. By concatenating RDNs along the
MIT path from the root to a given node, a unique distinguishing name
(DN) is obtained. This DN is used by CMIP to identify a node and access
its managed information.

Notice, however, that the DN is entirely dependent on the specific organization and contents of the MIT. The SNMP MIB path provides a static identifier, established at MIB design time, to leaves where data is stored. The only dynamic part of an identifier is the identification of table rows using data in key columns of MIB tables. SNMP access modes, accordingly, use direct GET to nontabular data whose location is statically known, and traversal of dynamic tabular data whose location is not known at MIB design time. The MIT access path, defined by a DN label, is entirely dependent on data that is unknown at design time and may even change during dynamic time as a result of creation/deletion of managed objects. As a result, the prime mode of accessing MIT is that of traversal of subtrees to locate information of interest. An MIT GET access typically results in bulk retrieval of all information of a subtree that meets some filtering criteria. Bulk and selective retrievals are (necessarily) the primary mode of MIT access.

The model of managed information pursued by OSI is also very different from that of SNMP. Managed objects include data attributes, in similarity to SNMP. Unlike SNMP, however, managed objects include actions (methods, operations) that may be explicitly invoked as part of management access. Managed objects can also provide event notifications to managers, unlike SNMP where event traps are associated with an agent rather than an object.

Management communications, too, are very different from the SNMP model. First, management communications are imbedded within the OSI application environment where they rely on OSI stack support, for example, association control service element (ACSE) and remote operations service element (ROSE) [14]. Second, management communications use a connection-oriented transport, unlike the datagram model of SNMP. Third, manager-agent communications are mostly confirmed. Fourth, constructs are provided to support synchronization of MIT transactions by multiple managers (even if this area is yet to be fully developed).

These additional features of the OSI model result in substantially greater complexity of communications interfaces, agent's structure, and the MIT, than their SNMP counterparts. In what follows we provide detailed descriptions of the different components of the OSI model and illustrate their application.

2.3.2 The Management Communication Model

OSI management communications require connection-oriented transport and rely on the OSI application layer environment (consult [14] for OSI application layer details). Agents (managed entities) and managers (managing entities) are viewed as peer applications that use the services of a Common Management Information Service Element (CMISE) to exchange managed information [18]. CMISE provides service access points (SAPs) to support controlled associations between managers and agents. Associations are used to exchange managed information queries and responses, handle event notifications, and provide remote invocations of MO operations. CMISE utilizes the services of OSI's ACSE and ROSE to support these services [14]. A typical structure of an agent communication environment is depicted in Fig. 2-13. A symmetric organization governs the structure of peer managing entities.

The top part of the figure describes the structure of the MIT. Managed objects instances and their attributes, operations, and event notifications are depicted as shaded rectangles at the top. The OSI agent provides selection functions to locate the MO records accessed by GET/SET/ACTION SAPs of CMISE. The agent also provides event detection and forwarding of notifications to managing entities enrolled (through MIT records) to receive them. A CMISE entity provides service access points (SAPs), depicted in the table below, to support communications with the agent. It dispatches/receives CMIP PDUs to/from other service elements such as ACSE and ROSE. These PDUs are exchanges through a connection-oriented transport.

Management communication services
— M-INITIALIZE: Establish management association
— M-TERMINATE: Terminate management association
— M-ABORT: unconfirmed termination
Management information tree operations
— M-CREATE   Creates an MO instance record in the MIT
— M-DELETE   Deletes an MO instance from MIT
Managed information manipulation services
— M-GET   Retrieve information
— M-CANCEL-GET   Cancel retrievals
— M-SET   Change an attribute value
— M-ACTION   Invoke an MO operation
— M-EVENT-REPORT   Generate an MO event report to a manager

The interactions pursued by management applications peers are typically confirmed through the standard OSI request-reply model. For example, an invocation by a manager of the M-INITIALIZE SAP results in a CMISE invocation of the ACSE through a CMIP PDU. The manager ACSE sends an association request to a peer. The peer ACSE at the agent passes the CMIP request to the agent's CMISE. A confirmation PDU will then propagate back from the agent's CMISE via an ACSE and back to the originating manager.

The core services of CMISE provide access to managed information. The GET construct provides a means for bulk retrieval and the agent's filtering of information. This is depicted in Fig. 2-14. To accomplish bulk

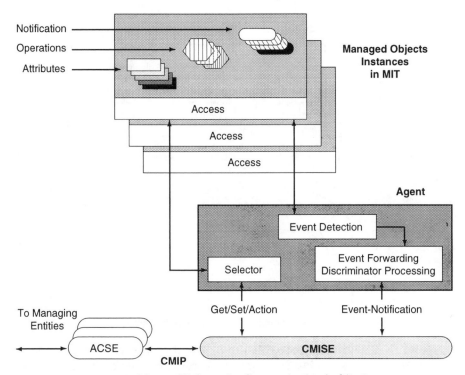

Figure 2-13 Managed Information Communication Architecture

retrieval, a GET need only specify a subtree of the MIT from which data is to be retrieved. This subtree is specified by its base node and the scope of the GET request. To specify a selection criterion a GET must provide a filter defined by a single language. Figure 2-14 illustrates retrieval of all port data on SalesNet whose error rates exceed some threshold. The GET request identifies the scope of the search and the filter and the agent performs the search and selection.

Remote invocation of operations is accomplished through M-AC-TION. It is necessary to specify the MO instance, the action to be invoked, and the parameters to be passed to it. The invocation is supported through the remote operation service element (ROSE). Event notifications are handled by enrolling appropriate records on the MIT using M-CREATE. A manager uses M-CREATE to place an event notification-managed object on the MIT. Upon detection of an event, the agent uses the MIT to identify subscribers for notifications. An M-EVENT-NOTIFICATION is generated for each such subscriber.

It should now be clear why the OSI model requires connection-oriented communications. The MIT structure, unlike the SNMP MIB, is not determined statically at design time. A software application or an

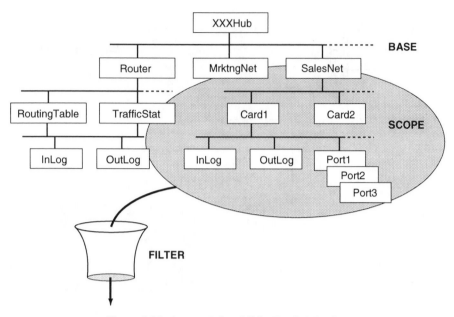

Figure 2-14   Aggregated and Selective Retrieval

operator may not know where certain information precisely located. As a result, the primary mode of information access depends on bulk/selective retrieval from subtrees. The amount of data generated through such access may exceed a single datagram. It is necessary to permit an entire stream of replies to be delivered in response to a query. To deliver such linked replies, a connection-oriented transport is required. This model of information retrieval is essentially identical to the organization of the X.500 directory services [14]. Indeed, the OSI model can be said to view the MIT as a directory service for managed objects information. A model of information retrieval that can result in a stream of responses requires a connection-oriented transport (or application-level handling of the stream).

It is useful to assess some of the difficulties implied by a connection-oriented management model. First, managers will require maintaining a connection to every agent (and vice versa—an agent must maintain a connection to every manager) in order to manage. This requirement has a serious impact on the complexity of agents and managers. Second, during stress time, when management is needed most, connections are likely to be lost. Managers and agents may spend their limited resources, at the most critical time, handling connection resetting and reestablishment rather than managing. Third, transport mappings of CMIP are rendered more difficult. In many LAN environments connection-oriented transport is rarely used by applications and does not benefit from the kind of efficient and wide support of datagram communications. Fourth, a GET retrieval may result in vast amounts of data that may not have been estimated a priori of issuing the GET; the manager may be clogged with linked replies. The CANCEL-GET command seeks to provide some remedy. One should contrast this, however, with the SNMP model where a manager is always in control of the amount of data retrieved. Platform-centered management would typically see its worst bottlenecks at the platform managers. Providing managers with control over their resources is an essential ingredient in handling the complexities of management.

To conclude this section we use the table below to summarize the CMIP primitives and the parameters passed by the respective PDUs. These PDUs are generated by CMISE entities in support of the respective SAP accesses by a manager/agent entity.

M-Create.Request: Invoke Id, MO Class, MO instance, Access control, Reference object instance, Attribute list

M-Create.Response: Invoke Id, MO Class, MO instance, Attribute list, Current time, Errors

M-Delete.Request: Invoke Id, Mode, Base object class, Base object instance, Scope, Access control, Synchronization.

M-Delete.Response: Invoke Id, Linked id, MO Class, MO instance,Current time, Errors.

M-Event-Report.Request: Invoke Id, Mode, MO Class, MO Instance, Event type, Event Time, Event Arguments

[M-Event-Report.Response: confirmation of report]

M-Get.Request: Invoke Id, Base object class, Base object instance, Scope, Filter, Access control, Synchronization, Attribute id list.

M-Get. Response: Invoke Id, MO Class, MO instance, Current time, Attribute list, Errors

M-Set.Request: Invoke Id, Mode, Base object class, Base object instance, Scope, Filter, Access control, Synchronization, Attribute list.

M-Set. Response: Invoke Id, Linked id, MO Class, MO instance, Attribute list, Current time, Errors.

M-Action.Request: Invoke Id, Mode, Base object class, Base object instance, Scope, Filter, Access control, Synchronization, Action info.

M-Action.Response: Invoke Id, Linked id, MO class, MO instance, Action type, Current time, Action Reply, Errors

2.3.3 The Structure of Managed Information Model

The structure of managed information (SMI) model plays a central role in the OSI standard. It is introduced in [19] and is elaborated in the guidelines for the definitions of managed objects (GDMO) [9]. This model is based on an extended object-oriented (OO) data model [20]. Managed Objects (MO), as OO classes, provide templates to encapsulate data and management operations (methods, actions) associated with managed entities. MO extends the class concept to include event notifications. Event notifications add a new dimension to the OO database model. Traditional OO software assumes a synchronous model of interaction between an object and its users (programs). Programs may invoke methods synchronously. Events, on the other hand, may occur independently and asynchronously with manager computations.

The MO model supports inheritance. An MO definition can include attributes, operations, and events of a more general MO. For example, a general MO describing an interface may be used to define specialized interfaces (e.g., Ethernet, Token Ring). The data, operations, and events associated with an interface MO will be inherited by these specialized subclasses. Inheritance is primarily a syntactic mechanism as one could simply include the definitions of the super-class in the subclass MO definitions to accomplish the same effect. Figure 2-15 depicts an MO defining a class of node objects. Ellipsoidal shapes describe data attributes. Rectangular shapes describe operations to test and start a node. Events are described by triangular shapes.

Consider now a specialization of a "node"—an IpSystem MO. An IpSystem can be defined as a subclass of node (Fig. 2-16). It inherits all the node attributes, operations, and events. The IpSystem may replace some of these inherited components (e.g., a new start operation) and add new attributes, operations, and events.

Relationships Are Significant in Management. Relationships among managed data items are of great importance in correlating information. In the storm example of Section 2.1, it was necessary to correlate observations of physical layer errors with those of an interface processor queue handling retransmission tasks. It would have been necessary to represent the relations among these objects to be able to correlate their behaviors. In general, traditional database models aim to provide representation of both entities and relationships to facilitate respective correlations of data.

The managed information model of OSI, in contrast to SNMP, provides explicit means to represent relationships. An MO may include relationship attributes with pointers to related MOs. For example, a port

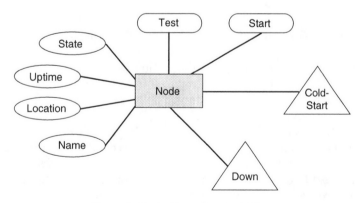

Figure 2-15 An Example of an MO

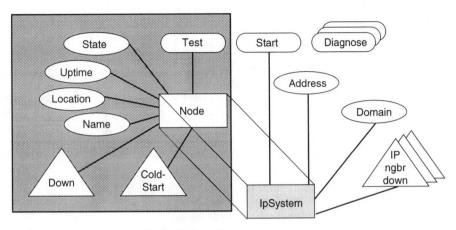

Figure 2-16 A Subclass of Node

attribute representing the relationship "contained-in" may include a pointer to the interface object that contains it. The pointer value is the Distinguishing Name (DN) path identifier of the interface object on the MIT.

Figure 2-17 depicts the OSI model of relationship. The MO boxes include relationship attribute cells. A relationship cell is identified by its "role" (e.g., the subordinate in a "contained-in" relationship). The value of a relationship attribute is a DN path identifier on the MIT, pointing to the related instance.

The OSI model includes a number of generic relationships that may be used in modeling MO such as "is-contained-in," "is-peer-of" (for protocol entities) and "is-backup-of" (for systems or components). The use of relationship attributes is similar to mechanisms used in the network model of databases [16]. It accomplishes great generality in representing, in principle, any entity-relationship model and in correlating different related data. For example, one can easily identify and retrieve information associated with all ports contained in a given interface by traversing the respective relationship pointers. The weakness of the model, however, is in the lack of mechanisms to facilitate efficient traversal of relationships. A manager will require retrieval of relationship pointers to the platform and then respective GET commands to retrieve data associated with the objects that are pointed at. This could have been simplified if CMIP included traversal primitives (GET-NEXT) to follow relationship pointers, similar to network databases.

Could SNMP MIB use a similar representation of relationship pointers? The MIB, unlike the MIT, has a static structure. If the relationship involves static objects, it would be known to the platform statically

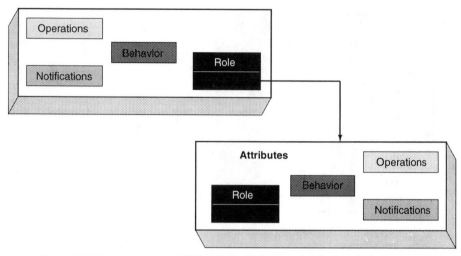

Figure 2-17   Representation of Relationship Using Role Attributes and Pointers

and require no query (and storage). Relationships among dynamic objects could, in principle, be represented by pointers to table rows of the respective instances. However, agents will need programs to maintain these pointers through changes. This requires significant changes in the agents. SNMP typically use relationship tables to accomplish a relational-database style model of relationships, however. In this case, traversal and correlations of data will require computations of the relational-join of relationship and entity tables in the MIB. Since this cannot be computed at the agent, a platform manager must retrieve and join these tables, a consuming task. In summary, while SNMP could, in theory, be used to represent and manipulate relationships, in practice this ability is limited.

Guidelines for the Definition of Managed Objects (GDMOs) Provide Syntax for MO Definitions.   The GDMO supports substantial extensions of ASN.1 to handle the syntax of managed information definitions. A new language structure, *template,* is introduced to combine definitions. Templates play a similar role as ASN.1 macro, except they do not lend themselves to simple extensions of ASN.1 compilers. The template structure is described below. Place-holders are enclosed in angular brackets, optional constructs are enclosed by rectangular brackets, and repetition of constructs is indicated via "*."

```
<template-label> TEMPLATE-NAME
CONSTRUCT-NAME[<construct-argument>];[CONSTRUCT-NAME  [<construct-
argument>];]*
[REGISTERED AS <object-identifier>];
[supporting productions  [<definition-label> -> <syntactic-definition>]*]
```

A sample template, depicted below, is the managed object class template. It is used to define MO structure and register the definitions on the ISO registration tree. It has a similar role to SNMP's OBJECT-TYPE Macro.

The <class-label> is a placeholder for an MO name. The "DERIVED-FROM" part describes the super-classes whose definitions are inherited by the MO. The "CHARACTERIZED BY" part includes the body of data attributes, operations, and event notifications encapsulated by the MO. "PACKAGES," "CONDITIONAL PACKAGES," and "PARAMETERS" are templates used to combine definitions of attributes, operations, and event notifications. The "REGISTERED AS" part registers the MO definition on the ISO registration tree.

```
<class-label> MANAGED OBJECT CLASS
[DERIVED FROM <class-label> [,<class-label>]*;]
[ALLOMORPHIC SET <class-label> [,<class-label>]*;]
[CHARACTERIZED BY  <package-label> [,<package-label>]*;]
[CONDITIONAL PACKAGES <package-label> PRESENT IF <condition-
definitions>
    [,<package-label>PRESENT IF <condition-definitions>]*;]
[PARAMETERS <parameter-label> [, <parameter-label>*;]
REGISTERED AS <object-identifier>;
```

An example of definition of an eventLogRecord [DMI], using the above template, is given below. An eventLogRecord inherits attributes of a general logRecord (its super-class). It includes definitions taken from a package for eventLogRecordPackage and a conditional-package event-TimePkg, and is registered on the OSI registration tree as the subtree labeled 5 under the label smi2MobjectClass. Attributes are followed with descriptors such as GET/REPLACE, denoting read/write access mode. Notice the informal statement of the condition under which the definitions by the conditional package are to be included. Thus, automated compilation of definitions, unlike SNMP MIBs, may be impossible.

The DMI defines a class hierarchy as depicted in Fig. 2-18. The boxes represent different MO classes, while the tree represents inheritance relations among them.

```
eventLogRecord  MANAGED  OBJECT  CLASS
DERIVED FROM  logRecord;
CHARACTERIZED BY  eventLogRecordPackage  PACKAGE
  ATTRIBUTES managedObjectClass GET, managedObjectInstance GET;;;
CONDITIONAL PACKAGES
  eventTimePkg PACKAGE ATTRIBUTES eventTime GET;;
  PRESENT IF the event time parameter was present in the CMIP event report;
REGISTERED AS  {smi2MobjectClass 5};
```

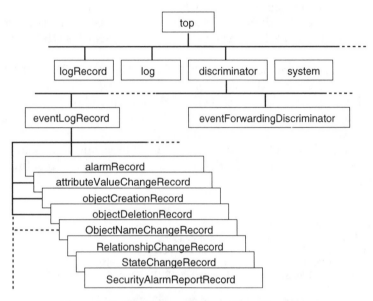

Figure 2-18 The Managed Objects Class Hierarchy

These generic MOs focus on definitions of various forms of management logs. The system MO is the main tool in building MOs associated
with a given system. It includes attributes to identify the system, represent its operational and administrative state, and provide generic notifications and packages of definitions for handling scheduled operations
maintenance. A schematic subset of its definition is provided below.

```
system  MANAGED  OBJECT  CLASS
DERIVED  FROM  top;
CHARACTERIZED BY  systemPackage  PACKAGE
        ATTRIBUTES  systemID  GET,  operationalState  GET,  usageState  GET,
        administrativeState  GET  REPLACE,  managementState  GET;
        ATTRIBUTE GROUPSstate,  relationship;
        NOTIFICATIONS  objectCreation,  objectDeletion,  objectNameChange,
        attributeValueChange,  state Change,  ......,environmentalAlarm;;;
CONDITIONAL PACKAGES
        dailyScheduling  PRESENT IF  both the weekly scheduling and
        external scheduler package  not present  in an instance
        ..........
        repairStatusPkg  PACKAGE
        ATTRIBUTES . . .
        PRESENT IF both the weekly scheduling and external
        scheduler package are not present in an instance
        ..........

REGISTERED AS {smi2MOBJECTClass 14};
```

Consider, for example, the problem of developing a class hierarchy to model typical networked systems, as a specialization of the system MO. A possible class hierarchy is illustrated in Fig. 2-19. This hierarchy considers two kinds of systems: complex systems, as on the left part of the tree, and simple systems (elements), as on the right side of the tree.

The DMI Provides a Large Number of Constructs. The structure of managed information (SMI) defined in [10] includes well over 100 definitions of managed information structures. Contrast this with the minimal set of managed information data types used by SNMP (under 10). These definitions are registered on an SMI subtree of the ISO registration tree. A section of this SMI tree is depicted in Fig. 2-20. Unlike SNMP, the OSI management model makes no use of the registration tree. Registration is merely pursued as a formal requirement of standardization. The MO definitions provide templates to model specific managed entities or their components. Instances of MOs are used to model specific managed entities, organized on the management information tree (MIT).

2.3.4 Putting It All Together: How to Build an OSI Model

This section seeks to complete the picture through a brief sketch of OSI modeling of a LAN hub. Typical hub components and their containment relations are depicted in Fig. 2.21.

Step 1: Identify Class Structure and Inheritance Relations Among Managed Objects. An MO must be designed for each managed component. The instances of these MOs will be organized on an MIT that reflects the containment relations. The first step is to identify similarities of managed elements and capture them in MO classes to take advantage of in-

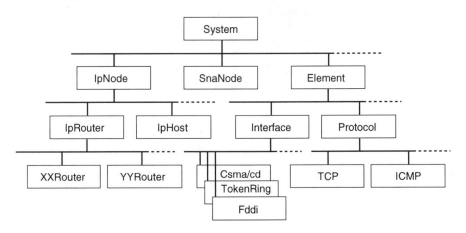

Figure 2-19   A Managed Objects Class Hierarchy to Represent Networked Devices

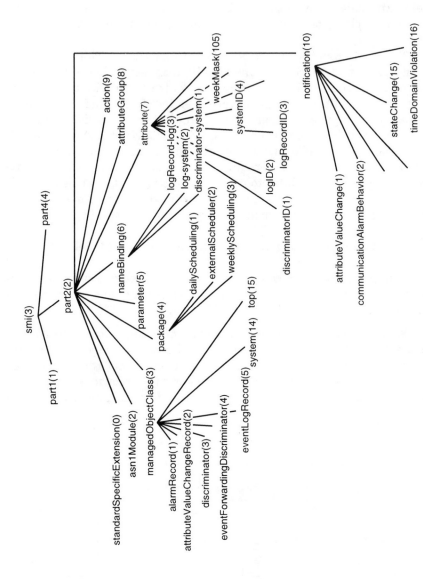

Figure 2-20 The SMI Subtree of the ISO Registration Tree

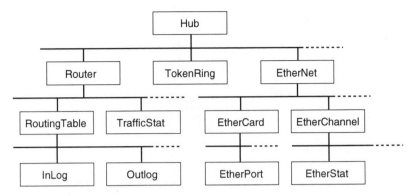

Figure 2-21 Typical Hub Components and Containment

heritance. For example, EtherPort and TokenPort components may share some attributes, operations, and events. These shared elements may be encapsulated in a generic port object. Similarly, a number of statistical logs may be generic for different types of channels. A possible partial class inheritance hierarchy is depicted in Fig 2-22.

Step 2: Design and Specify MO Syntactical Structures Using the GDMO. Using the GDMO, define managed attributes, operations, and event notifications for each of the MO classes needed. Class inheritance hierarchy is used to specialize generic MOs. An example of port MO definitions is depicted below. MO libraries defined by protocol committees (e.g., FDDI) can be used to capture standardized components.

```
port  MANAGED  OBJECT  CLASS
DERIVED  FROM  element;
CHARACTERIZED  BY  portPackage  PACKAGE
        ATTRIBUTES  portNum Get, portStatus GET,...;
        OPERATIONS  diagnose, disconnect, connect...;
        NOTIFICATIONS  portFailure, portInitialized..;
.......................................................
REGISTERED AS {......};
```

Step 3: Design a Generic MIT Structure for the Device. The design of the MIT follows the containment tree of Fig. 2-21. Each component is replaced by respective MO instances. For each MO, it is necessary to identify respective attributes forming a unique relative distinguishing name (RDN). For each device component the respective MO instance must be created on the MIT (using a CMIP CREATE primitive). Dynamic managed objects (e.g., different logs) may be created and deleted by managers during network running time. Relationship attributes values

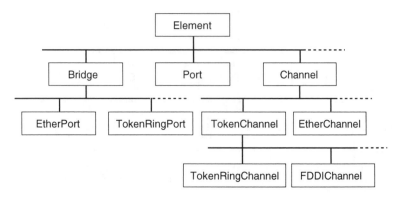

Figure 2-22   A Class Hierarchy for hub MOs

are set once the respective MO instances are located on the MIT. For example, a "contained-in" relationship may be associated with port-card pairs. An instance of an EtherPort MO may include a pointer to the EtherCard MO instance. Similarly, an EtherCard instance may point to an EtherChannel instance (subnet) to represent the relationship "attached-to."

2.3.5 Critical Assessment

The OSI model seeks to provide a comprehensive framework for handling management of arbitrarily complex systems. We devote this section to evaluate briefly some trade-offs associated with this generality and contrast the choices of OSI with those of SNMP.

The Managed Information Model. OSI provides an extended OO database framework to model managed information. It seeks to maximize the information modeling power to handle complex systems. Two central features distinguish this information model from that of SNMP. First, managed entities are modeled in terms of an extended OO model of managed entities that encapsulate data, relationships, actions, and event notifications. Of these, the primary area where OSI offers a distinct advantage over SNMP is the availability of explicit management actions. Data, relationships, and event notifications (under the improvements introduced by SNMPv.2) can be captured by MIBs similar to the OSI model. However, SNMP offers only limited invocation of agent procedures via side effects of SET. In contrast, CMIP enables remote invocation of rich and arbitrary agent procedures and parameter passing.

Abstraction and inheritance of the OO model can provide important benefits in developing effective management applications. For example, consider a troubleshooting application that analyzes a given network segment to determine problems. A failure of the segment will reflect in each node attached to it, regardless of the specific features of the node (e.g., workstation, bridge, server, printer, router). It is useful for such application to view these various heterogeneous components at a level of abstraction that captures their commonalty (e.g., nodes attached to a segment). This may be accomplished by defining an appropriate MO class to capture "node" objects and including an attribute to represent the "connected-to" relationship. An applicaiton designer could develop a troubleshooting application that uses only the properties of "node" of the various objects and ignores their details. With the increasing heterogeneity of managed elements and the level of details captured by MIB, such abstraction can be of great value in simplifying the development and deployment of management applications. It is interesting to note, however, that early GDMO defined by a number of standard committees often accomplish very minimal abstraction and utilize a shallow inheritance (1–2 levels deep). This raises the question as to the true value of selecting an OO modeling framework or of its application by committees. Either way, the practical value of the OO modeling framework is yet to be established.

Second, the MIT offers a dynamic agent database that can be changed by CMIP CREATE/DELETE accesses. A dynamic database can provide flexibility and efficiency in managed information access. Managing entities can control the contents and structure of the database and locate only information of interest there. The database may also be flexibly organized to reflect specific device configurations. A multiprotocol hub agent, for example, can dynamically include within a single MIT any information associated with new cards plugged in its chassis. This contrasts with the rigidity of static SNMP MIB.

A dynamic management database, however, presents significant implementation complexities. The resources required to store and process managed information cannot be predicted at design time. Managers may extend the MIT beyond available agent's resources. Changes in the MIT may result in corruption of the database. Deletion of MO records may result in orphaned relationship pointers, requiring complex garbage collection at agents. Multiple managers pursuing CREATE/DELETE activities can lead to inconsistent views of the MIT. Application software designers cannot share a single model of the MIT contents because each application must build and maintain its own MIT subset. Additionally, failure recovery is very difficult. Failure may cause the MIT contents to be lost. Upon booting the MIT will need to be reconstructed. Agent-created MIT entries may be rebuilt with ease whereas manager-created entries require complex interactions with managers to recover.

Additionally, the OSI model provides explicit representation of relationships among managed data items, affording it the full power to represent (in principle) any entity-relationship model. Relationships are of central significance in correlating related data. For example, recognizing which port at the physical layer serves a given connection at the network layer can be important in correlating symptoms detected by the layer management entities. The OSI model utilizes relationship pointers similar to those of the network database model. This means that agents must maintain these pointers through changes in the MIT. Additionally, no means are provided to traverse relationships at the agent. As a result, a platform must perform such traversal by retrieving the pointers and using them to guide its GET command. This renders the use of pointers slow and expensive (in polling). SNMP, in contrast, does not provide explicit representation of relations. MIB tables could be used to store relationships but correlating data would then require retrieval of entire tables and computing their relational join at the platform, a task that is too expensive (in polling).

The MIT provides, in effect, a general-purpose powerful OO database that extends traditional class structures by adding asynchronous event notifications. The cost of such generality, however, is in implementation complexities, substantial resource requirements, and decreased performance. Management information bases, however, need to balance conflicting requirements for functionality and real-time performance under resource constraints. The performance of OO databases, in general, is yet to be understood [20]. It is yet to be established whether the OSI design choices can strike such balance between generality and real-time performance.

The Managed Information Access Model. The OSI model introduces two important functions missing in SNMP: bulk retrieval and selective retrieval. Both capabilities are central to controlling the flow of management information. In the absence of bulk retrieval, SNMP managers are forced to perform a large number of polling requests to access managed information. In the absence of selective (filtered) retrieval, SNMP managers are forced to retrieve large amounts of data of no relevance. SNMPv.2 [13] recently introduced bulk retrieval capability. An SNMPv.2 GET-BULK request can retrieve a section of a table that fits within a single UDP frame. To retrieve an entire large table, multiple successive such GET-BULK requests must be issued. SNMPv.2 conducts its own transport of large tables using a window of one UDP frame at a time. In contrast, the entire retrieval can be handled via linked replies of a single GET request of CMIP. This gives CMIP a slight potential advantage in bulk retrievals, as its transport layer could pass large amounts of data more efficiently.

Neither SNMP nor SNMPv.2 include provisions for selective retrievals. As a result, managers must retrieve entire tables and filter the data at the platform. In the PBX example, the entire extension table may need to be retrieved periodically just to decide whether or not there are unhealthy extensions.

The Communication Model. The OSI communication model uses connection-oriented transport and relies on components of an OSI application layer environment, even though implementations are possible over other connection-based transport such as TCP [21]. The disadvantages of connection-oriented management communications may outweigh their value. Management interactions are needed most during the worst network stress times. During stress time, however, transport connections may be difficult to maintain. Management entities may spend significant time and resources in resetting and reestablishing lost connections, as connections may not be sustainable over a sufficiently long period to accomplish the management functions needed. Thus, connection-based transport may become an obstacle to accomplishing management interactions. In contrast, a communication model that assumes datagram transport can proceed to handle management functions even when the network is highly stressed.

2.4 CONCLUSIONS

The goal of this section is to provide a concluding critical assessment of both standards.

2.4.1 Managed Information Modeling

Platform-centered management requires a distributed database model of managed information. The SNMP approach has been to provide a minimal such database model. The limitations of the model were considered in previous sections. Some of these limitations are addressed by recent RMON extensions and by the successor, SNMPv.2. OSI designers, on the other hand, pursued a maximal database model. A minimal database model may be simple to implement and provide a controllable real time performance, but it lacks sufficient expressive power to model complex systems and relationships. A maximal database model may provide sufficient expressive power, but results in complex and costly implementations and intractable performance behavior.

A maximal database model raises an interesting question concerning the allocation of management responsibilities among the platform and agents. If agents are sufficiently resourceful to manage a complex database of managed information, they could use the resources to execute applications to interpret the data. Furthermore, if the primary goal of a platform is to provide a console access to device-managed data and applications, a maximal database model of agents should merely offer the capabilities to open a console remotely and attach it to the agent. The interactions of platform and agent could be reduced to handle display via one or another respective standards. If, on the other hand, the bulk of processing of managed data is to be conducted at the platform, why would an agent require a maximal database model? The proper balance between the roles of agents and platforms in management is yet to be understood. The simple agent model pursued by SNMP reflected the needs of devices of limited complexity and constrained management resources. It is unclear how useful this choice is, given the rapid changes in device and network complexity as well as their management processing resources. An agent may be often equipped with substantially more management processing resources than its respective share of platform resources. These ongoing rapid changes in complexity and availability of processing resources requires reconsideration of the allocation of functions among platforms and agents and, perhaps, new management models and protocols to support them.

It is interesting to note that both CMIP and SNMP designs chose to ignore the possibility of using a relational database model as a candidate. The relational model [16] offers well understood and standardized database structures. Relational databases have been used in the management of large telecommunication networks and computer systems. A minimal variant of structured query language (SQL) could have been used for a query protocol. Management platforms often utilize relational databases to store polled management information. Applications accessing these databases use SQL to query management information. A relational management model could permit these applications to access managed information, regardless of its location, at the platform or at device agents. This discussion does not seek to advocate SQL as a management protocol, only to indicate its a priori plausible candidacy.

2.4.2 The Curse of Heterogeneity

Both protocol standards provide a unified syntactic framework for organizing and accessing management information. Recall the problems of platform heterogeneity and of managed information semantics heterogeneity discussed in Section 2.1. Traditional software found it simpler to standardize program interfaces rather than shared global variables. SNMP MIBs may be seen as variables shared between internal agent programs and external management platform programs. A single platform program, furthermore, must access multiple MIBs of different devices. The semantic heterogeneity of these MIBs presents an obstacle in the development of such applications. CMIP managed objects, on the other hand, provide both data attributes as well as procedure interfaces. By shifting management interfaces to procedures rather than to variables, greater control of heterogeneity may be exercised. Managed objects may be designed to maximally hide information from management platforms applications by focusing on procedural interfaces. Platform applications will then view managed objects primarily as procedure libraries, not unlike the local platform libraries that they invoke.

Standard application program interfaces (API) can be developed to managed object procedures. Management applications could then be developed to depend only on the standardized managed object API, thereby resolving both the semantic and platform heterogeneity problems. CMIP thus provides a useful path to resolve the curse of heterogeneity.

The growth of multiple management standards seeds future problems of heterogeneity. While SNMP seems to have built its following within the LAN interconnect segment of the computing/communications industry, CMIP is building its following within the WAN and traditional telecommunication vendors segments. Within the computing industry, system management often uses private management systems and database servers often utilize relational databases to record management information. Throughout the decade, the boundaries between different segments and technologies will vanish rapidly, as accelerated integration of communication media and of computing and communication processes takes place. The availability of common standards to manage such integrated systems is of central significance to enabling these developments. It is likely that multiple standards will emerge and continue to coexist within various systems. Significant effort is required to find effective ways to bridge the heterogeneous management standards.

2.4.3 How Good Is Platform-Centered Management?

In this section we consider the very platform-centered management paradigm assumed by both standards.

Platform-centered management is unscalable. The rates at which device data must be accessed and processed typically exceed the network/platform capacity. For example, a platform managing 100 devices, each providing 100 managed variables sampled every 10 seconds, must execute 1000 data writes per second, well beyond the typical workstation's capacity. If the size (number of devices), complexity (number of managed variables), or speed of the network increases, the system becomes unmanageable. During failure times the platform will need to increase its interactions with devices, at a time when the network is least capable to handle these. Management response time, furthermore, will tend to stretch at a time when fast response is most needed.

Platform centered management leads to an unrealistic level of platform-agent interactions (micromanagement problem). For example, sup-

pose a platform needs to run a simple management program to handle
port failures, whose skeleton is described by:

```
{Upon   symptom_event
begin
    run diagnostic_test
    For i=1, num_ports Do if port_failure[i] then partition_port (i)
end}
```

The management events (symptom_event), procedures (diagnos-
tic_test, partition_port), and variables (port_failure), are all associated
with the device environment. The program itself, however, must reside at
the management platform. The symptom_event, describing potential port
failures, may either be built-in the agent or be detected by polling variables. Invocation of diagnostic_test, checking port_failure variables, and
invoking partition_port require a significant number of manager-agent
interactions. These interactions, furthermore, must be carefully
synchronized and confirmed to perform the procedure. Management platform applications are forced to micromanage agents through primitive
activities. This limits management processing of nontrivial tasks.

Platform centered management leads to barriers on development of
management applications due to semantic and platform heterogeneity.
Heterogeneity of platforms requires multiple versions of management applications to be developed. Semantic heterogeneity leads to restrictions on
devices that can be managed by a given application.

Platform-centered management derives from needs and constraints
of traditional networks. Device processing resources were limited and,
thus, management processing needed to be concentrated at platforms.
Organizations responsible to networks could, furthermore, afford the
costs of running management centers. Current and emerging networks
involve new tradeoffs and needs. Devices often involve substantial
processing resources. Some multiprotocol hubs and routers, for example,
include powerful RISC processing management platform workstations.
The value of moving management data from these devices to be processed
at a centralized workstation is unclear. At the same time, enterprises
often cannot afford the steep costs of running management centers; they
expect the network devices to maximally manage themselves.

Alternative management paradigms are needed to reflect the needs
and opportunities of emerging networks and resolve the limitations of
platform-centered management. Management should pursue flexible
decentralization of responsibilities to devices and maximal automation of
management functions through application software. Research toward
such distributed management is described in [22]. A management by
delegation (MBD) paradigm is used to distribute management applica-

tions to device agents dynamically. Management application programs are delegated by platforms to device agents that execute them under remote platform control. For example, the entire management program to handle the symptom_event (above) could be delegated to the device and executed there without platform intervention. MBD permits platforms to flexibly assign management responsibilities to devices and even program devices to perform autonomous management. Additional research efforts to develop distributed management are described in [3], [4], and [5].

In conclusion, neither platform-centered management nor its proposed standardization should be viewed as panacea. Network management needs and scenarios are likely to continue and change as new network technologies, new applications, new needs, and better management technologies arise throughout the coming decade. Significant research is needed to develop effective management technologies and to improve the standards to address emerging needs and opportunities. Standardization is likely to continue and evolve. The evolution of SNMP through incremental extensions and improvements provides a useful model for how standards can adapt to changes.

2.5 ACKNOWLEDGMENTS

The author would like to thank Dr. Ken Lutz of Bellcore for his careful review of the manuscript and suggestions for improvements, as well as Dr. S. Yemini, Ms. I. Levy, Mr. A. Dupuy, and Dr. S. Kliger from Systems Management Arts (SMArts) for their commentary. Research support by the National Science Foundation (NSF) under project number NCR-91-06127 is gratefully acknowledged.

REFERENCES

[1]. U. Black, *Network Management Standards The OSI, SNMP and CMOL Protocols.* New York: McGraw Hill, 1992.

[2]. A. Leinwand and K. Fang, *Network Management: A Practical Perspective.* New York: Addison-Wesley, 1993.

[3]. B. N. Meandzija and J. Westcott, presented at the First IFIP Int. Symp. on Integrated Network Management, Boston, MA, May 1989.

[4]. I. Krishnan and W. Zimmer, presented at the Second IFIP Int. Symp. on Integrated Network Management, Washington, DC, April 1991.

[5]. H-G Hegering and Y. Yemini, presented at the Third IFIP Int. Symp. on Integrated Network Management, San Francisco, CA, April 1993.

[6]. A. Kershenbaum, M. Malek, and M. Wall, *Network Management and Control Workshop,* Tarrytown, New York: Plenum Press, September 1989.

[7]. M. Rose and K. McCloghrie, "Structure and Identification of Management Information for TCP/IP-based Internets," RFC 1155, May 1990.

[8]. S. Waldbusser, "Remote Network Monitoring Management Information Base," RFC 1271, Nov. 1991.

[9]. OSI, I.S.O., 10165-4 Information Technology, Open Systems Interconnection, Guidelines for the Definitions of Managed Objects, 1991.

[10]. OSI, I.S.O., 10165-2 Information Technology, Open Systems Interconnection, Definition of Management Information, 1991.

[11]. K. McCloghrie and M. Rose, "Management Information Base for Network Management of TCP/IP-based Internets: MIB-II," RFC 1213, March 1991.

[12]. J. D. Case, M. S. Fedor, M. L. Schoffstall, and J. R. Davin, "A Simple Network Management Protocol (SNMP)," RFC 1157, May 1990.

[13]. J. D. Case, M. T. Rose, K. McCloghrie, and S. L. Waldbusser, "Introduction to the Simple Management Protocol (SMP) framework, draft," July 1992.

[14]. M. T. Rose, *The Open Book, A Practical Perspective on OSI,* Englewood Cliffs, NJ: Prentice Hall, 1990.

[15]. M. T. Rose, *The Simple Book: An Introduction to Management of TCP/IP-based Internets.* Englewood Cliffs, NJ: Prentice Hall, 1991.

[16]. J. Ullman, *Principles of Database & Knowledge Base Systems,* Vol I & II, Computer Science Press, 3rd ed, 1988.

[17]. A. Depuy, S. Sengupta, O. Wolfson, and Y. Yemini, "NetMate: A Network Management Environment," *IEEE Network* (special issue devoted to network operations and management), 1991.

[18]. OSI, I.S.O., 9595 Information Technology, Open Systems Interconnection, Common Management Information Services Definitions, 1991.

[19]. OSI, I.S.O., 10165-1 Information Technology, Open Systems Interconnection, Management Information Model, 1991.

[20]. E. Horowitz (Ed.), *Object-Oriented Databases and Applications,* Englewood Cliffs, NJ: Prentice Hall, 1989.

[21]. K. McCloghrie and M. Rose, "Common Management Information Services and Protocol Over TCP/IP (CMOT)," RFC 1189, March 1991.

[22]. Y. Yemini, G. Goldszmidt, and S. Yemini, "Network Management by Delegation," in The Second International Symposium on Integrated Network Management, Washington, DC, April 1991.

[23]. J. Galvin, K. McCloghrie, and J. Davin, "SNMP Security Protocols," RFC 1352, *MIT Laboratory for Computer Science,* July 1992.

[24]. OSI, I.S.O., 10040 Systems Management Overview, 1991.

[25]. OSI, I.S.O., 9596 Information Technology, Open Systems Interconnection, Common Management Information Protocol Specification, 1991.

CHAPTER 3

Telecommunications Management Network
Principles, Models, and Applications

Veli Sahin
NEC America, Inc.
Irving, TX 75038

3.1 INTRODUCTION

Telecommunications networks (TNs) are becoming more intelligent, distributed, and larger every day. Through the 1990s, it is expected that telecommunication networks will evolve toward software-based networks with the deployment of thousands of intelligent network elements (NEs) to support a wide range of information networking services, to generate new revenues, and to reduce network operations and management cost. It will be impossible to operate and manage these intelligent networks without the appropriate infrastructure in place. A management network with standard protocols, interfaces, and architectures is called a *telecommunications management network* (TMN) by the International Telecommunications Union—Telecommunications (ITU-T—formerly CITT) [1],[2] and it will enable the appropriate management infrastructure for the future networks.

3.2 SCOPE

The TMN provides a host of management functions and communications for operation, administration, and maintenance (OAM) of a telecommunications network and its services in multivendor environments.

TMN functions can be grouped into two classes—basic functions (BFs) and enhanced functions (EFs). BFs will be used as building blocks to implement EFs such as service management, network restoration, customer control/reconfiguration, bandwidth management, etc. TMN basic functions can be further grouped into three categories.

A. Management functions
   - Configuration management
     —provisioning
     —status
     —installation
     —initialization
     —inventory
     —backup and restoration
   - Fault management
     —alarm surveillance (analyze, filter, correlate)
     —fault localization
     —testing
   - Performance management
     —data collection
     —data filtering
     —trend analysis
   - Accounting management
     —collect accounting (billing) data
     —process and modify accounting records
   - Security management
     —provide secure access to NE functions/capabilities
     —provide secure access to TMN components [e.g., operations systems (OSs), subnetwork controllers (SNC), mediation device (MD), etc.]
B. Communications function
   - OS/OS communications
   - OS/NE communications
   - NE/NE communications
   - OS/workstation (WS) communications
   - NE/WS communications
C. Planning function
   This function is not specified in the ITU-T Recommendations M.30 and M.3010. It is, however, an important network management function.
   - network planning [including physical resource (facility, equipment, etc.) planning]
   - workforce planning

This chapter discusses TMN models, architectures, standard interfaces, communications function, and distribution of TMN functions. TMN
information model/architecture and details of TMN management functions can be found in other chapters of this book.

A TMN can be a very simple network connecting an OS to a single
NE or it can be a very large network interconnecting many different OSs,
NEs, and WSs. Figure 3-1 illustrates the relationship of a TMN to the
telecommunication networks that it manages.

Note that functionally, the TMN is a separate network that enables
management of a telecommunication network. However, it may use parts
of the telecommunications network to provide TMN communications [e.g.,
by using embedded operations channels (EOCs) in digital signals]. In
other words, some parts of the TMN can be an embedded logical network
within a telecommunications network.

3.3 BACKGROUND

The TMN project was started by the ITU-T in the fall of 1985. The ITU-T
M.30 includes the work of ITU-T Study Groups (SGs) IV, XI, and XV
during the 1985–1988 study period [1]. The industry and many standards
organizations (e.g., ITU-T, ANSI T1, etc.) all over the world are still working on TMN. The TMN architecture (functional/physical) models, definitions, and functions specified in the M.3xxx series might be modified.
Therefore, it is suggested that readers refer to latest ITU-T and ANSI T1
standards to get most current information [1],[2].

3.4 TMN MODELS, FUNCTIONS, AND INTERFACES

In this section, we discuss TMN definitions, models, functions, and standard interfaces as specified in the ITU-T M.3xxx [2].

3.5 TMN FUNCTIONAL ARCHITECTURE MODEL

TMN functionality provides the means to transport and process information related to the management of telecommunications networks and services. It is made up of operations systems function (OSF) blocks,
mediation function (MF) blocks, and data communication function (DCF)
blocks that are contained within the TMN, as shown in Fig. 3-2. The TMN
also contains that part of the network element function (NEF) blocks,
workstation function (WSF) blocks, and Q adaptor function (QAF) blocks
that support TMN management functions.

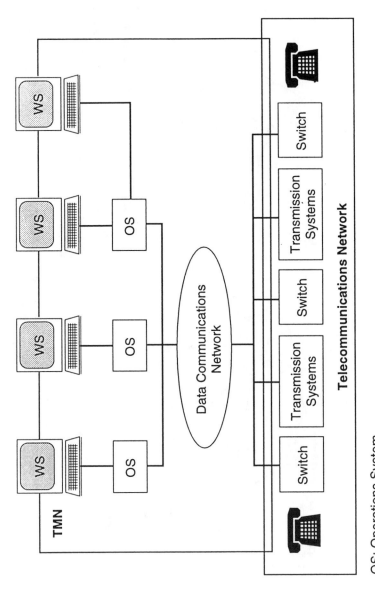

OS: Operations System
WS: Workstation

Figure 3-1 Relationship of TMN to Telecommunications Network

75

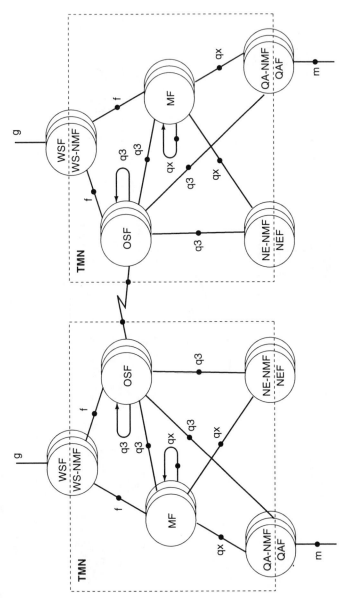

Figure 3-2 TMN Function Blocks and Reference Points

3.5.1 Operations Systems Function Blocks

The TMN management and planning functions defined in Section 3.2 are provided by OSFs. Many types of OSFs are needed to manage and plan today's telecommunications networks and services. Currently, the industry and standard organizations are working to specify the number of different OSFs and their implementations by physical systems. According to ITU-T M.3010, there are four OSF blocks to support the business management layer (BML), service management layer (SML), network management layer (NML), and network element management layer (NEML). The NEML is also called subnetwork management layer (SNML) or simply, element management layer (EML) [3]. Additional details on this subject can be found in Section 3.15 of this chapter.

3.5.2 Mediation Function Blocks

The TMN mediation function (MF) allows TMN function blocks to communicate with each other when they have different reference points or interfaces. In other words, the TMN MF block provides a set of gateway and/or relay functions. If we accept the MF definition in M.30/M.3010, then the MF block becomes a very confusing concept in the TMN. Since the MF definition in M.30/3010 includes information processing and information transport functions, the boundary between OSF and MF blocks becomes unclear. Some examples of MFs are:

- Information transport functions (ITFs)
    - —protocol conversion
    - —message conversion
    - —signal conversion
    - —address mapping/translation
    - —routing
    - —concentration
- Information processing functions (IPFs)
    - —execution
    - —screening
    - —storage
    - —filtering

3.5.3 Network Element–Network Management Function Blocks

Today's NEs can be viewed as networks of microprocessors and/or minicomputers. The intelligence of NEs has increased exponentially in the last decade. Therefore, many OSFs and MFs were integrated or are being integrated into NEs. These include switches, digital cross-connect systems (DCSs), add-drop multiplexers (ADM), multiplexers, digital loop carriers (DCLs), etc. Some examples of network elements network management (NE–NMFs) are:

- protocol conversion
- address mapping
- message conversion
- routing
- data collection and storage (e.g., performance, accounting, alarm, status)
- data backup
- self-healing
- self-testing
- self-fault localization
- NE level alarm analysis
- operations data transport via EOCs

3.5.4 Workstation Network Management Function

The workstation network management function (WS NMF) provides the means to interpret management information for the users by translating management information from "F" interface formats to "G" interface formats.

3.5.5 Q Adaptor Network Management Function Blocks

The Q adaptor network management function QA NMF provides translation/conversion between a TMN reference point and a non-TMN reference point (see Section 3.6.5 for the definition of m reference points).

3.5.6 Data Communications Function Block

The data communication function (DCF) will be used by the TMN communication function to exchange information among TMN function blocks. The prime role of DCF is to provide information transport for the OS/OS, OS/NE, NE/NE, WS/OS, and WS/NE communications. The DCF may therefore be supported by the bearer capability of different types of subnetworks. These may include point-to-point links, local area networks

(LANs), wide area networks (WANs), embedded operations channels (EOCs) in digital signals, etc.

3.6 REFERENCE POINTS

A reference point is a conceptual point of information exchange between nonoverlapping function blocks, as illustrated in Fig. 3.2. A reference point becomes an interface when the connected function blocks are embodied in separate physical pieces of equipment.

3.6.1 q Reference Points

The q reference points connect the TMN OSF, MF, NEF, and QAF function blocks with each other either directly or via the DCF. Within the class of q reference points, the q3 reference points connect NEF to OSF, MF to OSF, QAF to OSF, and OSF to OSF. The qx reference points connect MF to MF, MF to NEF, and MF to QAF.

3.6.2 f Reference Points

The f reference points connect OSF and MF function blocks to the WSF function block.

3.6.3 x Reference Points

The x reference points connect the OSF function blocks in different TMNs or between an OSF of a TMN and the equivalent of an OSF in a non-TMN environment.

3.6.4 g Reference Points

The g reference points are not considered to be part of the TMN even though it conveys TMN information. The non-TMN g reference points are located outside the TMN between the WSF and the human users.

3.6.5 m Reference Points

The m reference points are also located outside the TMN between the QAF and non-TMN managed entities or managed entities that do not conform to TMN recommendations. This will enable the management of non-TMN NEs through the TMN environment.

3.7 TMN PHYSICAL ARCHITECTURE AND INTERFACES

3.7.1 Physical Architecture and Its Components

TMN functions can be implemented in a variety of physical configurations. Figure 3-3 shows a generalized physical architecture for a TMN. A TMN physical architecture provides a means to transport and process information related to the management of telecommunication networks. As shown in Fig. 3-3, a physical architecture is made up of the following physical components:

1. Operations systems (OSs)
2. A data communications network (DCN)
3. Mediation devices (MDs)
4. Workstations (WSs)
5. Network elements (NEs)
6. Q adaptors (QAs)

It should be recognized that because of the complex nature of some telecommunication equipment, the increased use of microprocessors, and

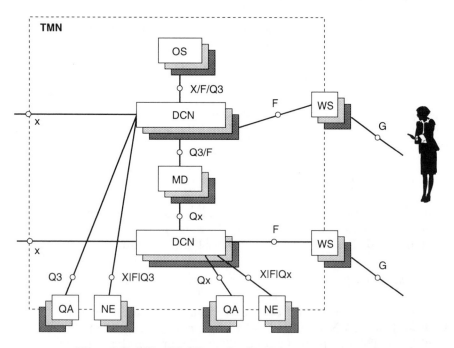

Figure 3-3 A Simplified Example of TMN Physical Architecture

the potential ability to distribute functionality within various network parts, these definitions may not completely cover every possible physical configuration that may be encountered. Depending upon the particular TMN implementation, MDs and/or QAs may not be included. Also, DCNs may be simple point-to-point connections or a more complex transport network such as a packet switched network. As shown in Fig. 3-3, a reference point becomes an interface (Q, F, G, X) when the connected function blocks are embodied in separate pieces of equipment.

3.7.2 TMN Interfaces

The interconnections of NEs, OSs, WSs, QAs, and MDs via a DCN are provided by standardized interfaces. These interfaces ensure interoperability of interconnected systems to accomplish a given TMN management/planning function. This requires compatible communication protocols, including compatible generic message definitions for all the TMN management and planning functions, independent of the type of device or of the supplier.

Two classes of Q interfaces (Qx and Q3) are defined in Fig. 3-3. These two classes are ordered in terms of the number of communications services that are provided and the variety and complexity of the TMN applications that can be supported over each interface. A Q3 interface will generally provide more complex services and protocols and will carry more functions for each NE.

Since TMN interfaces are based on the concepts of the OSI reference model (see ITU-T.X 200), the communications protocols specified for these interfaces shall conform to the ITU-T OSI recommendations of the X.200 series, unless economic considerations do not permit this. However, the protocol stacks (or suites) shall be capable of supporting an interoperable interface. The protocol stacks to be applied to the Q interfaces do not need to include all protocols of the OSI reference model. In such cases, the protocols shall be layered, with the services in each implemented layer equivalent to the corresponding OSI layer.

Qx Interface. A Qx interface that supports a small set of functions by using a simple protocol stack is appropriate for network elements that require few OAM functions and are used in large numbers. One appropriate application of the Qx interface with a simple protocol stack would be to carry bidirectional information flow related to simple events such as changes in alarm states in logic circuits, resets of alarms, and control of loopbacks. Protocols selected for such an application require some services from at least Layers 1 and 2 of the OSI reference model.

Qx interfaces with a complex protocol stack are required to support a larger set of OAM functions and require additional protocol services from Layer 3 or higher, up to the full 7-layer stack (e.g., the same protocol stack as Q3) to support the set. Qx interfaces with a complex protocol stack are generally used by the more complex NEs and MDs, which are required to support many OAM functions.

Q3 Interface. The Q3 interface supports the most complex set of functions and requires many protocol services to support the set. The protocol requirements for each set of OAM functions should be supported with protocol selections for Layers 1 to 7, as defined by the OSI reference model (ITU-T recommendation X.200). Nulling of individual service options may be necessary for economic and/or performance reasons. Currently, standardized specifications of protocols for OS to NE interfaces are given in ANSI T1.204-1989 and T1-208-1989.

X Interface. The X interface supports the set of OS to OS functions between TMNs or between a TMN and any other type of management network and requires many protocol services to support this set. The protocol requirements for each of the OAM functions should be supported with protocol selections for Layers 1 to 7 that are being defined in T1.217. The messages and protocol defined for the X interface may be suitable for use at the Q3 interface between OSs.

F Interface. The F interface supports the set of functions for connecting workstations to physical components containing the OSF or MF through a data communication network.

3.7.3 Interoperable Interfaces

In order for two or more TMN building blocks to exchange management information, they must be connected by a communications path and each element must support the same interface onto that communications path. It is useful to use the concept of an interoperable interface to simplify the communications problems arising from a multivendor environment. The connectivity of a TMN is confined by the message and information flow requirements. The interoperable interface defines the protocol stack and the message carried by the protocol stack. It is based on an object-oriented view of the communication. Therefore, all of the messages carried deal with object manipulations. The interoperable interface is the formally defined set of protocols, procedures, message formats, and semantics used for the management communication.

The message component of the interoperable interface provides a generalized mechanism for managing the objects defined for the information model. As part of the definition of each object, there is a list of the type of management operations that are valid for the object. In addition, these are generic messages that can be used identically for many classes of managed objects.

The scope of the management activity that the communication at the interface must support distinguishes one interface from another. This common understanding of the scope of operation is termed shared management knowledge (SMK). The SMK ensures that each end of the interface understands the exact meaning of a message sent by the other end.

3.8 TMN DESIGN STRATEGIES AND RELATED DEFINITIONS

In the past, dedicated and/or overlaid network architectures, NEs, and OAM plans were used to offer services. Furthermore, each service had different OAM plans for the loop and interoffice networks. Therefore, it was not practical to share network resources among operations functions as well as between operations and other applications such as services. These dedicated design approaches created traditional technical boundaries in all areas of telecommunications, such as transmission and switching, loop (access) and interoffice, switched and nonswitched services, and circuit-switched and packet-switched services. With the introduction of the digital NEs and transmission facilities in the access and interoffice networks, the access and interoffice networks were treated as two disjoint networks with respect to signal formats, signaling, and OAM plans. The main distribution frame (MDF) continued to serve as a convenient boundary that decoupled the access and interoffice networks.

Today, the introduction of digital cross-connect systems (DCSs) and local digital switches (LDSs) with digital transmission facility interfaces at the line and trunk sides in central offices is forcing the traditional boundaries to disappear. This digital integration of the digital loop and interoffice networks brings new opportunities to all areas of telecommunications, including network management (NM). One of these opportunities in the NM area is the sharing of digital facilities and intelligent NEs between network management and services. For example, portions of the digital transmission facilities and NEs can be used to provide cost-effective communications paths for the NM data collection and transport.

3.9 TMN DESIGN STRATEGY

A TMN design strategy greatly depends on the topology and capabilities of telecommunications networks (TNs) that the TMN must manage. In other words, we must first understand current TN architectures, TN architecture evolution, and the capabilities of NEs that make up a TN.

The traditional approach to provide communications between an NE and its supporting OSs is to design a standalone (i.e., dedicated) TMN. With modern intelligent NEs, that standalone approach may not be the most economical solution because there are many NEs distributed in a large geographical area. The strategy for designing a cost-effective TMN is to use the existing TN to also carry network management traffic as much as possible. When the TMN and TN physically overlap, network survivability becomes an important issue in designing the TMN. Diverse routes, diverse subnetworks, and alternate routing will be used to solve the network survivability problem. Today, this is possible because of synchronous optical network (SONET) architectures (e.g., self-healing rings) and SONET NEs [4]–[6]. Evolution to twenty-first century TN architectures is depicted in Fig. 3-4. Today's asynchronous networks at the digital signal 3 (DS3) and higher rates will migrate to synchronous networks in a few years, with the deployment of SONET products. In the next three to five years, asynchronous transfer mode (ATM) based products will be deployed to support broadband integrated services digital network (B-ISDN) services. Finally, at the end of the 1990s, TN architectures will migrate to information networking architecture (INA), which is discussed in [7].

The following are TMN design criteria and key features for twenty-first century networks:

- faster deployment of new services and technologies
- support multivendor environment with open architectures and standard interfaces
- reusable/portable software
- programmable platforms

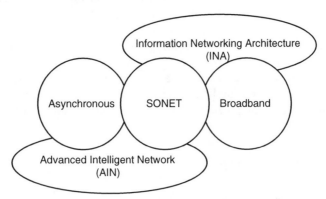

Figure 3-4 Evolution to the 21st Century Networks

- lower initial deployment cost
- lower life-cycle cost
- survivability
- distributed processing
- object-oriented design
- OSI 7-layer protocol stack and common management information service element (CMISE) messages
- modular functionality

Before we cover TMN design, we first must discuss and understand the capabilities of NEs and digital signal formats that can be used to design a TMN and/or to support TMN applications.

3.10 DEFINITIONS AND BUILDING BLOCKS

3.10.1 Embedded Communications Channels

Existing and proposed standard digital signals consist of service- and nonservice-carrying channels. Service channels are used to offer services whereas the remaining, or nonservice channels are not. The embedded communications channels (ECCs) are the nonservice-carrying channels in digital signals, as shown in Fig. 3-5 [8]–[11]. The word *embedded* indicates that the ECCs are included in digital signals and that the same physical transmission facility carries both the ECCs and service channels.

This section will focus on the parts of the ECCs that are used to support the network management applications. The ECCs are provided by using either the existing overhead housekeeping bits or reallocated bits in digital signals. The restructuring of the old DS1 and DS3 signal formats frees up capacity that can be used as ECCs for new applications. The proposed SONET and ISDN signal formats also include extra capacities (reallocated) to be used as ECCs. Note that a channel (e.g., DS0 time slot) in a digital signal can be used as an ECC when needed in the telecommunication networks.

3.10.2 Embedded Operations Channels

Embedded operations channels (EOCs) are those parts of the ECCs that are used for the TMN functions/applications. For the scope of this document, it is necessary to separate the EOCs from the remaining portion of the ECCs. Potential applications of EOCs are discussed in detail in Section 3.11. The proposed SONET, ISDN, synchronous DS3 transmission

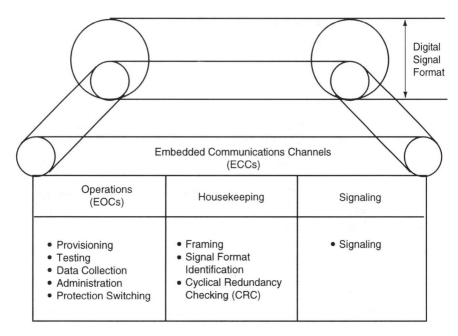

Figure 3-5 Embedded Communications Channel Concept

(SYNTRAN) signals, and digital signal 1 (DS1) with extended superframe format have at least one EOC in them (see Table 3.1).

TABLE 3.1 Examples of Major EOCs in Digital Signals

Digital Signal	Name of EOC	Speed (kbps)
ISDN Basic Access	DSL EOC	2
DS1 ESF	DS1 EOC	4
SONET	Section EOC (DCC)	192
STS-1	Line EOC (DCC)	576

Message-Oriented EOCs. When an EOC is used to carry messages, e.g., TL1 messages, it is called a *message-oriented EOC* (MO EOC). Messages are carried in the information field of the highest layer protocol of a protocol stack (PS). The MO EOCs applications may require thousands of bits; however, long messages are usually divided into small sizes, typically of 128 or 256 bytes.

Bit-Oriented EOCs. Bit-oriented EOCs (BO EOCs) are suitable only for simple messages not expressed in ASCII characters. Unlike the MO EOCs, BO EOCs require simple messages, language, and protocols and shall be used only for simple operation functions such as error and alarm indications. BO EOCs are encoded in bits (e.g., binary coded) and are typically used for the NE/NE communications.

The BO EOC applications usually have all or some of the following characteristics:

- They require near real-time processing.
- They require a few bits to encode information (therefore, a MO EOC, when used for these applications, will have low throughput and may not be efficient).
- They may require only a simple data link protocol; higher layers are not needed.
- They need simple rules to interpret information.
- They are usually restricted to NE/NE communications.
- They have a limited number of messages.
- They have limited flexibility.

For instance, an error or alarm indication requires one bit to indicate that an error or alarm has occurred. However, if there are many functions that each requires one bit, those bits can be encoded to reduce the number of bits needed to implement those functions. This is important when a digital signal format has limited capacity that is available for use as an EOC. For example, assume that a framed BO EOC has n information bits. If a position-dependent bit-oriented protocol (one bit allocated for each function) is used, then only n functions could be implemented over a BO EOC. Conversely, if an encoded bit-oriented protocol (i.e., not position dependent) is used, then 2^n functions can be implemented. To ensure that there is no error in the message, the message shall either be repeated or an error detection algorithm shall be used [e.g., cyclical redundancy checking (CRC)], depending on time limitations and applications.

3.10.3 Operations Channel

When a dedicated digital transmission facility (DTF) is used to provide an operations link/path for the NE/NE or NE/OS communications, the channel is called an *operations channel* rather than an EOC. The distinction between an EOC and an OC is needed because the physical level characteristics (i.e., electrical, mechanical, functional, and procedural) of their interfaces are different.

3.10.4 Shared Channel

A shared channel (SC) carries data for multiple applications (e.g., services, operations, signaling, etc.) by using the statistical multiplexing technique.

3.10.5 Network Element

An NE is any intelligent, typically processor-controlled network node that carries service traffic and may carry nonservice (e.g., operations and/or signaling) traffic. Neither an OS nor a WS is an NE because neither handles service traffic.

3.10.6 NE Types

Since large telecommunications networks include many NE types with different communications needs and capabilities, we attempt to map generic NE NM data networking functions for a typical NE in one of the three NE classes called *gateway NE* (GNE), *intermediate NE* (INE), and *end-NE,* as shown in Fig. 3-6. Table 3.2 lists some examples of NEs that can be used as GNE, INE, or end-NE.

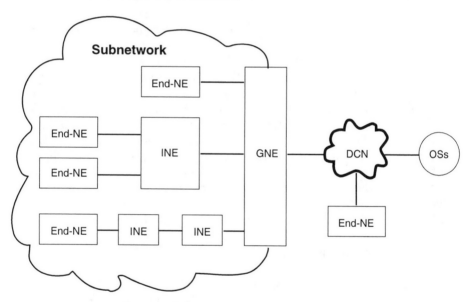

Figure 3-6   GNE, INE, and End-NE Concepts

TABLE 3.2 Examples of NE Types

	NE Type		
Network Element (NE)	GNE	INE	End-NE
Switch	X		
Digital Cross Connect System (DCS)	X	X	
Add-Drop Multiplex (ADM)	X	X	X
Remote Digital Terminal (RDT)		X	X

Gateway NE. Gateway NEs (GNEs) have the following characteristic functions or features:

- have direct access to a packet-switched network (PSN) or OSs or another GNE
- have a subtending subnetwork
- provide remote access to the subtending NEs via EOCs/OCs
- route tandem traffic
- provide statistical multiplexing
- provide gateway functions

The characteristic functions of a GNE are direct access to a PSN and/or to an OS, statistical multiplexing of the NM traffic, and routing of the tandem operations messages. A tandem message at an NE implies that the message is destined for another node (e.g., NE or OS) and generated by another node. GNEs can also have gateway functions such as protocol conversion, message translation, address mapping, and subnetwork flow control. The GNEs are the root of a subnetwork and they provide internetworking functions between a PSN and its subtending subnetwork. Thus, a GNE behaves either as a concentrator (CNC) or a packet switch.

Intermediate NE. The characteristic functions for INEs are:

- They do not have direct access to a PSN or OSs,
- They have subtending NE or NEs,
- They do statistical multiplexing and routing for the tandem traffic.

INEs also have subtending NEs and perform concentration and routing for the tandem operations traffic as do the GNEs. INEs may also have some gateway functions; therefore, they can be thought of as "semi-GNEs." The major difference between a GNE and an INE is that INEs do not have direct access to the OSs and/or PSNs.

End-NE. The characteristic functions for End-NEs are:

- They only handle their own traffic (local traffic); therefore, they do not need statistical multiplexing and routing functions, and,
- They can have direct access to any other type of NEs, a PSN, or OSs.

3.10.7 Operations Interface Module

An operations interface module (OIM) provides a set of generic functions and protocol architectures needed for network management. These functions and protocol architectures can be implemented by software, hardware, or both. They can be integrated into an NE(s) or standalone equipment, or both. The OIM functions are grouped into two classes:

1. OIM dedicated functions (OIM DFs)—these OIM functions cannot be shared with applications other than NM (e.g., services). Some examples of these functions are:
   - Parsing NM messages
   - Interpreting and executing NM messages
   - Providing access to OSs, craft, and NEs to support NM applications.

2. OIM common functions (OIM CFs)—These OIM functions can be shared among different applications such as NM and services. For example, statistical multiplexing and routing are the major service functions for a local digital switch (LDS) with built-in packet switch functions. Typical OIM CFs are:
   - Statistical multiplexing (SM)
   - Synchronous time division multiplexing (STDM)
   - Cross-connect
   - Routing
   - Flow control
   - Data collection
   - Gateway functions

An OIM can be accessed by shared and/or dedicated facilities. Since the OIM is a logical entity, the OIM access implies the access to the OIM functions and/or interfaces. Depending on applications and NE type, an OIM may include only the OIM DFs or a subset of all the functions mentioned above. Note that the OIM functions list is not intended to be complete.

3.10.8 Generic NE Functions for NM Data Networking

This section defines generic NE OIM functions that are needed for NM data transport and/or networking.

EOC Access. This function enables an NE to access an EOC (to locate bits that make up an EOC) in a digital signal. A digital transmission facility (DTF) can carry multiple digital signals, and each digital signal can contain multiple EOCs.

EOC Monitoring. This function allows an NE to monitor the content of an EOC without changing its content.

EOC Termination. This function enables an NE to terminate an EOC to process its contents. Examples of EOC content processing include:

- Parsing operations messages
- Interpreting operations messages
- Executing operations messages
- Protocol conversion
- Message translation
- Concentration (statistical multiplexing)

EOC Protection Switching. An EOC used to carry operations traffic under normal conditions is called a *primary EOC*, and an EOC used when the primary is unusable is called a *protection EOC*. Depending on the system architecture and communications needs, a primary EOC may have more than one protection EOC, or a protection EOC may protect more than one primary EOC. This function enables two NEs to use a protection EOC when a primary EOC is unusable because of, for example, failure of the digital transmission facility carrying the primary EOC, congestion in the network, or NE hardware/software failure.

The EOC protection switching does not always imply a facility protection switching. A system might not be required to do facility protection switching; however, it is desirable that a protection technique like one of the following be provided for an EOC that is used for the NE/OS communications:

- Protection via channel switching
- Protection via line switching
- Protection via alternate routing (see section, *Alternate Routing*, below)

To simplify administrative work and reduce the cost of EOC protection switching, the following guidelines are desirable for EOCs that are used for NE/OS communications:

- As a primary EOC, use an EOC in a digital signal that carries synchronization and/or signaling channels. This signal will be called the *primary signal*.
- As a protection EOC, use an EOC within a digital signal that is carried by the facility protection line.

• If there is no protection line, choose a protection EOC carried by another DTF.

EOC Status Indication and Report. An EOC can be in one of three states: (1) active (in use); (2) "standby" (for protection switching); and (3) available (unassigned). NEs shall store the status of active and standby EOCs and shall report any changes to the appropriate OSs.

EOC Idle Code Insertion. This function enables an NE to transmit continuous idle code sequences during idle conditions on all EOCs that the NE terminates. It is desirable that an EOC with a generic interface use the data link level flag (e.g., 01111110) for link access procedures balanced [LAPB] and link access procedures on the D-channel [LAPD] or all 1s as an idle code.

EOC-to-EOC Cross-Connect. EOC-to-EOC cross-connect is needed to make a cross-connection between two EOCs with the same data rate. It does not include concentration and speed adaptation (or matching). Some examples of this function include providing a common communication bus for multipoint polling or multiaccess [e.g., ALOHA, carrier sense multiple access (CSMA)] protocols [11], and various EOC intraoffice and interoffice applications.

EOC to Dedicated Port Cross-Connect. This function allows an NE to establish a cross-connection between an EOC and a dedicated port (e.g., RS-232-C, V.35) with the same data rate. This function also does not include speed matching and concentration. But it can be used for the EOC interoffice and/or intraoffice applications.

Statistical Multiplexing (Concentration). This function enables an NE to combine incoming messages and route them over the appropriate outgoing link(s) after any needed processing (e.g., protocol conversion) and buffering. Note that the incoming and outgoing link(s) can have different data speeds and can be the EOCs and/or dedicated facilities. Typical applications of this function include:

• NM data processing
• Routing
• DTF cost reduction
• Channel termination cost reduction

Buffer Status Indication and Report. This function enables an NE to detect that its OIM buffers is almost full. An example of the OIM buffer is the concentration buffer that is used to store local and/or tandem traffic.

Messages remain in the OIM buffer until an outgoing channel is available. An autonomous message shall be sent to the appropriate operations systems and/or network control center (NCC) when the threshold is reached (i.e., the buffer is 90 percent full). The GNEs and INEs shall be capable of rejecting all or some of the incoming messages (see section, *Flow Control,* below) when their buffers are full. In this architecture, OSs and NCC shall try to minimize data flow into these NEs.

Protocol Conversion/Interworking. Since NEs, OSs, and PSN nodes [packet switches or access concentrators (ACs)] support different generic interfaces for NM data transport among them, the protocol interworking function is required. This function enables two different protocol stacks to communicate with each other.

Address Mapping. This function allows an NE to route messages between two distinct interfaces that use different addressing algorithms and/or have address fields of different lengths.

Message Translation. When two nodes (NE and/or OSs) must communicate but use different messages, the message translation function is needed. Some operation functions require only simple, bit-oriented messages (e.g., error and alarm indications). An NE interprets these messages and sends a detailed report to the OSs using character-oriented messages (e.g., TL1 messages). Similarly, when a small NE uses a bit-oriented EOC to communicate with an OS, then these simple messages can be translated by GNEs and/or INEs into messages supported by the OSs for the NE/OS communications.

Multipoint Protocol Polling Function. The multipoint protocol polling function (MPPF) allows the use of a polling protocol for an NE-to-NE interface. A polling algorithm is needed to implement this function.

Routing. When an NE collects the NM traffic from its subtending NEs and performs statistical multiplexing (concentration) on that traffic, it requires a routing function to route the traffic (messages) that is destined for other nodes. A tandem message at an NE implies that the message is destined for another node (NE or OS) and generated by another node. End-NEs do not require this function. Routing is a Layer 3 function in the OSI reference model.

Alternate Routing. Alternate routing improves the reliability and performance of an NE. Performance is defined here as throughput (frame/packet per second) and queuing delay (waiting time in a buffer).

This function enables an NE to reroute all or some of the operations traffic from one link to another link (or links) for the following cases:

- Link or NE failure
- Overload (delay is close to threshold)
- Congestion (delay is larger than threshold)

The implementation of this function depends on the routing algorithm (e.g., fixed or adaptive) and the protocols in use.

Access to Routing Table. This function allows remote and/or local access to provision or change the routing table(s).

Flow Control

> 1—Message Rejection without Priority

This function enables an NE to reject incoming messages from subtending NEs and/or OSs during failure, overload, and congestion. For example, the GNE in Figure 3-7 shall reject all messages from OSs that are destined to an NE in the subtending subnetwork when the GNE detects a failure in the subnetwork that affects the NE in question. A message shall be sent to the operations systems indicating why the messages were rejected.

> 2—Message Rejection with Priority

These functions enable an NE to reject certain messages during the overload and/or congestion state (local or global). For example, an NE may

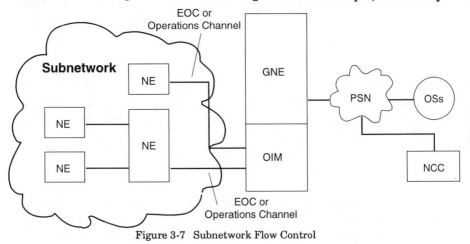

Figure 3-7 Subnetwork Flow Control

request an operations system to stop traffic and/or performance data collection from the subtending NEs when its concentration buffer(s) has a threshold alarm indicating that the buffer is full.

Message Duplication/Broadcasting. Message duplication enables an NE to receive a message from an OS or subtending NE and send it to multiple destinations by duplicating the message. As Fig. 3-8 shows, an OS may send a message to the GNE indicating intended destinations. The intended destination for the message may include all NEs in the subtending subnetwork or all NEs in a group in the subnetwork, such as:

- All DCSs
- All remote digital terminals (RDTs)
- All central office terminals (COTs)
- All add-drop multiplexers (ADMs)
- All ISDN network termination 1s (ISDN NT1s)
- Remote modules (e.g., remote switching module)

A GNE or an intermediate NE (e.g., an NE that has a subtending subnetwork but is not directly connected to a PSN node or to an OS) does not have to parse the message to implement this function. It is desirable that a multidestination address (a single address for a group of NEs) be put in the addressing fields of the proposed protocols. Message duplication must be used with care. If an OS requests performance data from the subtending NEs of a GNE, the GNE shall relay the message to its subtending NEs in a way that would avert overload or congestion at the GNE.

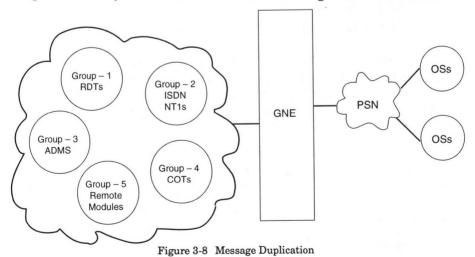

Figure 3-8 Message Duplication

OIM Access.   OSs and technician access to the OIM are required for
NE operations. The OIM shall be accessible via dedicated ports (e.g., RS-
232-C or V.35) and/or EOCs. For example, an OS may use one of the
dedicated ports on NE-i to have direct or remote (via PSN) access, or it
may use an EOC between NE-i and NE-j to have a remote access to the
OIM in the NE-i via the OIM in the NE-j. Figure 3-9 illustrates how the
OIM in NE-i can be accessed.

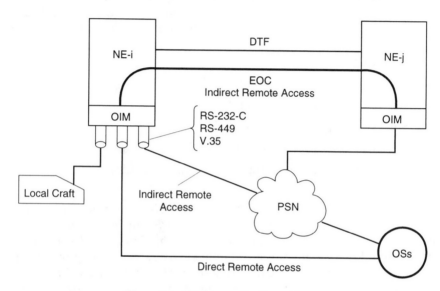

Figure 3-9   OIM Access Configurations

3.10.9 Mapping of NE Functions to NE Types

The generic NE NM functions to support the NM data networking
are defined. The key networking functions are concentration, routing,
protocol conversion, address mapping, and data transport (frame/packet
switching). Table 3.3 gives the mapping of these generic NE functions to
typical NEs in three NE classes.

TABLE 3.3 Mapping of NE Functions to NE Types

	Typical NE		
Functions	GNE	INE	End-NE
EOC Access	R	R	R
EOC Monitoring	R/O	R/O	NA

TABLE 3.3 (continued)

	Typical NE		
Functions	GNE	INE	End-NE
EOC Termination	R	R	R
EOC Protection Switching	R	R/O	O
EOC Status Indication and Report	R	R	R
EOC Idle Code Insertion	R	R	R
EOC/EOC Cross-Connection	R/O	R/O	NA
EOC/Dedicated Port Cross-Connect	O	O	O
Statistical Multiplexing	R	R	NA
Buffer Status Indication and Report	R	R/O	R/O
Protocol Conversion	R	R/O	NA
Address Mapping	R	R/O	R/O
Message Conversion	R/O	R/O	NA
Protocol Polling	R/O	O	NA
Routing	R	R	NA
Alternative Routing	R	R/O	O
Access to Routing Table	R	R	NA
Flow Control	R	R/O	NA
Message Duplication	R/O	R/O	NA
OIM Access	R	R	R

R=Requirement; R/O=Requirement or Optional; O=Optional; NA=Not Applicable.

3.11 TMN IMPLEMENTATION ARCHITECTURES

A large number of operations paths are needed to provide communications among intelligent NEs, OSs, and WSs in today's large digital telecommunications networks that support advanced services and network capabilities. It is not always economical to connect all NEs to a PSN or to an OS directly via dedicated transmission facilities. Therefore, an NE can be used to collect and transport NM data from other NEs. This can reduce the PSN access cost for the NEs that are located in the loop plant, on customer premises, and in central offices where PSN capabilities do not exist. The need for communications among NEs, OSs, and technicians requires that generic NE functions and interfaces be formulated to support multivendor's products in open architectures and to reduce the cost of network management. This section presents how the NE functions and EOCs in digital signals can be used to design a cost-effective TMN.

The data communication function of the TMN will be implemented by designing a data communication network (DCN). Since the DCN could be a very large network (at least for the North American networks), it will be divided into three subnetworks to simplify the design, as shown in Fig. 3-10. These subnetworks are a backbone DCN (B-DCN) and two access networks that connect OSs and NEs to the backbone DCN. The first access network is called the *OS-access network* and the second one is called the *NE-access network*. This section briefly discusses the OS-access network

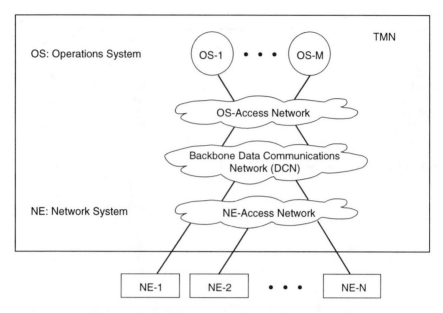

Figure 3-10   TMN Implementation Architecture Model

design and focuses on the design of backbone DCN and NE-access network. Figure 3-11 lists available technologies that might be used to carry
out each subnetwork.

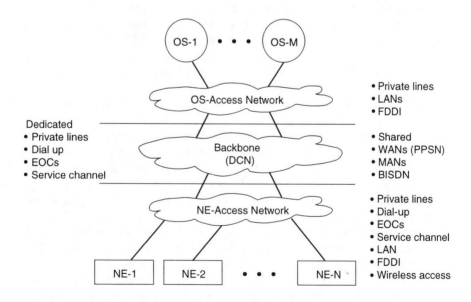

Figure 3-11   Examples of DCF Implementation Alternatives

3.12 OS-ACCESS NETWORK

The OS-access network provides data communications paths among OSs and connects appropriate OSs to the B-DCN. The simplest OS-access network architecture uses point-to-point private lines to provide needed connectivities for OS-to-OS and OS-to-NE communications. However, this simple architecture may not have the flexibility that would be needed for the OS-access network. Because it would be difficult to predict the exact capacity requirement for each link, a more complex architecture might be used, as illustrated in Fig. 3-12. A high-speed LAN or fiber distributed data interface (FDDI) can be used to connect OSs to the B-DCN as well as to provide a communication link between two OSs when needed. Also, more than one gateway can be used to increase the throughput and reliability of the OS-access network. Some OSs may still have direct connections to the B-DCN or a GNE to meet the requirements for a specific application (e.g., billing) and to further increase the reliability of OS-access network.

3.13 NE-ACCESS NETWORK

The NE-access network connects NEs to the B-DCN and provides connectivity among NEs. This access network uses EOCs to reduce the access cost of NEs to the B-DCN. The EOC concept is used to share network resources (transmission facilities, NEs, interfaces) among different applications, reducing the cost of NM data networking. The sharing of resources among different applications is one of the most important features of the integrated digital network access and ISDN architectures. The ISDN and integrated digital network access architectures eliminate voice frequency cross-connects and allow NEs to be accessible from a GNE via EOCs, as illustrated in Fig. 3-13. It is not always economically possible to connect all NEs to a B-DCN or to an OS via dedicated transmission facilities, especially when NEs are located outside of a central office (CO) and/or CO where B-DCN nodes are not available. Therefore, GNEs, INEs, and mediation devices can be used to design an NE-access network for the NEs that cannot afford to have a direct link to a B-DCN or their supporting OSs.

Mediation devices will be used to concentrate NM traffic from INEs and end-NEs that are not accessible via a GNE. Large GNEs (e.g., local digital switches) with mediation functions may have direct connections to the OS(s). A digital transmission facility would be a T1 line, fiber link, a digital radio link, etc.

These digital transmission facilities in the NE-access network are already installed to provide connectivity among NEs to offer services.

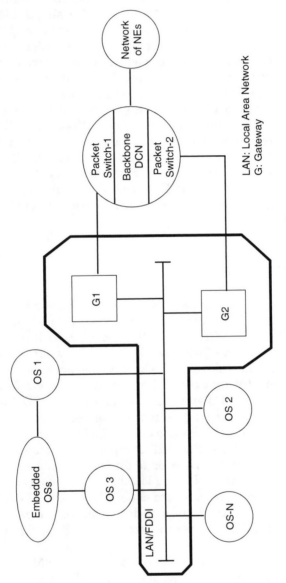

LAN: Local Area Network
G: Gateway

Figure 3-12   OS-Access Network

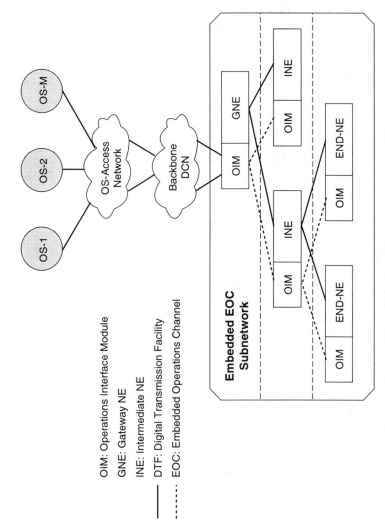

OIM: Operations Interface Module

GNE: Gateway NE

INE: Intermediate NE

DTF: Digital Transmission Facility

EOC: Embedded Operations Channel

Figure 3-13 An NE-Access Network Implementation via EOCs

They are capable of transporting digital signals with different rates. By allocating a porting of a digital signal capacity as an EOC, a cost-effective NE-access network can be designed. This access network can be an EOC-based private line network, a PSN, or both. INEs can be cascaded to form chain or ring configurations. Figure 3-14 illustrates an NE-access network implementation by using a LAN and EOCs.

3.14 TMN BACKBONE DCN DESIGN

The B-DCN may be a private line network, a circuit switched network, a PSN, or any combination of the alternatives mentioned above. Here we focus on the PSN portion of the B-DCN and discuss strategies to design a backbone packet switched data transport network. A standalone dedicated PSN would not be a cost-effective solution to our problem. In the following subsections we will discuss alternatives to implement a shared PSN.

3.14.1 Standalone Shared PSN

A public PSN (PPSN) can also be used to transport NM traffic as well as service traffic. This alternative has several disadvantages, however. First, most of the central offices in a local access and transport area (LATA) will not have a packet switch. Second, present packet switches and access concentrators (ACs) can terminate digital signals only up to a maximum of 56-Kbps rate. Today, this is forcing the use of a T1 multiplexer/demultiplexer in order to access a backbone T1 interoffice transport network, as shown in Fig. 3-15. Although digital cross-connect systems (DCSs) can be used to eliminate some of this unnecessary multiplexing/demultiplexing, the present PPSN architectures may not be as flexible as the other alternatives discussed below, unless packet switches and ACs can directly terminate T1 lines or fiber transmission facilities. This would not be cost-effective for the standalone packet switches or ACs, but it would be a cost-effective solution to our problem when packet switching and AC functions are integrated into large NEs, such as selected local digital switches, DCSs, or add-drop multiplexers, as discussed in the next section.

3.14.2 Embedded PSN

Because the OIM for the GNEs has the concentration and routing functions, the selected GNEs can be used to implement an embedded PSN (E-PSN), as shown in Fig. 3-16. These selected GNEs are called *backbone GNEs* (B-GNEs). The EOCs in interoffice digital transmission facilities will be used to provide communications links among B-GNEs. The GNEs

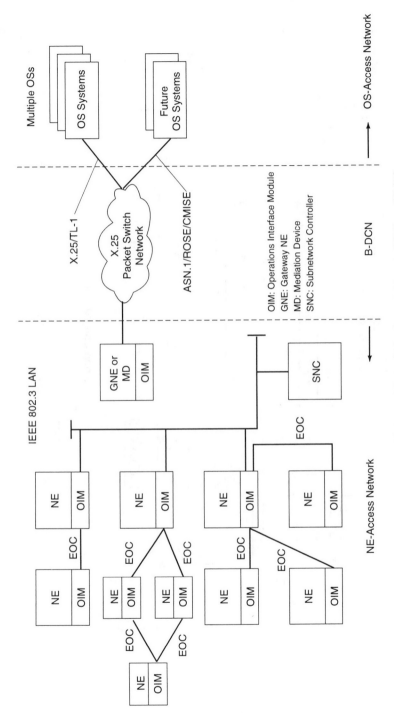

Figure 3-14   An NE-Access Network Implementation via EOCs and LAN

103

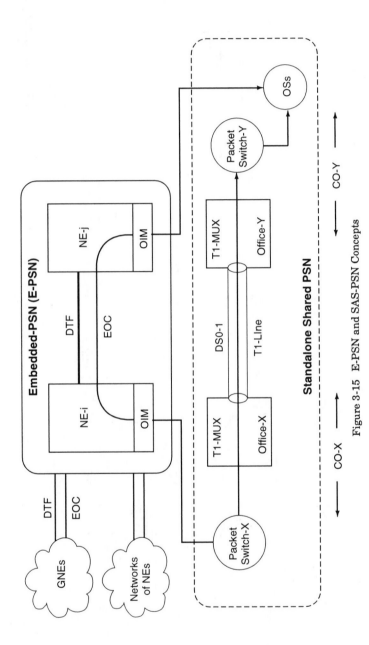

Figure 3-15 E-PSN and SAS-PSN Concepts

(or MDs) will collect and concentrate NM traffic from subtending NEs and provide internetworking functions between the B-DCN and NE-access network. As shown in Fig. 3-16, the OIMs for the B-GNEs will behave as a simple packet switch and OIMs for the other GNEs will behave as an access concentrator. As illustrated in Fig. 3-15, the embedded PSN architecture eliminates the disadvantages of the standalone shared PSN architecture because OIMs for the B-GNEs do not have to be connected to a multiplexer/demultiplexer to access a high-speed backbone interoffice transport network as in the standalone shared PSN architecture.

The embedded EOC-based PSN concept also has the following advantages:

1. Because a TMN shares intelligent transport NEs and/or switches, it will have advanced network capabilities and it will be automatically upgraded as the NEs are upgraded. Examples of these advanced capabilities are network reconfiguration and dynamic bandwidth allocation. For example, when additional capacity is needed, intelligent NEs can allocate additional 64-Kbps time slots as EOCs under the control of appropriate OS(s). Also, these NEs can reconfigure the network in case of failure, overload, or congestion in the network.

2. Currently, PPSNs use in-band signaling techniques. When NM traffic is also transported by PPSNs, the number of call set ups and the total amount of traffic will increase, and expected quality of service (e.g., transit delay) may be affected. Thus, appropriate traffic engineering criteria and network design principles must be considered. The embedded PSN concept has the opportunity to have a shared or dedicated out-of-band signaling network. A dedicated out-of-band signaling network can be designed by allocating a portion of EOC capacity in digital signals to carry signaling traffic. For instance, the SONET OC-N (optical carrier-N) signal has many available (unused) overhead channels (bytes) that can be used to design a dedicated signaling network for the embedded PSN. Or, since the embedded PSN concept will share large transport NE (e.g., DCSs) and/or LDSs, it might be possible to share a common channel signaling network to also carry the embedded PSN signaling traffic.

3. Because the number of large NEs (e.g., LDSs, DCSs) in a LATA would be much higher than the number of standalone PSs and ACs, an embedded PSN will reduce the NE access cost to the B-DCN.

4. There will be many alternatives for reconfiguration and/or rerouting because large GNEs are connected to many other GNEs via multiple facilities. This will also increase the reliability of the TMN.

An embedded PSN can be a dedicated PSN for network management or it can be a shared PSN. It would also be possible for a TMN to use any

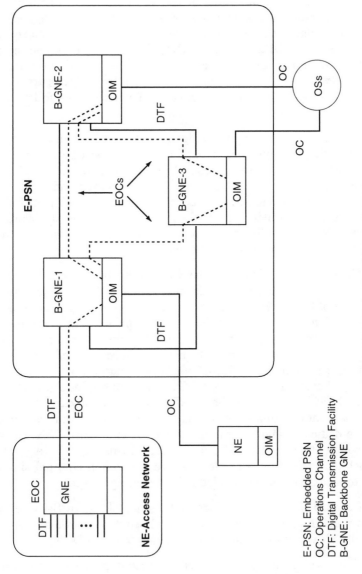

Figure 3-16 Embedded PSN via B-GNEs

E-PSN: Embedded PSN
OC: Operations Channel
DTF: Digital Transmission Facility
B-GNE: Backbone GNE

combination of embedded-dedicated PSN, embedded-shared PSN, and a standalone-shared PSN. This embedded PSN concept would be much more economical when metropolitan area network (MAN) and B-ISDN concepts are implemented, because those concepts would be introducing packet switching technology to large transport NEs (e.g., MAN NEs) and LDSs (e.g., B-ISDN switches).

3.15 TMN LAYERS AND DISTRIBUTION OF TMN FUNCTIONS

A management network designer must consider physical and logical views separately when designing a TMN. Physical view deals with the management of physical entities (e.g., NEs, facilities, power supplies, air conditions, etc.) and logical view deals with the logical entities such as services, network planning, work scheduling, accounting, security, contracts, etc. The total TMN functions are partitioned into five functional management layers, as discussed below, to simplify the design of a TMN.

3.16 MANAGEMENT LAYERS

TMN management functional/logical layers as defined in M.3010 are:

- Business management layer (BML),
- Service management layer (SML),
- Network management layer (NML),
- Network element management layer (NEML), and
- Network element layer (NEL).

NEL TMN functions are provided by NEs. These functions include basic TMN functions such as performance data collection, alarm collection, self diagnostics, address translation, protocol conversion, etc. Some NEs can also do data screening and analysis. Basically, NEL is responsible for management of an NE.

NEML is also called *subnetwork management layer* (SNML). NEML TMN functions are provided by SNCs. This layer manages a collection of NEs (a subnetwork). Some functions in this layer are the same as NML functions, but have a more limited span of control.

NML functions are usually provided by OSs. NML functions are used to support TMN applications that require an end-to-end view of a TN. This layer receives aggregated or summarized data from NEML and creates a global view within its scope/domain. This layer provides applications based on standards and it is not responsible for the management of non-standard products or features. It communicates with other layers via standard interfaces.

SML is responsible for the management of services that are provided to customers. This layer is not concerned with the management of physical entities. It provides the basic point of contact with customers for all service transactions, including opening a new account, service provisioning, service creation, providing information about quality of service (QoS), service contracts, etc.

BML has responsibility for the total enterprise and is the layer at which agreements between operators are made. Strategic planning and executive actions are supported through this layer.

3.17 PHYSICAL IMPLEMENTATION OF TMN FUNCTIONS

In current telecommunication networks, network management intelligence mainly resides in the operation systems (OSs). The network elements have little intelligence. Because the current management network does not have a subnetwork management layer, low-level, nonservice-related equipment management becomes a time-consuming job for OSs. This prevents the OSs from managing the network and service effectively. As the telecommunication network increases its global size and provides new services, network management functions become more complex. Without partitioning and distributing network management functions into SNML (or NEML) and NEL, the situation will become worse. Figure 3-17 depicts how the emerging TMN physical architecture will change from a traditional two-layer hierarchy (centralized intelligence and management) into a three-layer hierarchy (distributed intelligence and management) in the future. OSFs and MFs of a TMN will be implemented by OSs, SNCs, and NEs in a distributed architecture environment.

TMN functions can be grouped into two classes called global and nonglobal functions. Nonglobal functions are those that do not require an end-to-end network view and/or a very large span of control. Some examples of this function are NE provisioning, data aggregation (performance and alarms), protocol conversion, message translation, graphical user interface, etc. Global functions use nonglobal functions as building blocks to support more complex TMN applications such as service management, network restoration, network reconfiguration, dynamic bandwidth management, workforce administration, network planning, etc. Since the global functions require a large span of control and an end-to-end network view, they should be provided by OSs. As shown in

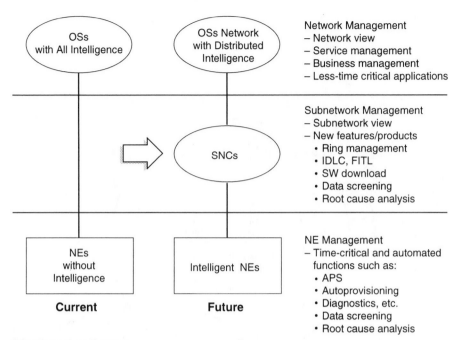

OSs: Operations Systems
SNC: Subnetwork Controller

Figure 3-17   Physical Implementation of TMN Management Layers

Fig. 3-18, global functions usually require interworking between different OSI management domains.

In general, nonglobal functions and nonstandard functions (vendor/technology specific) will usually be implemented by SNCs and NEs. Cost and performance should be also considered in selecting a physical system to implement a TMN function. The boundary between NEs, SNCs, and OSs with respect to network management is illustrated in Fig. 3-19.

3.18 NEED FOR AN SNC AND ITS APPLICATIONS

Currently, the development and deployment of OSs capabilities lag at least two or three years behind the availability of the new technologies, services, products, and architectures. This is because OSs and NEs are usually developed by different vendors. SNCs provide practical, economical, and timely solutions to the current problem. Below are the major reasons why SNCs are needed:

• To support faster introduction of new services, products, features, and architectures. This is possible because SNCs are developed by NE vendors.

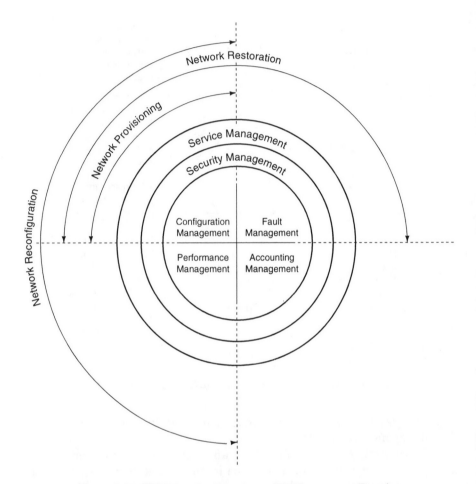

Figure 3-18   TMN Function Mapping to OSI Management Domains

- Because SNCs use NE network management capabilities/functions and support distributed management, it is possible to provide cost-effective solutions.
- To support OSI migration and upgrade.
- To provide mediation of different interfaces for interworking.
- To solve and/or improve bottlenecks and performance problems for existing architectures and products.

SNCs can be developed by using standard hardware platforms, software platforms, and a standard application program interface (API) as shown in Fig. 3-20 to support multivendor products and to reduce the cost of developing an application. Major SNC applications are:

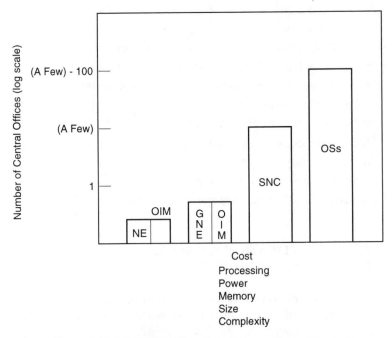

Figure 3-19 NE/SNC/OS Boundary (Scope or Span of Control)

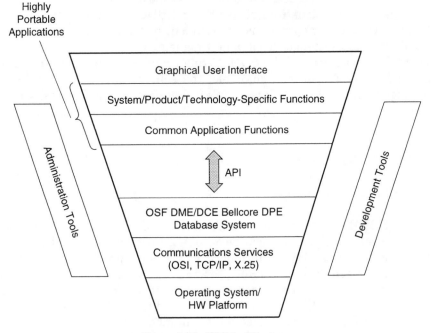

Figure 3-20 SNC Architecture

- NE management (provisioning, installation, data collection)
- Ring management
- Integrated digital loop carrier (IDLC) system management
- Fiber in the loop (FITL) system management
- Management of access networks as a whole system
- Software download to NEs
- Software release management
- Data screening and analysis
- Alarm filtering and correlation in its span of control
- Support of new services and technologies
- Bandwidth/path management in its span of control

3.19 A SPECIFIC APPLICATION EXAMPLE: SONET RING MANAGEMENT

SONET rings are required to provide and/or support self healing, autoprovisioning, quick response to network management requests, software/firmware download, etc. The current centralized OSs are not adequate to meet these requirements. These requirements can be satisfied in a timely and a cost-effective way by moving some of the network management functions into ring NEs and SNCs. Figure 3-21 shows a typical SONET ring operations environment in North America. Based on the distributed system concept, a new management method is suggested. Due to the regularity of the SONET ring architecture and powerful embedded SONET overhead channels, the solution for the SONET ring management is relatively less complex; therefore, it could be easily realized.

3.20 DEFINITIONS

3.20.1 SONET Rings

A SONET ring is a collection of SONET add-drop multiplexer (ADM) nodes forming a closed loop where each ADM node is connected to two of its adjacent nodes [5],[6]. Since each ring always provides a protection channel for each working channel, services will be restored when there is a failure. Rings are categorized as unidirectional or bidirectional rings according to the traffic flow under normal conditions. Rings are also called two-fiber or four-fiber rings based on the number of fibers that make them up, as shown in Figs. 3-22–3-24. From the point of view of ring protection

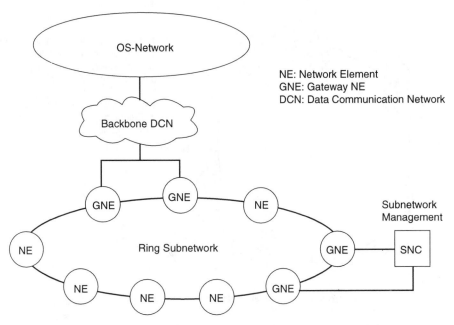

Figure 3-21 A Typical SONET Ring Management Environment in North America

switching, the SONET rings are called either *line switched* or *path switched* rings. Line switching is triggered by the line-level indications and path switching is triggered by path-level indications. In the event of a failure, line switched rings loop back the failed (nonworking) channels to the protection channels and the path switched rings recover from the failure by operating based on 1+1 protection switching principles using SONET path overhead information.

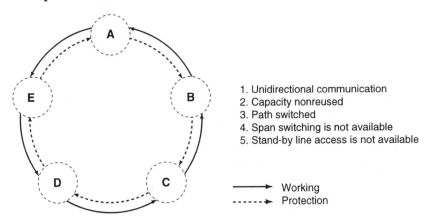

Figure 3-22 Two-fiber Unidirectional Path Switched Ring

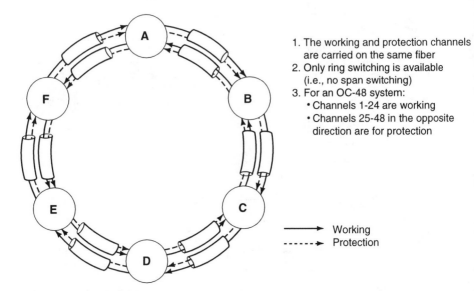

Figure 3-23 Two-fiber Bidirectional Line Switched Ring

3.20.2 Ring Autoprovisioning (Self Provisioning)

Ring provisioning includes time slot assignment (TSA), time slot interchange (TSI), and all cross connections from line and add-drop sides for each NE on the ring. Traditionally, rings are provisioned by a provisioning OS using hundreds of cross-connection commands. On the contrary, ring autoprovisioning is a new method that allows NEs on the ring to automatically provision themselves according to a distributed ring knowledge base.

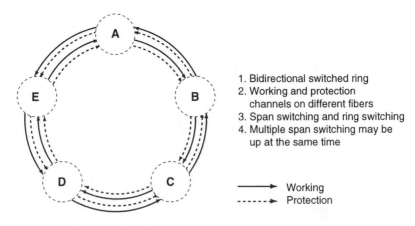

Figure 3-24 Four-fiber Bidirectional Line Switched Ring

3.20.3 Automatic Protection Switching (APS) and APS Map

Automatic protection switching (APS) provides survivability when there is a fiber cut and/or a node failure. Each ring NE has a map that stores required provisioning information to perform the APS function. This map is called the *APS map*. Currently, the T1X1.5 standards committee is working on the APS requirements and protocols for the bidirectional line switched rings (BLSR). To avoid misconnections during the protection switching, some ring provisioning information must be stored in all NEs on a ring by using APS maps. In general, each NE only needs to store the provisioning information for its adjacent nodes. Therefore, the APS map for each NE on the ring will be different.

3.20.4 Ring Table and Managed Ring Object

The ring table is a knowledge base for every NE on a SONET ring. This knowledge base is a data structure that contains complete ring provisioning information and additional information that will be used to support TMN applications other than provisioning (e.g., bandwidth management). Therefore, the ring table supersedes the APS map. It is stored in every NE on the ring and used by NEs to perform autoprovisioning and self healing operations. The ring table could be further expanded to support other ring network management functions such as performance monitoring, alarm surveillance, path management, bandwidth management, etc. Ultimately, the knowledge presented by the ring table and its associated algorithm will likely evolve to a high-level managed object. The ring knowledge base (RKB) and the ring table are used interchangeably.

3.21 SCOPE

SONET ring table could be used to support many TMN applications. Here, we only discuss ring autoprovisioning and APS by using the suggested ring table. Also, we focus on NE ring management requirements. OS/SNC ring management requirements will not be discussed.

3.22 SONET RING PROVISIONING METHODS

There are three possible methods for provisioning a SONET ring. The first is the channel-based method, which treats each cross-connection as a managed entity. With this method, NEs do not have any information about the status of connections. The second method is called *node-based provisioning* and it treats each node (each NE) as a managed entity and provides a complete bandwidth view for the node by using a node map. All cross-connect information associated with a node is stored in a node map in each NE. However, the node map does not provide a completed ring (subnetwork) view. Both methods require that an external management system must provision each NE on the ring. These two methods are not adequate to manage a ring because ring restoration must be completed in less than 50 ms.

Figure 3-25 explains the channel- and node-based provisioning methods. With the first method, Channel 1 is dropped and continued at Nodes A and B. This requires that two cross-connect commands must be sent to Nodes A and B. Therefore, another four commands must be sent to these two nodes in order to provision Channels 2 and 3. The second method requires that a node map, which depicts all the cross-connects in a node, will be sent to each node.

The third method is a ring-based method that treats a ring as a managed entity. It distributes the ring management intelligence into ring NEs and automates the ring management functions. This method uses a ring table that includes all provisioning information for all nodes on the ring. This is a more systematic way to manage a SONET ring as whole rather than many individual nodes or channels within the ring subnetwork. With this method, ring NEs will perform ring management functions with minimum OS and SNC involvement. The ring table will be first loaded into one of the NEs on the ring only once by an external management system (OS or SNC) or by using the local craft interface. Then, the ring table will be distributed to the other NEs on the ring via broadcast.

3.23 METHOD COMPARISON

Table 3.4 illustrates the differences among the three provisioning methods. It is obvious that the third method provides a systematic network management method for SONET ring management. This method allows NE vendors to distribute the subnetwork management functions into ring NEs. It will minimize the future OS interoperability problem by increasing the NE independence. Also, this method simplifies NE/OS interface by using a very limited command set for SONET ring operations. For instance, an OS or an SNC need send only one table retrieval command to obtain the current status (e.g., provisioning, bandwidth usage, etc.) of the ring.

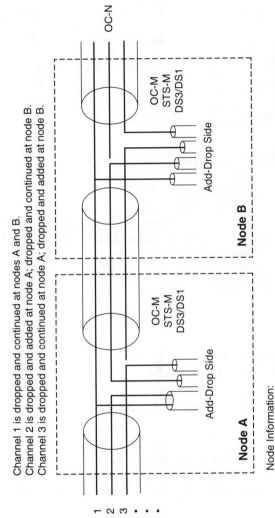

Channel 1 is dropped and continued at nodes A and B.
Channel 2 is dropped and added at node A; dropped and continued at node B.
Channel 3 is dropped and continued at node A; dropped and added at node B.

Node Information:
Node A: Channel 1 and 3 are dropped and continued; Channel 2 is dropped and added.
Node B: Channel 1 and 2 are dropped and continued; Channel 3 is dropped and added.

Figure 3-25 Channel- and Node-based Provisioning

TABLE 3.4 Comparison of Possible Provisioning Methods

Comparison/Methods	Channel-Based Provisioning	Node-Based Provisioning	Subnetwork (Ring) Based Provisioning
Scope	Single cross-connect	Complete node proviing	Complete ring proviing
OSS/Ring Manager Dependency	Very high	Less	Much less
Ring Autoprovisioning	Very difficult	Difficult	Good
Number of Commands Needed to do Ring Provisioning	Hundreds of single cross-connect provisioning commands	One node map download command for each node on the ring	One ring table download command
Ring Subnetwork Management Support	Very Complex	Complex	Simple
Provisioning Data Confirmation	Tedious—each single cross-connect provisioning command requires an acknowledgment	Each node could confirm with a node bandwidth view.	Easy—only one piece of information must be confirmed.
	NE lacks a way to know if a mistake is made	Difficult to know the status of other nodes	
Operation Error	Very high	Medium	Much less

3.24 GUIDELINES FOR THE RING TABLE AND THE ASSOCIATED ALGORITHM

The design of this ring table and the associated algorithm shall meet the following guidelines:

1. It should contain provisioning information for the entire ring. This includes ring type, ring ID, number of nodes on a ring, node IDs, node sequence, and all termination points and their status.

2. It must include all required information to support the SONET ring standard APS application.

3. Standard TSA and TSI are supported using the path ID. The path ID could be used to provide path traceability. The TSI support is important to support interlocking rings.

4. Expendability must be allowed in the ring table format and the algorithm.

5. Extra traffic carried by protection channels must be supported. In case of failure, some of the extra traffic may not be dropped.

6. Table size must be minimized while the functions supported with the associated algorithm are optimized.

7. Ring table verification must be easy and guaranteed. Broadcasting/checksum can be used for ring table verification.

8. Initially, the ring table can be generated by an external management system.

9. A standard ring object should be specified to support the multivendor environment.

10. Use of a ring-based provisioning method does not exclude channel-based and node-based provisioning. In fact, the ring-based method is supplementary to both channel- and node-based provisioning methods. The ring table shall be designed in such a way that each row of the ring table provides detailed channel information and each column, with two subcolumns, provides detailed node provisioning information. All management information carried by the three methods is convertible by using a simple algorithm.

11. Ring-based provisioning is not only limited to the ring provisioning function. By expanding the contents of the ring table, ring NEs, OSs, and SNCs could perform other network management functions.

3.25 CONCLUSIONS

TMN allows multivendor environments via standards and a common language (set of standard definitions) to communicate with each other in a way that we can understand each other. Through TMN standards and architectures, NE, OS, and TN planning processes are integrated. Network management requirements play a major role in designing current and future network architectures, services, and NEs.

REFERENCES

[1]. CCITT Recommendation M.30, "Principles for a telecommunications management network," Blue Book, vol. IV, Fascicle IV.1, 1989.

[2]. CCITT TMN Draft Recommendations M.3xxx series, 1992.

[3]. "Generic requirements for element management layer (EML) functionality and architecture," Bellcore, TA-TSY-001294, issue 1, December 1992.

[4]. "Synchronous optical network (SONET) transport systems: Common generic criteria," Bellcore, TR-NWT-000253, issue 2, December 1991.

[5]. "SONET add-drop multiplex equipment: A unidirectional dual-fed, path protection switched, self-healing ring implementation," Bellcore, TR-TSY-000496, issue 2, September 1989/Supplement 1, September 1991.

[6]. "SONET bidirectional line switched ring equipment—Generic criteria, Bellcore, TA-NWT-001230, issue 2, April 1992.

[7]. "Cycle 1 Initial Specifications for INA, Bellcore," SR-NWT-002268, issue 1, June 1992.

[8]. V. Sahin, "Generic NE functions and protocol architectures for operations data networking," Bell Communications Research, Special Report, SR-TSY-000657, issue 1, December 1986.

[9]. V. Sahin, "NE functions and OSI protocol architectures for operations data networking," *Proc. IEEE* ICC '87, pp. 1643–1647, June 1987.

[10]. V. Sahin, "A telecommunications management network (TMN) for intelligent digital networks," *Proc. GLOBECOM '87*, pp. 1273–1277, November 15–18, 1987.

[11]. V. Sahin, C.G. Omidyar, and T. M. Bauman, "Telecommunications management network (TMN) architecture and interworking designs," *IEEE JSAC*, vol. 6, no. 4, pp. 685–696, May 4, 1988.

[12]. V. Sahin, "TMN principles, models, and applications, tutorial No. 5," *SUPERCOMM/ICC '92*, Chicago, Illinois, June 15, 1992.

[13]. T. K. Lu and V. Sahin, "Distributed SONET ring management," Proc. IFIP/IEEE International Workshop on Distributed Systems: Operations and Management (DSOM '92), Munich, Germany, October 12–13, 1992.

[14]. D. Evans and V. Sahin, "SONET sub-network controller (SNC)," Proc. Eastern Communications Form, Session OPS-08, Rye Brook, NY, May 5, 1992.

[15]. V. Sahin and S. Aidarous, "TMN and subnetwork management, tutorial #6," ICC '93, Geneva, Switzerland, May 27, 1993.

[16]. D. Subrahmanya and V. Sahin, "Getting the Myth Out of SONET Interoperability," Proc. of 9th Annual NFOEC, June 13–17, 1993, p. 23–32.

[17]. V. Sahin, "TMN, tutorial #1," NOMS '94, Kissimmee, Florida, February 14–17, 1994.

[18]. V. Sahin and S. Aidarous, "TMN standards, implementations, and applications," SUPERCOMM/ICC '94, New Orleans, LA, May 1–4, 1994.

CHAPTER 4

Domestic and International Open Systems Interconnection Management Standards

Philip A. Johnson
AT&T Bell Laboratories
Holmdel, NJ 07733

4.1 INTRODUCTION

The growing complexity of the telecommunications enterprise, as new technologies have arisen and communications reach has spanned the globe, has created a tremendous need for open systems technologies in the management of telecommunications. Telecommunications standards bodies, both regional and international, have worked on open systems interconnection (OSI) management and their work is briefly reviewed. These standards-setting arenas have set the stage by defining terms and specifying functionality in terms of interface requirements. This work, employing an object-oriented paradigm with the client-server concept, has led to the development of OSI messages across interfaces.

This chapter discusses the work generated in the standards arena and the various standards organizations themselves. From Chapter 3, we review general principles, including the client–server process. Because of the interest in functionality across interfaces, this chapter briefly discusses the principles of the telecommunications management network (TMN) in order to relate functionality to management standards.

The aspects of layering within the TMN are potential future work in the standards community; current status is discussed in this chapter. Other emerging technologies, such as broadband ISDN and intelligent networks, will continue to interact with and impact TMN.

4.2 GENERAL PRINCIPLES USED IN DEFINING THE TMN

Much of the TMN can be implemented on general-purpose computers and general-purpose software systems, which is unlike the traditional telecommunications devices that have relied on specialized hardware and software for optimal performance and reliability. Because of this thrust, general principles underlying open systems development are used in defining the TMN and its related standards. Because Chapter 3 has presented a complete discussion, a brief overview only is presented in this chapter.

4.2.1 Information Architecture

In generic OSI terms, network management is a distributed application that involves the exchange of management information between management processes for the purpose of monitoring and controlling the various physical and logical networking resources (real network, circuits, lines, equipment, etc.). TMN architecture is considered from two perspectives:

1. The management information model that describes management resources and related management activities. This activity takes place at the application level and involves a variety of subfunctions such as storing, retrieving, and processing information.
2. The management information exchange that involves a communications network and the message communications function that allows particular physical components to attach to the telecommunications network at a given interface. This level of activity only involves communication mechanisms such as protocol stacks.

Object-Oriented Approach. In order to allow effective definition of a growing set of managed resources, the TMN information architecture makes use of OSI management principles and is based on an object-oriented paradigm. These principles are those defined for OSI management by the International Organization for Standardization (ISO) (ISO 10040 and ISO 10165-4). A detailed presentation of the concept of objects is given in Chapter 5.

It must be noted that object-oriented principles apply to information modeling, i.e., to the interfaces over which communicating management systems interact, and should not constrain the internal implementation of the telecommunications management system.

A managed object is defined by:

- the attributes visible at its boundary;
- the management operations that can be applied to it;
- the behavior exhibited by it in response to management operations or in reaction to other types of management—related stimuli. These can be internal (e.g., threshold crossing) or external (e.g., interaction with other objects) or the physical resource modeled by the object; and
- the notifications emitted by the object.

Manager/Agent. Management of a telecommunications environment is an information processing application. Because the environment being managed is distributed, network management is a distributed application that involves the exchange of management information between management processes for the purpose of monitoring and controlling the various physical and logical networking resources (e.g., switching and transmission resources) (ISO 10040).

For a specific management interaction, the management process will take on one of two possible roles.

1. A manager role: the part of the distributed application that issues management operations requests and receives notifications.
2. An agent role: the part of the application process that manages the associated managed objects. The role of the agent will be to respond to directives issued by a manager on its behalf. It will also reflect to the manager the data view of these objects and emit notifications reflecting the behavior of these objects.

A manager is the part of the distributed application that, for a particular exchange of information, has taken the manager role. Similarly, an agent is the part that has taken the agent role.

Manager/Agent/Object Relationship. All management exchanges between manager and agent are expressed in terms of a consistent set of management operations (invoked via a manager role) and notifications (emitted by the agent role). These operations all can be realized through the use of object-oriented protocols such as the common management information protocol (CMIP). The way agents interact with the resources they are in charge of is an implementation issue and not subject to standardization.

Interworking Between Management Entities. These management entities use the manager–agent (M–A) relationship described above to achieve management activities. The manager (M) and agent (A) are part of management subfunctions and, as such, are part of the TMN.

Shared Management Knowledge. In order to interwork, communicating management systems must share a common view or understanding of at least the following information:

- Supported protocol capabilities
- Supported management functions
- Supported managed object class
- Available managed object instances
- Authorized capabilities
- Containment relationships between objects (name bindings)

The mechanisms implementing the shared management knowledge (SMK) are being defined by ISO at this writing.

When two management entities exchange management information, it is necessary for them to understand the SMK used within the context of this exchange. Some form of context negotiation may be required to establish this common understanding within each entity.

TMN Naming and Addressing. For the successful introduction of a TMN (within an OSI environment) into an administration, a logical, integrated naming and addressing scheme for identifying and locating the various communications objects within a TMN is critical. In order to locate TMN systems and identify various entities within each system, unambiguous naming and addressing methods are required.

4.3 FRAMEWORK OF TMN STANDARDS

The basic goal of a TMN is to provide an organized network structure to interconnect various types of operations systems (OS) and telecommunication equipment using standardized protocols and interfaces.

To organize data and perspectives about the TMN, different architectures have been defined: functional architecture, physical architecture, and the generic object model. Some in industry have added a layered architectural concept, which is in the process of emerging in the standards.

4.3.1 Functional Architecture

TMN functionality provides the means to transport and process information related to the management of telecommunication networks. It is made up of operations systems functions (OSF), mediation functions (MF), and data communication functions (DCF) that are wholly contained within the TMN. The TMN also contains the part of the network element functions (NEF), Q adapter functions (QAF), and workstation functions (WSF) that support TMN management functions.

Three types of OS function blocks have been identified. They are:

1. The basic OSF that performs the TMN application functions related to network elements.
2. The network OSF that covers the realization of network TMN application functions by communication with basic OSF.
3. The service OSF that performs specific TMN application functions for managing individual services. A service OSF may connect to another service OSF that has no direct relationship to a service but is needed for administration of the telecommunications business.

Other types of OS function blocks may be identified in the future.

The mediation function (MF) block routes and/or acts on information flowing over the TMN Q interfaces. The MF may threshold, buffer, store, filter, condense, adapt, or process this information. The data communication functions (DCF) will be used for exchanging information.

The prime role of DCF is to provide information transport mechanisms, including routing and relaying functions. The DCF may, therefore, be supported by the bearer capability of different types of subnetworks. These may include X.25, LANS, or the DCC of the synchronous optical network (SONET) [T1.105] or SDH.

When the DCF is implemented using interconnected DCN subnetworks based on different technologies (e.g., LAN, MAN, SONET DCC), any necessary interworking functions will be part of the DCF.

When DCFs have to be inserted at a given reference point, they are implemented at every point of attachment. (A reference point is a service boundary between two function blocks.)

The DCF provides a means to transport information related to telecommunications management between management function blocks. The data communications are typically provided by layers 1 through 3 of the OSI mode.

The TMN reference model in Fig. 4-1 shows an example of each pair of function blocks that can be associated by a reference point. Figure 4-1

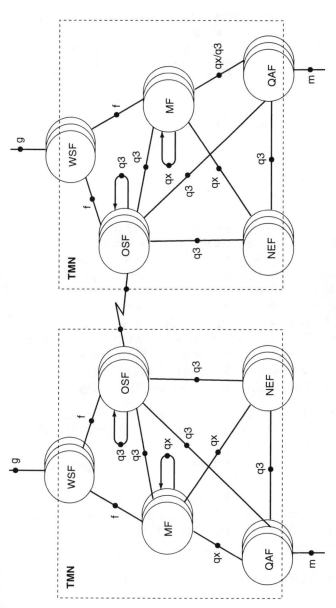

NOTE: This figure shows the valid reference points between pairs of function blocks. It does not imply that all possible configurations or paths between multiple blocks are valid.

Key:

MF – Mediation Function	TMN – Telecommunication Management Network
NEF – Network Element Function	WSF – Workstation Function
OSF – Operation System Function	q3, qx,q, f – Reference Points
QAF – Q Adapter Function	m, q – Non-TMN Reference Points

Figure 4-1 A Generalized Functional Architecture for a TMN

127

also illustrates the typical flow of information between function blocks and between two TMNs.

4.3.2 Physical Architecture

TMN functions can be implemented in a variety of physical configurations. A TMN physical architecture provides a means to transport and process information related to the management of telecommunication networks. As shown in Fig. 4-1, a physical architecture consists of the following physical components:

1. Operations systems
2. Data communication networks
3. Mediation devices
4. Workstations
5. Network elements
6. Q adapter

It should be recognized that because of the complex nature of some telecommunication equipment, the increased use of microprocessors, and the potential ability to distribute functionality within various network parts, these definitions may not rigidly cover every possible physical configuration that may be encountered.

As shown in Fig. 4-1, a reference point becomes an interface (Q, F, X) when the connected function blocks are embodied in separate pieces of equipment. The non-TMN G interface provides connection between the workstation and the human user. The non-TMN M interface provides connection between QA and the non-TMN equivalents of OSF and NEF.

The following interfaces are defined within the TMN concept:

1. *Q Interface.* Connection of network elements (NEs) to a mediation device (MD), NE with MF to another NE, MD to MD, OS to OS, and OS to MD or NE.
2. *X Interface.* Connection of a TMN to other management-type networks, including other TMNs.
3. *F Interface.* Connection of TMN function blocks to workstation support of a *Q* interface. If an NE contains mediation functions, the degree of mediation functionality within the NE determines whether the *Q* interface is supported.

4.3.3 The Generic Object Model

The goal of being able to universally exchange information concerning TMNs has led to the creation of the generic network information model. This model defines a set of object classes that are generic and are used to describe information exchange across all TMN interfaces. As a generic model it is intended to be applicable across different technologies, architectures, and services.

The model is described using the object-oriented techniques developed for OSI systems management. Thus the object classes are defined formally using abstract syntax number 1 (ASN.1) while entity–relationship diagrams give a pictorial impression of some of the relationships.

4.3.4 The Idea of Layered Architecture

The logical layered architecture (LLA) is a development concept based upon hierarchical principles in which the architecture can be thought of as being based on a series of layers. The scope of each layer is broader than the layer below it. In general, it is expected that upper layers will be more generic in functionality while lower layers will be more specific.

The LLA implies the clustering of management functions into layers. It is only when a specific LLA instance is defined that functions, or groups of functions, may become processes that are separated by interfaces.

The LLA uses a recursive approach to decompose a particular management activity into a series of nested functional domains. Each functional domain forms a management domain under the control of an operation system function. Such a domain may contain portions dealing with other operation system functions to allow further layering and/or it may represent resources (logical and physical) as managed objects within that domain.

All interactions within a domain take place at generic q reference points. However, interactions between peer domains can take place at a q or x reference point depending upon the business strategy applicable for that interaction. For the provision of network services, it is common for network management issues to cross the boundaries of a network provider (e.g., AT&T, Bell Atlantic, etc.).

It is the flexibility of the layered architecture, together with generic q reference points, that give the LLA the ability to be used as the basis for many different types of architecture. In all cases, it is the scope of the model for each domain that dictate the layering and the interdomain interactions required.

In consequence, the LLA architecture defines a TMN as being the overall management domain of a network provider. The network provider has business objectives and operational strategies. These can dictate the general composition of the network provider and the services marketed, and thus lead to the management requirements.

The LLA thereby provides a logical view of the management components and their control authority, which can be put together to create the management solution (i.e., those particular TMN). Another aspect is the partitioning of management components based on abstraction level (e.g., service rather than supporting resources).

Mapping between layers is thought of as an information model containing objects within an operation system function's control plus the objects made visible to the OSF by the subordinate domain at the next lower layer. Mapping provides a way of hiding objects in a subordinate domain to the next higher layer. Hence, through the recursion, the idea is to manage the real resources.

4.4 LISTING OF STANDARDS AND ASSOCIATED FORUMS

Documentation of TMN and TMN-related information is voluminous. Further, due to increased interest and activity, work is currently being done in many standards-setting and implementators' agreement forums. What now follows is a description, as of August 1992, of the more active groups.

4.4.1 TMN-Related Groups

First, the ISO/IEC Joint Telecommunications Committee (JTC1) provides international standards on a number of technical areas related to TMN. Such areas include protocols and services, and rules for application of specific services. These are framework standards, whose principal intent is to set perspectives and to give guidance on how more detailed interface standards are done. Committee X3 is the United States participant. It is also noted that this ISO work is oriented toward data applications, not telephony. For this reason, this ISO work supports, but does not replace, TMN. The systems management area of ISO has been very active in deriving principles and procedures for management, which have been used in driving the TMN information models. The International Telecommunications Union—Telecommunications (ITU-T—formerly CCCIT) is the international telecommunications standards body, chartered under treaty through the United Nations. The joint ITU-T/ISO activity within ITU-T Study Group VII is also involved with TMN information models.

Study Group IV of ITU-T has the TMN and TMN-related standards responsibility. Concerning the modeling, the current work allocation is to do the generic work within Study Group IV, with service- and technology-specific needs being addressed and negotiated between the Technology-Specific Study Group and Study Group IV. A list of the currently available

ITU-T TMN standards appears later in this section. ITU-T Study Groups XI and XV also are actively involved in protocol and modeling. Study Group XI has produced much work on protocols and management models for SS7 and various ISDN applications. Study Group XV deals with transmission-related aspects of modeling.

Committee T1 is the ANSI-accredited developer of telecommunications standards in North America. Technical Subcommittee T1M1 has the responsibility for internetwork operations, administration, maintenance and provisioning standards, and technical reports. TMN and the resulting information model work is the responsibility of T1M1 within Committee T1. This work results in interface standards defined for the TMN interfaces. This work began in 1986 and continues. At present, the intent of T1M1 is to take the ITU-T TMN work as the basis (where such work exists) and extend this basis to regional-specific needs. Where no ITU-T work exists or there is no agreed-to ITU-T direction, T1M1 supports ITU-T by providing information on T1M1's views concerning TMN and TMN-related details. A list of the currently available T1M1 documents appears later in this section.

The European Telecommunications Standards Institute (ETSI) is the European counterpart to Committee T1. The NA4 Technical Subcommittee within ETSI has responsibility for TMN and related work. ETSI NA4 mirrors T1M1 in that it produces European standards and also provides input to the ITU-T process. ETSI and T1M1 are currently reviewing procedures that will allow both to meet both their regional and common global needs in an optimum way. A partial listing of the ETSI standards currently available appears below. It should also be noted that ETSI TM2 is involved in traffic management.

TTC is the Japanese counterpart to T1 and ETSI. At this writing, TTC is not producing separate TMN standards for Japan. TTC desires currently are communicated to ITU-T for review.

Although not a standards body, the Network Management Forum (formerly the OSI/NM Forum) is a consortium of more than 100 organizations from all parts of the globe. Their focus is on implementors' agreements to implement network management solutions from an open systems context. The Forum is interested in this model and messages of the network management work, compatible with a TMN. When standards are available, the Forum uses this information; when standards are immature or otherwise unavailable, the Forum will draw up a "straw man" model for implementor agreement.

4.4.2 Lists of Standards

ITU-T. TMN Recommendations from Study Group IV have been allocated the M.3000 number series. This series provides a generic description of TMN. In this context the term *generic* means independent of the type of network to be managed.

Within the M. series the TMN Recommendations are split into a number of general subcategories:

M.3000 series on TMN overall principles and framework,

M.3100 series on TMN models and object definitions,

M.3200 series on the management services TMN might support,

M.3300 series on aspects for workstations connected to the TMN,

M.3400 series on management functions supporting the TMN services

The set of recommendations that will soon be published are briefly identified below. Each will be considered in more detail later in this chapter.

M.3010   Principles of a Telecommunications Management Network—An updated version of the previous M.30. It lays out the general structure for TMNs.

M.3020   TMN Interface Specification Methodology—This documents the procedure for specifying TMN.

M.3100   Generic Network Information Model—The set of Managed Objects representing resources that are applicable to all types of telecommunication networks.

M.3200   TMN Management Services: Overview—Gives a description of some conceptual management capabilities that a TMN could be expected to support.

M.3300   TMN Management Capabilities presented at the F Interface—Describes capabilities of workstation interfaces to the TMN.

M.3400   TMN Management Functions—Identifies the functions that need to be supported at TMN interfaces.

M.3180   Catalogue of TMN Management Information—Catalogues all the TMN managed objects that are documented in the ITU-T recommendations.

It should be noted that the following ITU-T standards have been produced for application-specific modeling: G.773, G.774, Q.811, Q.812, and Q.821.

T1M1. The set of standards and other information currently available on TMN and TMN-related issues include:

T1.204   Lower Layer Protocols for OS/NE Interfaces

T1.208   Upper Layer Protocols for OS/NE Interfaces

T1.210	Architecture and Functions for TMN
T1.214	Generic Network Model for Interfaces between NEs and OSs
T1.215	Fault Management Messages between OSs and NEs
T1.224	Protocols for OS/OS Interfaces
T1.227	Trouble Administration Object Modes
T1.228	Trouble Administration Messages
T1.214a	Performance Management Object Model
T1.229	Performance Management Messages

These documents are available through ANSI. A technical report, TMN methodology, is available through the Committee T1 Secretariat, the ECSA in Washington, DC, as Technical Report No. 13.

ETSI. The set of standards includes:

NA432	General Principles of TMN
NA43203	TMN Vocabulary
NA433	Modeling of TMN; Specification of Interface Requirements.

ISO. OSI management is discussed from a framework view in ISO 7498-4. This document discusses concepts and modeling principles. Based on this, a series of systems management standards is currently available:

Title	Standard Numbers	
Systems Management Overview	ISO/IEC 10040	CCITT X.701
Management Information Model	ISO/IEC 10165-1	CCITT X.720
Definition of Management Information	ISO/IEC 10165-2	CCITT X.721
Guidelines for the Definition of Managed Objects	ISO/IEC 10165-4	CCITT X.722
Object Management Function	ISO/IEC 10164-1	CCITT X.730
State Management Function	ISO/IEC 10164-2	CCITT X.731
Attributes for Representing Relationships	ISO/IEC 10164-3	CCITT X.732
Alarm Reporting Function	ISO/IEC 101-64-4	CCITT X.733
Event Report Management Function	ISO/IEC 10164-5	CCITT X.734
Log Control Function	ISO/IEC 10164-7	CCITT X.735
Security Alarm Reporting Function	ISO/IEC 10164-7	CCITT X.736
CMIS	ISO/IEC 9595	CCITT X.710
CMIP	ISO/IEC 9596-1	CCITT X.711

4.5 WORK PROGRESSION IN REGIONAL FORUMS

This section will discuss current issues for work progression of TMN and TMN-related areas.

4.5.1 T1M1

At this writing, T1M1 is working on the International Standard digital network (ISDN) service profile management/service profile verification objects. Further work is anticipated on state change management for ISDN, local loopback (ISDN) management, and generic managed objects for ISDN.

Other modeling areas are also being considered for T1M1 for broadband ISDN management, management for intelligent networks (IN), and personal communications management. Security management and testing are also candidates.

4.5.2 ETSI

The ETSI customer administration service profiles management/service profile verification and traffic management standards. The management of broadband ISDN, intelligent network, and personal communications are active candidates.

4.5.3 TTC

At this writing, the plans of TTC are unknown.

4.6 WORK PROGRESSION IN INTERNATIONAL FORUMS

Because of timing, both ISO and ITU-T work on TMN are in a lull at the moment. Many documents discussed earlier have or are just going through their respective approval cycles. Anticipated new work for 1993 and beyond include extensions and minor enhancements to the capabilities just described.

ITU-T Study Group XI is currently completing work on performance management objects and messages. Study Group XV is enhancing SDH objects and messages. Because of the interest in modeling, the areas of M.3400 should be very active in 1993 and beyond.

Work is also progressing in the Network Management Forum. OMNIPoint 1, introduced in August 1992, includes trouble, testing, and security management in advance of international standardization. The *raison d'etre* for OMNIPoints is that of snapshots across standardization activities so that implementors would use OMNIPoint references. These documents include additional implementation detail, which the standards do not.

4.7 FUTURE DIRECTIONS FOR TMN

One major effort, anticipated to begin in the next ITU-T study period, is to reconsider and expand the management services and components. These ideas have their basis in M.3200.

M.3200 includes a description of a "template" tool that is designed to help identify how the functional areas of each service map into the hierarchical layering of management.

The expansion is anticipated to include a fuller description of each service and a collation of all TMN management components into one document. This will allow commonalities to be identified, leading to easier rescue within implementations.

The second major effort, which is ongoing at this writing, is a potential integration of the TMN with the intelligent network concepts. However, much further work is needed to understand this tentative mapping, its implications, and ensuing directions. It is also an open question whether TMN and IN are separate and distinct or whether one is part of the other (or perhaps an earlier view of the other). This promises to be an exciting area of research.

CHAPTER 5

Applying Object-Oriented Analysis and Design to the Integration of Network Management Systems

Raymond H. Pyle
Bell Atlantic
Beltsville, MD 20705

5.1 INTRODUCTION

What is network management? At an intuitive level, network management can be defined as the ability to monitor and change the state of every physical and logical component of a telecommunications network. These components could range from an object as complex as a digital switch to a relatively simple modem connected to a PC. The *state* of a component can be thought of as a collection of associated indicators that provide information about the component, e.g., an LED on a modem that indicates that the power is on.

What is a network management system (NMS)? A technician who notes a flashing alarm signal on a switch and takes corrective action is one example of a network management system. A computer program on a PC, minicomputer, or mainframe that stores information about the network or interacts with it is another example. An NMS for a modern digital switch may well consist of a very complex piece of software that provides information on literally hundreds of aspects of the state of the device. Each NMS has its own functionality, interfaces, and operations requirements.

Given that a modern telephony network consists of thousands of components of different types, spanning generations of technology and made by different manufacturers, an obvious question arises: What if every component came with its own NMS? NMS operators would have to learn to use a thousand different systems, learn the different interfaces used and, perhaps more important, have to learn that "ALARM LEVEL 5" on component A means *regular maintenance due* and on component B means *complete system failure*! Unfortunately, far from being hypothetical, such a situation is a fairly accurate picture of the realities facing network managers today.

Integrating network management systems (INMS) is an attempt to bring order to the chaos described above. The premise is simple: if every network component was built to work with a standard generic INMS, then this INMS can replace the profusion of disparate systems in existence. Most network managers and vendors accept this premise, and also accept that INMSs must be built around commonly accepted standards so that products from multiple vendors can coexist and interwork. Beyond this, however, are sharply diverging views on how standardized INMS should be built. For instance, should the standards be evolutionary, built around simple, proven, limited functionality technologies (the Internet model), or should standards be designed comprehensively on paper first, be built around complex, sometimes unproven, full function technologies that have the promise of surviving well into the future (the open system interconnection model)?

We will begin by describing how NMSs have evolved from managing physical objects and events to managing *information* that abstracts physical details. We will show how integration must begin at the information and schema (i.e., the way we describe the information) level. We then describe requirements at the schema level that must be met in order to achieve integration. This discussion will set the stage for a tutorial introduction to object-oriented design, its salient features, and its application to INMS.

5.2 EVOLVING NETWORK MANAGEMENT

Network management has had—and, in many places in the world, still has—a very local and physical orientation. Resources such as multiplexers, switches, routers, etc., are managed locally with physical actions. A technician connects a meter or pushes a button to monitor or change the state of the resource. This means that either someone must be located at the site of the resource or they have to be dispatched to the site.

Most often the action to be taken is motivated by a physical alarm or a user trouble report. In other words, network management strategy has been reactive and dependent on failure—not the best strategy for providing reliable network service to users. Where mostly electromechanical network elements exist, it is, possibly, the only strategy.[1]

---

[1]In many areas of the world, and even in some North American networks, this is the situation today.

5.2.1 A New Paradigm—Remote Surveillance

As the design of network technology advanced with the evolution of electronics, especially the microprocessor, the potential for a new strategy arose and a new paradigm began to take hold—*surveillance-based network management*. The health of the network is monitored on a scheduled or, if possible, real-time basis. Systems are used to anticipate problems. This is a significant paradigm shift from traditional network maintenance and administration. The emphasis moves from managing physical things (routers, bridges, etc.) with a local perspective, to managing logical things (information) remotely. At first, the distinction between physical and logical may seem trivial to a programmer or systems developer since the ultimate purpose of the system is to manage the physical resource. But to a network manager, it is significant. The shift is from local and on-site observations and actions (different perceptions, local and varied interpretations, decisions, and actions) to predictable observation and interpretation of well-defined information over a wide area.

If we extend the paradigm one step further and say that when accomplished through mapping (applications) of the state and behavior of resources onto a well-defined set of data values, the range of functions that have to be performed by a remote system can be simplified to a set of database management functions [i.e., common create, read, update, delete (CRUD) functions]. This paradigm promises a relatively simple platform for remote systems-based surveillance and control. Figure 5-1 provides a conceptual model for the paradigm.

As public and private voice and data networks have grown and become more complex, the remote surveillance and control paradigm has become the primary tool for dealing with increasing size and complexity. But the way the paradigm has evolved has been the cause of many problems for the network manager. As mentioned before, what is needed is the ability for one NMS to remotely monitor and control many network elements that are not necessarily made by the same manufacturer. For various reasons, that is not what has evolved.

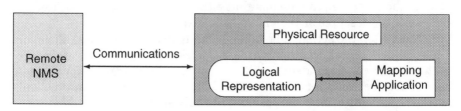

Figure 5-1 Remote, Systems-based Surveillance, and Control Model: Remote systems communicating with applications in resource that map data values onto the resource behavior.

Until very recently, network surveillance and control systems tended to be technology-, product-, and/or manufacturer-specific. Network equipment manufacturers developed proprietary systems to manage their own technology and specific design. Functions and data varied from manufacturer to manufacturer. Network management information has been specified in highly optimized proprietary schemata and manipulated with custom functionality. Attempting to integrate different systems under these circumstances is a nightmare and often results in two or more terminals on a single network manager's desk: the integration is provided through the human operator. Still, where it has occurred, it has been a significant improvement over local and physical management.

5.3 THE SYSTEMS MANAGEMENT MODEL

Standards committees dealing with information systems and telecommunications management have developed a model for the remote surveillance and control paradigm. It is called the systems management model and is intended as a fundamental guide for developers. Figure 5-2 presents that model.

The systems management model makes clear the goal of managing resources remotely across a communications interface. It also introduces the agent process into the paradigm and the concept of managed objects with defined management operations and notifications. These terms will be further developed as we begin applying object-oriented analysis principles to the realization of this model. The important point at this time is the picture of remote processes communicating for the purpose of network management.

5.3.1 The Model Has Recently Begun to Be Realized

The systems management model has begun to be realized in recent years. This is largely because certain standards groups have attacked the problem at the most elemental point in development. They have shown that when integration of function and information is accomplished at the

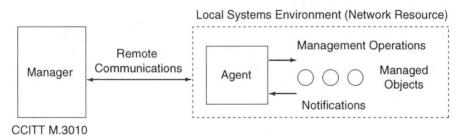

Figure 5-2

schema level, there exists a platform for integration of network elements (NEs) and systems, regardless of manufacture.

5.4 DESIGNERS HAVE CONCENTRATED ON INFORMATION

It must be remembered that, in this new NM paradigm, it is information about a resource that is important. At the design level, the physical resource itself is important only as the source of the information. For information exchange to occur across a communicatiõns interface between two applications, both applications involved in the exchange must view the exchanged information identically. Therefore, to realize our model, a common, shared information schema is essential for integrating NM systems [1].

5.4.1 The Schema Provides a Common Reference

The purpose of the schema (or information model) is to provide a common frame of reference for integration, not to specify internal data structures in a particular system. The schema must specify the information that can be communicated over an interface [2].

The systems management model and the problem of a shared NM information model have been approached in two similar but fundamentally different ways, as discussed in Chapter 2. The Internet Engineering Task Force (IETF) developed the Simple Network Management Protocol (SNMP) and its accompanying management information bases (MIBs) for managing TCP/IP internets. ANSI X3 and T1, the International Telecommunications Union—Telecommunications (ITU-T—formerly CCITT) and the International Organization for Standardization (ISO) have been working on the Common Management Information Protocol (CMIP) and its accompanying MIB(s) for managing OSI networks. The two approaches are similar in that they each are realizations of the remote surveillance paradigm and systems management model and they attempt to integrate network management at the schema level. As such, they each provide a standard set of information management functions (e.g., read, update, etc.), a standard management information specification using a standard abstract syntax, and a standard protocol for communications. They differ significantly in complexity and design paradigm. Both make use of object-oriented (OO) principles, but to different extents. Since the OSI CMIP approach makes more complete use of the object-oriented paradigm, it will be used for examples through the rest of the chapter.[2]

---

[2]For a look at the internet model, please refer to Chapter 2.

5.5   THE OO PARADIGM IS HELPFUL IN MODELING NM INFORMATION

The object-oriented paradigm, applied to analysis and design, was especially attractive to analysts who began the efforts to develop systems management principles for OSI because it met several requirements that other analysis and design methods did not. Those analysts in ITU-T and ANSI T1 who followed with development of models for telecommunications network management came to the same result.

5.5.1   Algorithmic versus Object-Oriented Decomposition

The information model decomposes the system problem space into manageable and understandable pieces. This decomposition normally takes one of two well-known forms—algorithmic or object-oriented. Traditional top-down structured analysis is the most widely used example of algorithmic decomposition. In this case the problem space is decomposed by successively breaking down some major process into subprocesses and dealing with the lesser complexity of each subprocess [3], [4]. Object-oriented decomposition breaks out key abstractions in the problem domain rather than steps in a process [3]. Figure 5-3 gives examples of the two methods applied to the decomposition of the problem of updating an alarm record file in a NM system. As Booch [3] points out, both designs solve the same problem but in distinctly different ways. Whereas the structured method concentrates on the ordering of events in a process, the object-oriented method emphasizes the agents that either cause action or are the subject of actions [3]. If we were to use the analogy of the vocabulary of the problem space, then algorithmic decomposition is concerned with the verbs while the object-oriented method is concerned with the nouns [2].

5.5.2   OO Design Is a Powerful Tool for Managing Complexity

What is a good design methodology? Entire books can be written (and have been written) that address that question, but we shall restrict ourselves to a simple answer: If it works well for your problem domain, it's a good design methodology. The operative phrase is *"works well."* In the INMS domain we can equate works well to the following fundamental criteria:

1. *The resulting system should be capable of supporting different perspectives or views of the same object or group of objects.* This is motivated by the reality that the INMS will be perceived very differently by different users, depending on their perspectives. For instance, a net-

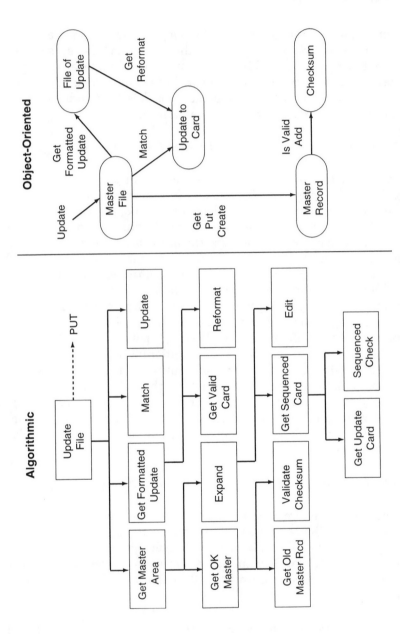

Figure 5-3 (Adapted from Figs. 1-2 and 1-3 of Object-Oriented Design with Applications by Grady Booch [3].)

142

work planner may see it as a tool for collecting usage statistics for future capacity planning; an accountant may see it as a tool for calculating capital depreciation costs for the year; a security analyst may see it as a tool for intrusion detection and, yes; a network manager may see it as a tool for managing the behavior of the network.

2. *The designer should not be forced to make any assumption about the implementation of the design.* Any modern network will consist of components of many different types, made by different manufacturers with very different technology perspectives. Requiring these disparate manufacturers to agree to a common schema at the information level requires a tremendous amount of commitment and work. Asking for a common implementation may be impossible simply because of the physical limitations of the different devices. Consequently it is critical that a design methodology for an INMS be completely divorced from the implementation.

3. *The design should proactively exploit what commonalities exist, i.e., it should avoid* reinventing *the wheel.* So far we have repeatedly emphasized the heterogeneity of network elements. However, it is also true that there exists a great deal of commonality among the features and functionality of different elements. For instance, an electromechanical switch and a digital switch may be separated by a generation of technology and functionality, yet both perform the same basic functions and share common features that are characteristic of all switches regardless of implementation. The design must be capable of explicitly capturing these common features, thus avoiding repetitious modules. A related facet of exploiting commonality is what can loosely be termed *extensibility.* For instance, if a vendor develops a new model of a modem that has six new features and has omitted two old features, then a good design should be able to handle the new component without extensive rework.

4. *A good design will allow common actions to be taken on different objects.* For example, the INMS might have to query different devices for their time of day, without caring about whether the component consults a digital clock, a sun dial, or a remote time service to answer the question. In the network management world, this concept is essential to permit version control [5].

The criteria listed above are not exhaustive, and their relative importance to others we have not listed might be questioned. Furthermore, more than one design methodology exists that, to different degrees, help meet these criteria. But, for the rest of this section we will focus on object-oriented design, a methodology that analysts in many standards bodies have found to be better suited than most, to help meet the above criteria.

Booch [3] states that software systems are inherently complex. By utilizing OO analysis and design techniques, the complexity of the software system can be simplified through the use of decomposition, abstraction, hierarchy, and associated principles. As we will see, the specification of a managed object, when coupled to a complete information model using object-oriented principles, is a powerful tool for managing complexity and integrating systems.

5.6 OBJECTS

What is an object? A person, an apple, a switch, a modem, a cable, a terminal, this book . . . all these are objects. Objects have attributes: an apple may be red in color, a cable may be cut, this book may be deserving of your attention, a person may have an attitude problem. Objects can perform actions (or have actions performed on them): a cable can transfer data, a book might let you turn a page, an apple can be eaten (and thereupon cease to be an object), a person can lend you a dollar. Objects can consist of other objects: a person has limbs, a torso, and a head; a digital switch has thousands of different ICs and wires. Objects can have common properties: both a terminal and this book can be read from, the apple and the person have common biological characteristics. Objects can be operated upon by common actions: both an apple and this book can be digested. So far, *nothing* that has been described is in any sense new or not common knowledge to any person three years old and older. It is precisely this simplicity that makes object orientation a powerful design paradigm. Humans intuitively perceive the world as a collection of objects that interwork. This is in contrast to a process-oriented world view that is often inherently less intuitive, more complex, and less likely to meet the design criteria listed above [5].

The concept of an object in the OO paradigm allows the analyst to concentrate on the information view of a resource without losing sight of the resource itself. In modeling terms, an object might be defined as a representation of some logical or physical entity of interest in the design of a system. It is a grouping of characteristics that are important in the construction of some view of the entity. More formally, Smith and Tockey [6] define an object to be a representation of an individual, identifiable item, unit, or entity—either real or abstract—with a well-defined role in the problem domain. This definition makes clear another characteristic of the OO technique that makes it important to NM. It does not restrict the analyst to the "real" or concrete world. The NM analyst may wish to specify information about a real world object (physical or logical) such as a circuit pack plug-in unit in a network router or an event register in some network element. But there are also entities of interest to NM that are purely conceptual abstractions. Services, circuits, connections, and information paths are all examples of these. Each is an abstraction that may consist of an ensemble of logical and physical entities but is managed as an entity itself.

5.6.1 Managed Objects

In modeling for network management, logical and physical resources of interest are defined as managed objects and structured within an OO information model. A managed object is defined in terms of the attributes it possesses, the operations that may be performed upon it, the notifications it may issue, its relationship with other managed objects, and other forms of behavior it exhibits [2]. The object model is constructed based on the principles of encapsulation, abstraction, inheritance, and allomorphism. It is these principles that make the OO paradigm an attractive tool for modeling integrated systems.

Before discussing the four principles directly, let us look at the components of a managed object. Looking ahead to Fig. 5-5, each of these components is displayed in an example managed object specification. We discuss them only briefly here as they are dealt with in much more depth in other chapters of this book.

Attributes.   Attributes are the object's characteristics, expressed as data, that we are interested in managing (using in some fashion). In the example, the attribute *networkElementId* is shown. Further down the specification, two more important pieces of the object specification are shown. First the specification of the attribute is shown and then the abstract syntax definition for the data values that this attribute can take on is shown.

Operations.   Operations define the management actions that can be carried out on the characteristics (attributes, etc.) of the object. In our example you will notice the notation to the right of the attributes: GET and GET-REPLACE. In the case of the *networkElementId*, the specification allows only the GET (or read) function, but for the *systemTitle* attribute, both GET and REPLACE (read and change) operations are allowed.

Notifications.   Notifications are the issuance of information by the object without prompting. The designer can specify certain events that the object must be capable of recognizing and notifying the managing system about in some standard form. In the example, the notification called *attributeValueChangeReporting* (which is further defined in the same fashion that the attribute was) requires the object to report to the managing system whenever the value of an attribute changes.

Relations and Behavior. The format used in the example for defining managed objects defines the behavior of the object in plain English. The behavior can describe the resource, and explain relationships to other objects in the model and any other behavior the designer feels appropriate to explain to developers. Behavior is also a part of the specification of attributes and notifications.

5.7 OBJECT-ORIENTED ANALYSIS AND DESIGN PRINCIPLES

A note of caution is in order at this point. More than one perspective can be taken in regard to the principles discussed here. Different perspectives can derive different implications about the application of the principles. So it is important that the discussion continue from a single perspective. It is especially important to distinguish between the perspective of an implementer/developer and that of an analyst/designer. The analyst/designer, who would attempt to integrate systems of different manufacture, is concerned with decomposing the problem and creating a design that can be widely understood and implemented in many ways. The developer/implementer normally views the problem from the perspective of some particular implementation method. This chapter proposes that OO analysis and design principles are effective tools for developing a design platform for integration of NM systems but stops short of suggesting that OO programming and DBMSs are the best way to implement the design. We shall use examples from the C++ object-oriented programming language to illustrate the major OO design principles so that a language perspective can assist our understanding. However, we want to make it clear that the basic OO design principles can be implemented in several different languages, both object-oriented (e.g. C++, Smalltalk, Eiffel, etc.) and non–object-oriented languages (e.g. C, PASCAL, etc.). Naturally, OO programming languages provide better facilities for implementing OO design than their non–OO counterparts.

There are many so-called principles of good design. Their relative importance within the design process can depend upon the designer's perspective, goals, and background. The principles discussed here are particularly representative of, or supported by, the OO approach.

5.7.1 Abstraction

Our first design criterion was that the system should be capable of providing different views to different people. The OO principle of *abstraction* provides a method for achieving this. Any object in the system can be *abstracted* by its interface. In other words, every other detail of the object can be (and is) hidden from the view of the user. For instance, when you walk up to a teller in a bank and withdraw money, you abstract away the incredible complexities of the teller object into a simple interface wherein the teller talks English, accepts your withdrawal slip, and gives you money. The same teller object has a very different interface to her or his six-month-old baby who does not know what money is, and has another very different interface to the bank manager. The bank manager, the baby, and you used the OO principle of abstraction to narrow your focus to the interface and to the properties of relevance. In the network management world, such abstraction is critical to managing complexity. Large complex objects interface with the INMS and the user through a simple abstract interface.

The designers of integrated network management systems use this principle in two ways. First, abstraction is used to construct a hierarchial decomposition of the problem by backing away from particular brands and types of resources to find classes of commonality. As Meyer puts it, "rather than describing individual objects, you should concentrate on the patterns that are common to a whole class of objects" [4]. Second, it is used to construct managed objects that present characteristics that represent some widely held view of a network resource. The task is to choose a view that is truly useful to a wide audience and to craft the abstraction so that the view of interest is fully represented. We will see that this task is supported by the other principles, especially that of inheritance.

Figure 5-4 is an example of the use of the abstraction principle to construct a hierarchical decomposition. This figure was taken from the early analysis work done in the ANSI T1M1.5 committee as they began the work of developing a North American standard network model. At this point we will concentrate on the abstraction process used rather than the details of the model itself. Notice that rather than objects such as routers, servers, switches, terminal multiplexers, etc. (specific types), the model presents objects that are higher-order abstractions. In the case of the resources mentioned, a single object (network element) captures a view that is common to all of them. This is the first use of abstraction. The standards committee saw that the resources mentioned fell into a class (elements of a network) and that identifying the characteristics common to the class is an important step in decomposing the problem of managing the network. Once the class has been identified and abstracted at the highest usable level, specialization, as described in the next section (inheritance), can be used to accommodate unique instances of the class.

The second use of abstraction is demonstrated in Fig. 5-5. Here we see an example of a specification for a managed object in the model. The committee has specified the information that provides a particular view of the object. Again we will concentrate on the attributes of the object and leave other facets of the object for later. The network element specification

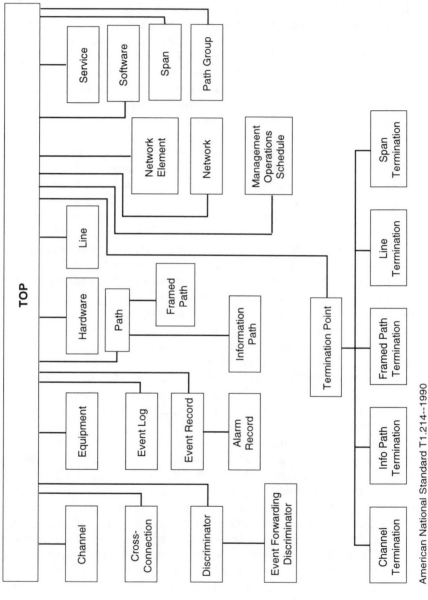

American National Standard T1.214–1990

Figure 5-4

```
                        GDMO Data Dictionary


network Element                 MANAGED OBJECT CLASS
   DERIVED FROM                 "ISO/IEC 10165-2":top;
   CHARACTERIZED BY             networkElementPkg              PACKAGE
      BEHAVIOR DEFINITIONS      networkElementBehavior;
      ATTRIBUTES                networkElementID              GET,
                                systemTitle                   GET-REPLACE;
   NOTIFICATIONS                eventReporting
                                attributeValueChangeReporting,
                                stateChangeReporting;


                                    .

                                    "

                                alarmIndicationPkg


                                    .

   CONDITIONAL PACKAGES
                                    ,
                                                              PRESENT IF an
                                alarmSeverityAssignmentPkg    instance supports
                                                              it
- - - - - - - - - - - - - - - - - - - - - - - - - - - - - - - - - - - - - - - -
networkElementBehavior    BEHAVIOR
   DEFINED AS   The Network Element object class is a class of managed objects that represent
                telecommunications equipment (either groups or parts) within the telecommunications
                network and performs network element functions, i.e., provide support and/or service
                to the subscriber. Network Elements may or may not additionally perform mediation
                functions. A Network Element communicates.....

                A Network Element contains equipment that may not be geographically
                distributed. The various equipment that comprise a geographically distributed Network
                Element may themselves be interconnected using other Network Elements.
- - - - - - - - - - - - - - - - - - - - - - - - - - - - - - - - - - - - - - - -
networkElementId ATTRIBUTE
   SINGLED-VALUED
   WITH ATTRIBUTE SYNTAX     NameType
   MATCHES FOR               Equality
   BEHAVIOR                  {--see section 6.2.49--}
::= {attribute 49}
- - - - - - - - - - - - - - - - - - - - - - - - - - - - - - - - - - - - - - - -
NameType::= CHOICE {
   number    INTEGER,
   pString   PrintableString
}
--The familiar reader will note that the original ASN.1 macro format has been respecified in GDMO form --
```

Figure 5-5 (Note to Figure 5-5: The specification is noted in an abstract syntax ASN.1 for Abstract Syntax Notation One and the format follows the Guidelines for Definition of Managed Objects (GDMO) issued by the International Organization for Standardization (ISO). Although it is not the purpose of this chapter to deal with formal specification languages, it is appropriate to mention that these tools are also important concepts in creating a common schema. Note that the syntax allows us to specify the object without worrying about encoding or data values. The value of each attribute is important only to the syntax and range level. ASN.1 is an admittedly complex language to work with, but is also very powerful as a specification language. For further information on this widely used syntax and specification template, please refer to the bibliography.)

should identify information that would be of interest about every network element regardless of type. Quickly note the kinds of information provided in this object. Name, operational state, administrative state, etc. are items of information we would expect to need, regardless of the resource.

As an example of how the abstraction principle influences an object oriented language,[3] consider the simple example of a modem-object (which we shall create for illustration purposes) shown in Fig. 5-6. Using the C++ syntax for class declaration, we have defined a class called modem. (In C++, a class is a description of a user-defined type, which when instantiated, is termed an object.) This object has a private (internal) portion, which we shall discuss shortly, and a public (external) portion. The public portion is the interface of this object to the outside world. For instance, a network manager's abstract view of this object may be limited to simply the functions baud_rate(), error_rate(), and set_param(). That is, the manager does not care about the modem's cost or the model number. On the other hand, the accountant's abstraction of this modem is restricted to the interface function cost(). Notice, for now, that when we were using the abstract syntax to specify objects in the design model, we were only concerned with the public portion—what is seen on the interface. This concept of a public interface that conveniently provides a simple, abstract view of what may be a very complex object is central to OO and is the method we use to satisfy our first INMS design criterion of being different things to different people. In summary, use of the abstraction principle means that a managed object will provide a specification of information, in some form, that abstracts the essential characteristics of an object relevant to a particular purpose (or view).

5.7.2 Encapsulation

Hand in hand with *abstraction* goes the second OO design principle of *encapsulation*. If abstraction provides us with a simple interface to a complex object, encapsulation hides the complex and proprietary implementation details.

Returning briefly to the systems management model (Fig. 5-2), we note that the overall complexity of the system being managed is hidden. In fact, the complexity is a major stumbling block in developing remote management. Each element in a network has its own unique complexity. There are many different types of resources and many different manufacture designs within each type. Encapsulation is a principle applied to cope

---

[3]Although the C++ class declaration public interface specifies function calls within a system and therefore differs from the network management interface specified in GDMO form (which is a message passing interface between systems), the OO principles can be more fully understood by seeing how they apply in both cases.

```
/class Generic_Modem{
  public
    int return_baud_rate(void);
    int return_status(void);
    int last_activity_time(int port_num);
        void set_parameters(int num_bits,char  parity,int
  stop_bit);
  private
    .
    .
    .
    .
}
```

Figure 5-6

with network architecture, technology, and functionality that is continual-ly changing. Coad and Yourdon [7] emphasize encapsulation as the prin-ciple of information hiding and formally define it as: "A principle, used when developing an overall program structure, that each component of a program should encapsulate or hide a single design decision. . . . The in-terface to each module is defined in such a way as to reveal as little as possible about its inner workings." (Hence the public and private portions of our C++ example.) Booch [3] defines encapsulation to be the ". . . process of hiding all of the details of an object that do not contribute to its essential characteristics." This principle suggests that an object has two parts—the view that is presented at the interface (public) and the details of the implementation (private). Encapsulation suggests that the object specification in the design model need be concerned with the view on the interface only and leaving the details of the implementation hidden. The goal of applying the encapsulation principle is to specify modules that are independent and, therefore, are resilient to change and can be convenient-ly reused without modification to solve new problems [8]. This is an impor-tant principle for integration of systems. If the model is modularized so that each module is independent of all other modules and independent of implementation details, then we can hope to build managing systems that share the same understanding of information with a variety of agent ap-plications built on the same model.

The analysts in the T1 committees (who are concerned only with the external view of the object) encapsulate objects much as Kim and Lochovsky [8] describe for encapsulation of programming languages. The visible interface is a set of standard operations on the object enacted by message passing. The behavior and syntax of the object's attributes are specified but the internal data structures and implementation of the

```
class Generic_Modem{
public
  int return_band_rate(void);
  int return_status(void);
  int current_cost(void);
  void set_parameters(int num_bits,char parity,int stop_bit);
private
  /* Private Data */
  char baud_rate;/* A=300,B=600,C=1200 D=2400 E=9600 */
  int status;
  int initial_cost,age;
  struct byte_type{
    int number_of_bits;
    int parity; /* 0: Even 1: Odd */
    int stop_bit; /*0: Yes 1: No*/
Rs232_interface interface_ports[2];/* The modem object HAS-A two
                    Rs232_interface objects (composition) */


  /* Private Functions: Can only be called by functions within this object */
  int calc_depreciation(int initial_cost,int age);}

int Modem::return_baud_rate(void) {
  case baud_rate
    {
     'A': return(300);
     'B': return(600);
     'C': return(1200);
     'D': return(2400);
     'E': return(9600);
    }
}

int Modem::return_status(void){
  return(status);
}

void Modem::set_parameters(int num_bits,char parity,int stop_bit){
  byte_type.number_of_bits=num_bits;
  byte_type.parity=parity;
  byte_type.stop_bit=stop_bit;
}

intModem::current_cost(void){
 return(initialcost-calc_depreciation(initial_cost, age));
}
```

Figure 5-7

operations is not. Communications is via a standard communications protocol in accordance with the systems management model.

To illustrate how this principle is realized in an OO language, let us see a more fully developed view of our modem object. The details of the syntax in our simple example are not important. The points to note are:

1. The private area contains the data and functions that are internal to the object. These *cannot* be accessed by any external entities.

2. As long as the interface definition is frozen, the internals of the object, as represented in the private areas, can change as much as needed without impacting the whole system or systems that must interact with it; i.e., the internals are encapsulated and invisible to the outside world.

3. Observe (line X) that objects can be composed of other objects.

4. Observe (line Y) that the functions in an object can "call" functions present in other objects.

5.7.3 Hierarchy/Inheritance

Abstraction is greatly assisted by the principle of hierarchical decomposition and the concept of inheritance (specialization). First an abstraction that accommodates a class of objects is identified and then specialized by successively subclassing for more and more specialization. Using the analogy of vocabulary, the modeling methodology from T1M1 describes the process similar to the hierarchy of nouns. ". . . the sequence of nouns: animal, mammal, carnivore, and cat could all be used to refer to a pet" [2]. Each successive term, though, is a further specialization that better defines a particular kind of pet. The result is a hierarchical decomposition that presents a very understandable abstraction that is also extensible. As an example, we can come back tomorrow and subclass carnivore with an additional subclass "dog" without modifying the rest of the hierarchy.

In the model in Fig. 5-4, notice the hierarchy of the termination point (TP) object class. At the highest level of abstraction we identify the top object from which all other object classes are subclassed. It contains certain information we wish to make common to all objects (e.g., objectClass, which can be used for filtering). The TP object class comes next, which is then subclassed into specific types of TPs. Looking at Fig. 5-8, which is an annotated view of the object specification, the inheritance or specialization principle is demonstrated. The TP object specifies attributes, behavior, operations, and notifications common to all types of TPs. The subclass specifications do not repeat those characteristics, but instead specify only the differences that make them specializations of the super class and indicate which superclass they are derived from (the more

terminationPoint   MANAGED OBJECT CLASS
    DERIVED FROM   :"iso/iec 10165-2":top;
    CHARACTERIZED BY   terminationPointPkg   PACKAGE
        BEHAVIOR DEFINITIONS   terminationPointBehavior
        ATTRIBUTE   tpDirection   GET;
        NOTIFICATIONS   eventReporting,
                                    objectCreationReporting,
                                    objectDeletionReporting,
                                    attributeValueChangeReporting,
                                    stateChangeReporting,

    CONDITIONAL PACKAGES   isoStatePkg   PRESENT IF......,
                                    alarmIndicationPkg   PRESENT IF......,
                                    alarmSeverityAssignmentPkg   PRESENT IF......,
                                    associatedObjectPkg   present if......;

- -

terminationPointBehavior   BEHAVIOR
    DEFINED AS   The Termination Point Object class is the superclass of managed
                  objects that delimit entities such as framed paths, lines, spans or
                  channels. This object class is only used for inheritance purposes and no
                  managed object instance of this object may be created.

spanTermination   MANAGED OBJECT CLASS
    DERIVED FROM   terminationPoint;
    CHARACTERIZED BY   spanTerminationPkg   PACKAGE
        BEHAVIOR DEFINITIONS   spanTerminationBehavior;
        ATTRIBUTES   spanTerminationId   GET;

- -

spanTerminationBehavior   BEHAVIOR
    DEFINED AS   The Span Termination object class is the superclass of classes of managed
                  objects that represent the physical connections where a signal enters or exits
                  equipment or NE. Instances of this object class may provide functions such as
                  reshaping, regeneration, optical-electrical conversion, and alarm performance
                  monitoring on the span. Such monitoring may include inserting or extracting
                  overhead information.

Figure 5-8

abstract object from which they are specialized). It is understood that they inherit the characteristics of the superclass. The implication is that if we craft each abstraction in the hierarchy carefully, then an object-oriented model can be a long-lived design indeed, even in the face of quickly evolving technology, well into the twenty-first century.

The model developed by the T1 committee illustrates the principle well for the analyst, but to further understand this principle let us look again at our C++ example of the modem object (Fig. 5-9).

Now the original parent class of Generic_Modem has been specialized to include two children called FAX_Modem and DATA_Modem. Each offers additional attributes that define its particular specialization and each inherits the attributes of Generic_Modem, the parent class. Should a new classification of FAX_Modem arise in the future (for example, Copper Cable FAX_Modem and Fiber Optics FAX_Modem), then the Generic_Modem and FAX_Modem modules can be reused with the additional specialization provided by the new subclasses.

Subclasses can be instantiated or uninstantiated. Uninstantiated objects may be specified as object classes used specifically as building blocks for other classes (note the behavior description in the TP object class in Fig. 5-8). They may be used when certain characteristics are not intended to be implemented without further specialization but ensure that all subclasses of a particular class have certain characteristics. In Fig. 5-4, TOP is a ready example of an uninstantiated object class. Every class in the model is a subclass of TOP and it is not intended that the TOP object ever be implemented as an independently managed object. It contains certain attributes (such as objectClass) that all objects must have. The concept of multiple inheritance is also supported by uninstantiated object classes.

Two refinements to the principle of inheritance should also be mentioned—strict inheritance and multiple inheritance. To ensure continuing clarity in the decomposition of the problem domain, most NM analysts have found it helpful to restrict specialization to strict inheritance. In other words, we can specialize by adding characteristics, not by removing them. A subclass must inherit the characteristics of all of its superclasses. This rule would imply that if a subclass does not possess one or more of the characteristics of its parent, then it should be treated as a new object class. This preserves the modularity of the class hierarchy.

Multiple inheritance refers to the ability of a subclass to inherit characteristics from more than one superclass. An important use of this facet of the inheritance principle is the ability to create an instantiated subclass from clusters of uninstantiated objects that were created for just this purpose [9]. The NMF principles offer the object-oriented graphics interface where window, text, and tree objects are combined as an example of the use of multiple inheritance.

```
class Fax_Modem:public Generic_Modem{/* This says that the
                    Fax_Modem class inherits from the Genric_Modem */
public
  void send_data(bit_map_object doc);/* bit_map_object is a 'type' which
                    represents a document that can be faxed. */
  bit_map_object receive_data(void);
private
       .
       .
}

void Fax_Modem::send_data(bit_map_object doc){
       .
       .
}

bit_map_object Fax_Modem::receive_data(void){
       .
       .
}

class Data_Modem:public Generic_Modem{/*This says that the
                    Fax_Modem class inherits from the Genric_Modem */
  public
    void send_data(ascii_data_object doc); /* ascii_data_object is a 'type'
                    which represents a document that can be sent over a
                    data modem */

    ascii_data_object receive_data(void);
  private
       .
       .
}

void Data_Modem::send_data(ascii_data_object doc){
}
       .
       .

ascii_data_object Data_Modem::receive_data(void){
}
```

Figure 5-9

5.7.4 Allomorphism

Reuse of objects without modification through encapsulation and extensibility through abstraction and inheritance are important qualities of a model that would provide a platform for integrating NM. But if the integration is to last, these principles are not enough. Evolution of technology will add the complication of vintages to the problem. Over time, resources and managing systems will be deployed that will have different versions (and therefore different behaviors) of the model implemented. So, even though the model can be extended for new classes and new specializations of existing classes, some existing resources and systems may not implement the extensions. Allomorphism (called polymorphism in older literature) is a principle for managing the complexity introduced by coexisting versions. It requires an allomorphically related object class to be capable of being managed as though it were an instance of another object class (called its allomorph) in its inheritance hierarchy. This occasion would arise, for example, when a system implemented on an earlier version (before current extensions) attempted to manage some facet of the behavior of the allomorphic object. From the modeling point of view, allomorphism is a principle that allows two applications built on different versions of the same model to *move back up the inheritance tree* until they find an object they have in common and then operate from that common point.

This principle is very important for version control in network management. Here we have an environment where the network elements (and their agent processes) evolve quickly while the manager processes may be part of large centralized systems that evolve much more slowly (the familiar "dot release").

Returning once again to Fig. 5-9, we observe that the function send_data() is defined twice, once in the Fax_Modem subclass and once in the Data_Modem subclass. However, the actual function definitions will be completely different, as the manner in which a fax bit map image is transmitted is entirely different from the way in which an ASCII data file is transmitted. As can be observed from the function definitions:

- void Data_Modem::send_data(ascii_data_object doc)
- void Fax_Modem::send_data(bit_map_object doc)

the calling arguments are different. When an operations system makes a function call to:

- Generic_Modem::send_data(X)

the actual function called will depend on the "type" of X. If it is of type bit_map_object, then the send_data(+) function in the Fax_Modem subclass will be called, and if it is of type ascii_data_object the function in the Data_Modem subclass will be called. The advantages of this behavior are:

1. The manufacturer can add new subclasses that implement the send_data(+) object *without* having to change any existing code in the Generic_Modem class. For our simple example this may not seem like a tremendous advantage, but for any realistic programming effort, a principle that lets you *add* new code without *changing* old code is of immense value.
2. From the operations system perspective, older systems do not have to change the way in which they communicate to the now extended modem.

It must be observed that if the object-oriented model is implemented in a non–object-oriented programming language, then many of the advantages to the software engineer disappear.

These four object-oriented principles, when applied to create a complete design, offer the analyst a powerful array of tools for specifying a flexible, reusable, and long-lived model. It is these properties that brought the OO design method to the attention of standards analysts and institutionalized it as the method of choice in American national and international standards bodies developing network management standards.

REFERENCES

[1]. Bellcore, "Transport configuration and surveillance, background of operations interfaces using OSI tools, Tech. Rep. TR-NTW-000836, Issue 1, July 1992.
[2]. ANSI T1M1, "A methodology for specifying telecommunications management network interfaces," Tech. Rep. T1M1/90-087, 1987.
[3]. G. Booch, "Object oriented design with applications," Redwood City, CA: The Benjamin/Cummings Publishing Company, Inc., 1991.
[4]. B. Meyer, *Object-oriented software construction*, Englewood Cliffs, NJ: Prentice Hall, 1988.
[5]. R. Ganesan, "Objects and entities, a perspective," Research Notes, Bell Atlantic, 1992.
[6]. M. Smith and S. Tockey, *An integrated approach to software requirements definitions using objects*. Seattle, WA.: Boeing Commercial Airplane Support Division, 1988.
[7]. P. Coad and E. Yourdon, *Object-oriented analysis,* Yourdon Press, Englewood Cliffs, NJ: Prentice Hall, 1991.

[8]. K. Won and F. H. Lochovsky, *Object-oriented concepts, databases and applications,* Chap. 1 "Survey of OO Concepts," Oscar Nierstrasz, ACM Press, 1989.

[9]. Network Management Forum, J-Team report on modeling principles, June 1990.

CHAPTER 6

Modeling and Simulation in Network Management

Victor S. Frost
Electrical Engineering and Computer Science
University of Kansas
Lawrence, KS 66045

6.1 INTRODUCTION

Users of today's communications networks want them to be 100 percent available, exhibit ideal performance (i.e., be transparent to end users), provide ubiquitous connectivity, and be smoothly improved to accommodate additional users, new services, applications, and technologies. The goal for network management is to meet these customer expectations. Thus, network management [1] must deal with:

- Operational control of the network
- Basic administration
- Network monitoring
- Network planning

The operational control of the network is often on a short (e.g., minute-by-minute) time scale to maintain system availability. Basic administration (e.g., access control and change management) occurs on a longer time scale, i.e., days. The network must also be monitored to measure performance and identify usage trends that can become part of longer-term growth planning. Network bottlenecks are also identified through monitoring. In addition, network management is concerned with the long-term planning of the evolution of the system to maintain performance in the presence of traffic growth and technological advances.

Network performance prediction plays an important role in network management. Clearly, long-term planning requires the evaluation of topological and technological alternatives based on both cost and performance [2]. The elimination of bottlenecks can be accomplished through performance analysis. Waiting until a performance problem arises and then simply adding hardware (e.g., facilities between network elements), may not necessarily be cost effective. Anomalies in the operational characteristics can also be identified using performance analysis. The operation of networks in the presence of failures can also be determined using performance prediction.

From the perspective of the network manager, performance analysis provides metrics that include delay, throughput, and call and packet loss probabilities, among others. Performance prediction transforms an assumed level of the demand for network services and resources and a network description into the desired performance metrics. There are several alternative methods used to predict the performance of communications networks, including analytical techniques, computer simulation, projections from existing experience, and experiments (i.e., build it and measure the performance).

Analytical techniques use a mathematical model that is typically derived from queueing theory [3–15] to represent the network. Often the average resource holding time (e.g., packet lengths and link capacities) and the frequency of user demands (e.g., packet arrival rate) are used to summarize the network load. The desired performance metrics are expeditiously calculated from the load and model. Speed is a major advantage of analytical models; results can usually be obtained within minutes. However, these models are often based on a set of assumptions regarding the nature of the load and the operation of the network. For example, many models assume that the time between requests for network resources is a random variable with an exponential probability density function and that the requests are served in a first-in, first-out (FIFO) order. Even though modeling assumptions can limit the validity and accuracy of these models, analytic techniques are quite useful in the early stages of a project to develop a quick view of system behavior. Often the theoretical approach can provide an initial approximate prediction of system performance. (See references [3–15] for a detailed discussion of analytical techniques used in telecommunications networks.)

Performance can also be predicted by extrapolating past experience. For small changes from a base network, this technique may be useful. However, the relationship between the network load and the performance metrics can be highly nonlinear. It is common for such systems to work in almost a binary mode, i.e., below some threshold of network demand, the system operates satisfactorily—small changes in load cause small changes in performance. Above such a threshold, a small change in load can cause a very large change in performance. Above the threshold, the system is essentially broken. For example, using the simplest queueing model (a M/M/1 [9]) to represent a statistical multiplexer with an output link rate of 56 kbps and an average packet length of 1024 bytes, we can see that a change in load from 20 to 25 percent only causes about a 7 percent increase in delay, a change in load from 70 to 75 percent results in a 20 percent increase, and going from a load of 85 to 90 percent causes a 50 percent increase in delay. Care must be taken when projecting future network performance based on past history.

Performance can also be determined from measurements obtained from operating networks. In most cases this is not a practical approach. Networks are too large and expensive to build to then determine their performance. Increasing the load or experimenting on an operating network to predict its response will disrupt users.

A valuable tool in science and engineering is computer simulation. Simulation involves developing a computer program that "acts" like the system under study. This chapter is not intended to be a detailed tutorial on computer simulation or a monograph for experts in discrete event simulation. There are several excellent texts [16–24] and overviews [25–27] on this subject. Rather the important aspects of simulation and modeling environments that are relevant to network management tasks will be discussed. It is important for network managers to understand the complexities involved in computer simulation of telecommunications networks. Profitable application of simulation technology requires much more than executing a computer program. Its use demands a clear awareness of the network operation and desired performance metrics. This translates into a need to know how to model the network under study as well as the identification of the appropriate statistics to collect. As we will see, some existing simulation tools help the users with these complex tasks. A goal of this discussion is to provide a background for network managers to enable them to acquire and productively use simulation technology.

The execution of the computer simulation model is comparable to conducting an experiment on the target system. Communications networks are amenable to representation by computer programs; the protocols in networks are algorithms. The random demands for network resources are modeled in simulation using pseudo random-number generators. Such models are characterized as stochastic [16]. Simulation can be used to closely model reality. Such models contain the immense detail of network operation and the simplifying assumptions involved in analytical models are thus avoided. Detailed models may require enormous amounts of computational power, however, it is not uncommon for network simulations to require days of processing time on modern workstations. Unfortunately, in many cases there is no viable alternative to simulation.

Several factors have combined recently to increase the use of simulation in network management. First, the computational power exists to perform extensive simulation studies of realistic networks. Second, a new generation of communications network simulation environments has become available. These new tools provide an integrated setting to graphically describe the target network, execute the simulation, and analyze results. Some of these tools provide models for common networks. The analyst no longer needs to write detailed simulation programs. These tools can be directly used by network managers who are not necessarily experts in writing simulation software. Flexible plug-and-play modules are the basis for these tools. The ultimate goal is to make these tools as easy to use as spreadsheets. Care must be exercised in applying simulation, however, as it is easy to draw erroneous conclusions from the inappropriate use of these environments. A goal of this chapter is to guide the network manager toward the legitimate and productive use of simulation because such tools are increasingly applied in the long- and short-term planning process.

The highly dynamic nature of demands for network resources combined with rapidly changing technology require detailed planning and performance prediction. The need for highly reliable networks is also leading to the use of simulation to determine the system response to component failures. Simulation environments will increasingly be used as they become more tightly coupled to other network management tools. For example, several simulation tools now accept actual traces of network data measured by network analyzers. These data are used in the simulator as input traffic to drive models of future system configurations. Measured performance data can also be compared to simulated results to tune and validate the models. It is projected that many tools will soon have interfaces to network management systems, e.g, the simple network management protocol (SNMP). Such interfaces will allow the tool to automatically configure simulation models based on interactions with the management system, relieving the user from the sometimes tedious task of describing the base model. These interfaces will encourage the integration of simulation into the short-term operational management of communications networks.

The purpose of this chapter is to introduce simulation in a network management context, discuss how networks are represented for simulation, and introduce some criteria for selecting a simulation paradigm for specific applications. The goal is to give network managers who are unfamiliar with simulation technology a basis for evaluating not only the limitations of simulation but the myriad evolving simulation products. An overview of the specific capabilities of each of these products is beyond the scope of this chapter (a discussion and listing of current simulation products can be found in [28]). A representative case study will be presented to illustrate the use of simulation in determining the performance of a campus network.

6.2 WHAT IS COMPUTER SIMULATION OF COMMUNICATIONS NETWORKS?

The goal of executing a computer simulation of a communications network is to obtain some notion of how a system will perform under a given set of conditions. A comparison between two or more alternative systems is often desired. From this goal we can deduce the three elements involved in computer simulation of a communications network: (1) modeling the demands on network services and resources, (2) modeling the mechanism the system uses in processing those demands, and (3) instrumentating the model to collect statistics to form the performance predictions. In this section each of these elements will be discussed, along with the limitation of simulation technology.

6.2.1 Characterizing the Demands for Network Resources

Applications (e.g., simple phone calls, electronic mail, and file transfers) generate demands for network services and resources. These demands can be modeled on several time scales [29]—subscription, call, burst, packet, and cell—as shown in Fig. 6-1.

Analysis is often done at only one time scale. In some cases multiple levels of time scales must be considered. However, in each of these instances the resource demands form a stream of requests and resource-holding times. For example, consider a LAN-LAN file transfer application operating over the future broadband integrated services digital network (B-ISDN). The information transport in B-ISDN is based on asynchronous transfer mode (ATM) using small, fixed-length packets called cells, of length 53 bytes. In this case a request for a file transfer of 1 Mbyte must be broken up into about 122 Ethernet packets, each 1024 bytes in length. Each of these packets are segmented into about 22 ATM cells for transmission. These 2,684 cells are then transmitted through the B-ISDN. Note such file transfer traffic can be modeled as a burst of 122 packets onto an Ethernet or 2864 cells on the B-ISDN, depending on the level of modeling required and the desired performance metrics. Not only are different time scales involved in modeling demands for network resources but telecommunications traffic is often time-variant. Typically, there are peak demands in the mid-morning and mid-afternoon.

Fundamentally, the modeler must obtain characteristics for the demands on the target network and then translate those demands into a

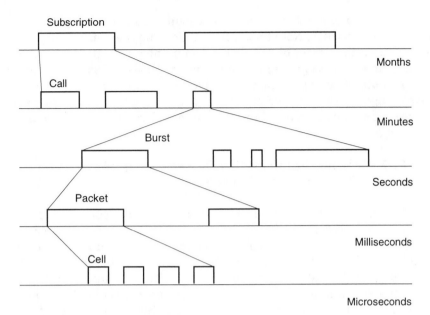

Figure 6-1 Time Scale of Demands for Network Resources

traffic model, which is a difficult task. Network traffic is dynamic and changes with the applications running over the network and the technology. Different applications (e.g., E-mail, credit card validation, data entry, and computer-aided design) will generate different demands on network resources. These demands will also be a function of the presence of an interactive human interface in the application. Some studies have reported traffic measurements [30–33]. Unfortunately, there is often a lack of correlation between the measurements and the higher layer traffic sources. For example, measurements of only packet length and interarrival times on a LAN do not help in characterizing the end-to-end application-to-application traffic. In addition to length and interarrival times, source, destination addresses, and application are needed to complete the picture. Time and thought must go into modeling the traffic demands on the system for accurate modeling. The quality of the results can be no better than the input.

Traffic sources are often characterized by two random processes—one describes the interarrival time while the other models the holding time. In queueing theory these are usually assumed to be Poisson processes, i.e., the interarrival and holding times are independent exponential random variables. Simulation allows for arbitrary specification of these processes through the use of pseudo random numbers. (See [16] for a detailed discussion of pseudo random numbers.) When developing simulation models, the analyst selects the time scale (e.g., call, burst, or packet) for the model and constructs (or uses prebuilt) modules to represent the traffic characteristics. Note that segmentation and reassembly frequently are required. Sequences of pseudo random numbers are used to generate holding times, message lengths, interarrival times, message/packet errors, and other random elements involved in communications networks. Developing accurate models of network traffic is an evolving task; such models are elusive because of the rapidly changing nature of communications networks.

6.2.2 Modeling the Information Transport Mechanism in Communications Networks

Information is originated by users to be transported over communications networks. The desire to move this information creates demands for network services and resources as described in the previous section. The communications network processes these demands and delivers the information to the destination. Thus, a major element of simulation of communications networks is the development of a computer program that mimics this network processing. Simulation allows for the modeling of the system in minute detail. In the extreme, the system could be represented from the application through the physical layer of the protocol stack, including the waveform and noise aspects of the transmission medium. Obviously this is not practical in most cases. The analyst must be aware of the trade-off between the detail in the model and the time required to execute the simulation. It is important to include in the model only those network functions that have an impact on the desired performance metrics. For example, queue fill statistics are often desired under different network loading conditions—in this case steady-state traffic flows would be considered. Including the details of call processing (i.e., session setup and tear down) would not be needed. Fortunately, tools have been developed that ease the modeling tasks and in some cases include plug-and-play modules of common network elements. Network simulation "experts" have built these components at the appropriate level of abstraction for the desired performance studies. However, as pointed out in [16], care must be taken in reusing simulation software. A reverification and revalidation process is recommended to confirm that the models operate as expected. That is, simulation software that is to be reused should be considered suspect until the user has confirmed that it performs the desired functions and that those functions represent the system under study. The purpose of this section is to introduce how the network modeling is accomplished in simulation. This basic understanding is needed to appreciate the relative merits of simulation for specific network management tasks. Furthermore, this background will be useful in evaluating alternative modeling tools.

Networks process the demands for resources using sets of communications protocols. These protocols are implemented in algorithms (for lower layers in hardware while in upper layers, software is used). From a network performance perspective, typical functions of the protocols include error control, flow control, rate control, encapsulation, segmenting/reassembly, routing, contention resolution, blocking, and transmission scheduling among others. The network simulation mimics the target system by first generating events to represent the arrival of user demands and then uses a model of the protocol to mimic the network processing. Note the actual protocol is not necessarily implemented, rather an abstraction of the process is often used. Usually, only the elements of the protocol that effect performance are considered.

Protocols govern the response of the network to user demands; i.e., for a demand event that occurred at time t, the network processing of that event depends on the *state* of the system at time t. For example, a simple routing algorithm may send packets to the output link with the shortest buffer. This event-based processing lends itself to a method known as *discrete event simulation* (DES) [34]. Discrete event simulation is the underlying methodology used in most of the high-level simulation tools. An alternate approach is activity scanning. In activity scanning the analyst specifies the activities that each element (packet) in the system encounters and the conditions that cause each activity to start and end. Because of the need to scan each activity at each event, the activity orientation is inefficient and not widely used as a modeling framework [35]. When digital computers were first applied to simulating queueing systems, DES was implemented using a general purpose language, such as FORTRAN. Even today some analysts prefer to develop DES code in a general purpose language and C is often used. It was recognized that some efficiencies could be obtained through the development of "modeling languages." As discussed by Kurose [36], these textual based languages (e.g., GPSS [37], SIMULA, SLAM II [35], and SIMSCRIPT [23]) were aimed at general queueing problems and freed the modeler from low-level details of the simulation. They formed the first generation of modeling languages. Further improvements could be made by focusing on a specific class of systems, e.g., communication networks. The second generation of modeling languages were constructed especially for these systems. Included in this set are RESQ [22, 38, 39], PAWS [40], and Step-1. These were still textual based languages. Modern workstations provided the technology for the third generation of communications network modeling tools; included in this group are Q+ [41], SES/Workbench [42], OPNET [43], Genesis [36], TOPNET [44], NEST [45], and BONeS [46]. A summary of such tools can be found in [28] and [47]. These tools are graphical based and provide a complete environment for performance evaluation. That is, the network model is constructed using pop-up menus and a set of user-connected icons on a workstation screen. The simulation is executed and controlled and the results are analyzed in a common environment. Some of these tools can be thought of as graphical programming languages tailored to communications network simulation. Another feature of some of these tools is their ability to animate the network traffic flows and protocol processing. Such animation is a valuable aid in debugging the models as well as communicating the nature of the system to nontechnical personnel. We are now seeing the fourth generation of tools. These tools enhance the previous generation through the addition of interfaces to network monitors and management systems [48]. The fundamentals of DES will be presented next.

We have seen that the input stimulus to telecommunications networks is a sequence of demand events occurring at random times. The processing of an event depends on the state of the system, e.g, queue fill. A DES has the same structure. The state of the simulated system is stored in a set of system-state variables. Pseudo-random events cause this state to change and the scheduling of other events using event routines. These event algorithms describe the specific state variables to be modified as well as the timing of future events. An event calendar stores the time-ordered sequence of events to be executed. Event routines can change the event calendar. Network simulation is usually event driven, i.e., time increments to the event with the earliest time on the calendar. For example, in Fig. 6-2 a message arrives at time t with the state of the system shown as Queue Fill = 0.

In this case the arrival event schedules the departure of the arriving message at time t plus its length and schedules the arrival of another message at t plus some random time, RN. Note that the system state will be different upon the arrival of the second message. In this simple ex-

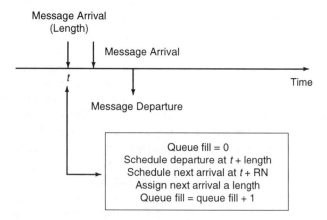

Figure 6-2   DES Example

ample the next event to be processed would be an arrival. It is clear that the DES must have an initialization mechanism to establish the starting system state; statistical collection routines; a post processor to transform the collected statistics into the desired performance estimates (to be presented in the next section); and a coordinating program to control the event calendar, post processor, and the initiation/termination of the simulation.

The entity acted upon by the event routines form another element involved in DES. In this example the entity is the message. Here the message has a simple data structure, i.e., its length. In more realistic models this data structure would contain other information. For communications network simulation, the data structure used in DES is often closely coupled to the message/packet format. For example, a standard logical link control (LLC) packet contains a destination service access point, source service access point, control information, and data. In a DES of an LLC, the data structure for the entities would contain these elements in addition to simulation specific information, e.g., a packet creation time stamp that would be used for statistics collection. The data field could contain a length indication or, in more detailed models, a pointer to another data structure that represents a network layer packet data structure. Encapsulation of data structures is a common feature in communications networks and DES tools for these systems often have an encapsulation component.

DES is the combination of:

State information
Event calendar
Event routines
Entities (data structures)
Coordinating program
Initiation/termination processes
Post-processor

and is used to mimic the operation of the target system.

Thus, modeling the processing of user demands for network services and resources entails the specification of data structures to represent the entities flowing through the system and the construction of algorithms that represent the manipulation of those data structures within a DES framework. The underlying DES foundation provided in modern network simulation tools allows the user to focus on high-level network modeling issues rather than lower-level event scheduling and event manipulation tasks. The difference in the current generation of tools is how the user specifies the data structures and algorithms. Various approaches currently used will be presented later. Given a DES model of the communication system, the user must have a mechanism to obtain the desired performance predictions. The issues involved in acquiring those predictions will be addressed next.

6.2.3 Instrumentation of Simulation Models to Collect Statistics and Form Performance Predictions

It is convenient to view the simulation model as a replica of the target system, and just as with the target system, instrumentation is needed to collect statistics to form performance predictions. Further executing the simulation is analogous to doing an experiment—the results from a simulation must be treated as random observations. However, the synthetic environment provided by the simulation model allows for the gathering of information typically unattainable from the target system, e.g., accurate time stamping is easy in simulation but often difficult with real systems. The purpose of this section is to illustrate how to obtain performance predictions and the associated potential pitfalls.

Here we will consider obtaining a message delay estimate as an example. Assume that an element in the message data structure is its creation time, i.e., at the instant the message is generated the current simulation time is obtained and inserted into the message data structure. Let $T_c(m)$ be the creation time for the mth message. The message departs the system at time $T_d(m)$. The time in the system or the message delay is

$$T_s(m) = T_d(m) - T_c(m)$$

During the course of the simulation, M messages are created and depart the system. The estimate for the average delay is then given by

$$T_{ave} = \sum_{m=1}^{M} T_s(m) / M$$

It is often important to estimate the deviation in the delay estimator, which can be related to the variance of the random variable T_{ave}. Because the samples $T_s(m)$ are correlated, direct calculation of the variance must be done with care. That is, using the simple sample variance formula on the delay samples will lead to incorrect results. The correlation in the sequence of delay samples can be intuitively seen by noting that if message k has experienced a large delay, it is likely that message $k+1$ will also find the system in a loaded state, causing it to also have a large delay. The correlated nature of the samples obtained from simulations (or real measurements) complicates the formation of performance predictions. In theory the variance of the random variable T_{ave}, $Var[T_{ave}]$ can be found from

$$Var[T_{ave}] = \sum_{n=1}^{M} \sum_{j=1}^{M} COV \{T_s(n), T_s(j)\} / N^2$$

Where $COV\{T_s(n),T_s(j)\}$ is the covariance between the nth and jth message delays. Now the problem is estimating $COV\{T_s(n),T_s(j)\}$ from the sequence $T_s(m)$. Advanced spectral analysis techniques are sometimes applied to this problem [16]. It is common, however, to take other approaches to deal with the correlation problem. Basically, the desire is to obtain identically distributed statistically independent, identically distributed (iid), samples of delay. The common approaches to this problem are: 1) simple replication, 2) batch means, and 3) regeneration.

Simple replication entails executing the simulation model K times. Simple replication is illustrated in Fig. 6-3. In this figure $N(t)$ represents the number of packets in the system at time t.

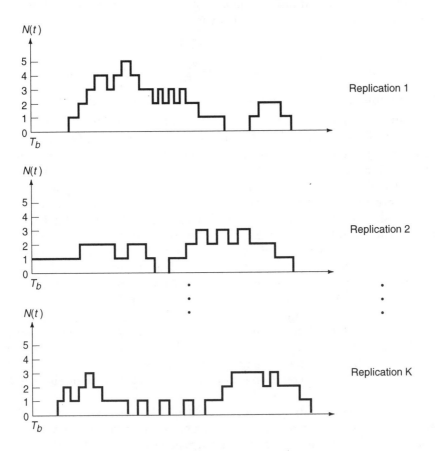

Figure 6-3 Simple Replication Method

Note that a different starting point in the pseudo-random sequence is needed for each run, i.e., a different seed is used in the pseudo-random number generators for each run. Re-executing a simulation model with the same seeds will produce the same results. The $T_{ave}(k)$, $k=1 \ldots K$ obtained from this set of runs are assumed to be iid. The sample variance, S^2 of T_{ave} is directly estimated from the K samples of T_{ave}; i.e.,

$$\hat{T} = \frac{1}{K} \sum_{j=1}^{K} T_{ave}(j)$$

$$S^2 = \frac{1}{K-1} \sum_{k=1}^{K} \left(T_{ave}(k) - \hat{T} \right)^2$$

Confidence intervals can now be easily formed. Making a normal probability density function assumption (from the central limit theorem this will be a good assumption as long as K is large enough) the confidence interval for the expected value of delay is given by

$$\hat{T} \pm z_{1-\alpha/2} \sqrt{\frac{S^2}{K}}$$

This statement says that the probability the expected value of delay is between $\hat{T} - z_{1-\alpha/2} \sqrt{S^2/K}$ and $\hat{T} + z_{1-\alpha/2} \sqrt{S^2/K}$ is $1 - \alpha$. The parameter is $Z_{1-\alpha/2}$ is the $1 - \alpha$ quantile of a unit normal random variate. For example, a 95% confidence interval corresponds to $\alpha = 0.05$ and $Z_{1-\alpha/2} = 1.958$. Thus if the confidence intervals are formed 100 times, we would expect the true value of average delay to lie within these intervals 95 out of the 100 times. Confidence intervals provide an indication of the quality of the performance estimate; the smaller the confidence interval, the better the estimate. However, there is a trade-off between this quality and the cost of the simulation. A smaller confidence interval can be obtained by increasing K, which increases the execution time of the simulation model.

In the method of batch means, results from a single simulation run are divided into batches, as shown in Fig. 6-4. Statistics obtained from each batch are assumed to be iid and the above equations applied. Note that the iid assumption will become increasingly appropriate as some samples in between each batch are ignored, i.e., not used in statistical calculations. Because there are no ignored samples between Batch 1 and Batch 2 in Fig. 6-4, a correlation may exist between the statistics obtained from those batches.

Regeneration techniques are based on the observation that the memory in a sequence of delay measurements is reset every time the system empties (there are other possible regeneration points). That is, if message m arrives right after the system empties, then its delay will be

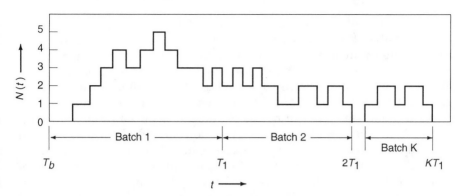

Figure 6-4 Batch Method

independent of the delays of messages 1 . . . $m-1$. See [49] for a formal discussion of regenerative processes. Figure 6-5 illustrates the regenerative simulation. The statistics obtained from different regenerative cycles (see Fig. 6-5) are iid. One complication with the regenerative approach is that the time between regeneration points is not the same. Furthermore, not all systems are regenerative and even if the system is regenerative, the regeneration time may be very long in some cases, which diminishes its advantages.

The discussion above points out the problem associated with the correlation in network measurements. Upon examination of Figs. 6-3 to 6-5, two other issues arise. First, how does the user know when to stop the simulation? If the simulation is too short, then the performance predictions will be unreliable, too long, and computational resources will be wasted. As already noted, simulation can be a computationally intensive task. Stopping rules are formally based on the confidence intervals given above. That is, the simulation is run until a predetermined level of confidence [19] is obtained.

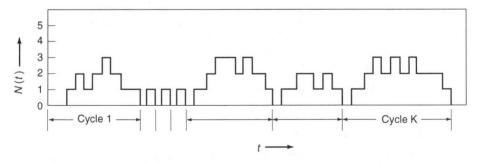

Figure 6-5 Regenerative Method

In a DES of a network, as in the real system, the response of the system evolves from its initial state. For example, if all the buffers are initially empty, then the delay of the first messages entering the network will be lower, on average, than those arriving later. Often DES is used to characterize the network in steady state, that is, after some time in a transient state, the effect of the initial condition disappears and the system reaches a stable condition. To obtain the steady-state performance, the effect of the transient period must be removed. Transient removal typically involves the deletion of data generated at the beginning of the simulation. In this case the simulation is executed from some initial condition and the measurements obtained during the transient period are not used in the calculation of the performance metrics. Determining the end of the transient period (shown as T_b in Fig. 6-3 and 6-4) is difficult in general; several different procedures have been developed. (See [16] and [24] for details.) Note that this problem is automatically handled in regenerative simulation. In some cases, however, the transient response is the target of the simulation study. For example, examining the response of the network to component failures to determine how long it takes to recover from a failure, and the system performance after the failure, would require a transient analysis.

This simple discussion shows that care that must be taken when building performance estimates from simulated data. Correlation, stopping rules, and transient effects must be taken into account. Some existing network modeling tools aid the user in considering these complex issues, for example, post-processors that automatically use the method of batch means are available. The degree to which a modeling environment helps the network manager deal with the unique statistical aspects on simulation should be carefully considered when selecting a tool.

6.2.4 Elements of a Successful Simulation Study for Network Management

It is clear from the above discussion that successful application of simulation technology to network management tasks requires a degree of sophistication. The purpose of this section is to present elements that are needed to successfully use simulation in the context of communications networks. Pitfalls and common errors involved in DES are discussed in general in [16] and [24].

Elements of a successful simulation study include:

| Know the customer |

There must be clear objectives of what is expected from the simulation study. Simulation should be used to address specific, well-identified performance questions and issues. The customer of the simulation results, whether the technical network manager or higher-level decision makers, should be aware of the goals and limitations of the simulation study; it is important to have a match of expectations. The performance metrics must be agreed to and clearly defined. Simulation studies take resources, i.e., manpower, capital (computers), and computer time. There should be a match between the goals of the simulation study and the allocated resources. Note that better results usually can be obtained by allocating more resources to the study through using longer runs and/or more detailed models. It takes longer for the analyst to develop more detailed models. A common error, like in many software projects, is the underestimation of the time and effort required to conduct a simulation study. The customer (and management) should be part of the process throughout the lifetime of the simulation effort. They should be aware of the status of the effort and be encouraged to comment on its evolution.

Know the System/Network

A detailed understanding of the network under study is needed. This knowledge is required to develop an appropriate system model. A common error is the inclusion of too much detail in the simulation model. This wastes manpower during the model development as well as computer time during the production runs. Comprehension of the operation of the network will aid in focusing the effort on the critical components. As suggested by [16], sensitivity studies as well as network experts can be employed to determine the appropriate level of modeling detail. It is critical to understand the random nature of the demands on the network—traffic modeling is essential to the success of the study. Use of arbitrary models can lead to erroneous results. Inappropriate models, e.g., assuming a Poisson process, can in some cases lead to missing the actual system performance by orders of magnitude. Knowledge of the system operation will also ease the verification and validation process. Verification is confirming that the simulation algorithm is doing what you told it to do. Validation is confirming that you are telling it to execute the right process. Clearly, the more you know about the system, the easier it will be to verify and validate the model of that system.

Know the Important Performance Metrics

Model detail as well as run length are dependent upon the desired performance metrics. For example, simulations targeted to estimate low packet loss probabilities will typically require significant computations, while studies aimed at obtaining average delay and throughput will be shorter. As the simulation execution time may be quite long—i.e., hours or days—it is important to collect as much information as possible during each run for post-processing. Again there must be a match between the customers' desires and the simulation results. For example, it is not uncommon for a system specification to be in the form of *the delay must be less than X ms 99% of the time*. This specification clearly indicates the nature of the performance metric expected from the simulation study. Simulation is often used to generate performance curves (e.g., delay vs. load) and not a single point. In these cases it is recommended to have the nature of these parametric studies defined before executing the model. Simulation studies are usually aimed at exploring a part of the system response characteristic, i.e., the network performance is a function of a number of variables. Understanding the critical segments of the response characteristic prior to the study enhances its likelihood of success. Application of response surface methodology [50] and standard experimental design techniques [16] can be helpful.

> ## Know How to Interpret the Results

Simulation is analogous to conducting an experiment—the results are observations of random phenomena. The significance of the variability in the results must be appreciated. A team member with statistics training will be useful. The application of the sample variance or confidence intervals to characterize this variability is important to establish the credibility of the results with the customer. Results from a single replication of a model are often meaningless and the inappropriate use of the iid assumption can lead to significantly erroneous results. State-of-the-art simulation environments provide an animation capability; that is, the flow of the packets through the network can be visualized. This is a valuable tool for debugging the model and demonstrating the system operation. Typically, however, short animation runs cannot be relied on to provide accurate performance estimates.

> ## Know How to Establish a Credible Model

A model is called credible if it and its results are accepted by the customer as valid and used in the decision process [16]. As simulation becomes tied into network management systems, the decision process may deal with the operational aspects of the network as well as the traditional planning activity. A process to establishing a valid and credible simulation model is presented in [16]. An important step is to make sure the model has high *face value*, i.e., the results should *look* reasonable to knowledgeable people. The results must pass sanity checks. Favorable comparison of the results to theory (i.e., change some of the model assumptions to match available theory) or similar studies will increase its credibility. Obviously, favorable comparison to measurements obtained from a real network will validate the model.

| Expect the Model to Evolve |

A major advantage of simulation is that it can be used to address performance issues as the network traffic as well as the technology evolves. It is important to realize from the beginning that the simulation model will not be static. The customer will hopefully like the results, often leading to requests for modifications to the model, e.g., a change in traffic, topology, or protocol. Planning for the evolution of the model is important. Thus modularity and extendibility of models within a simulation environment is an important factor.

| Know How to Apply Good Software Management Techniques |

Simulation models are implemented as large computer programs. Even if simulation tools are employed, the network processing must be represented in some form (as will be discussed in a later section) and the model implemented within some framework. Thus, standard software development and management techniques should be used. This process begins with the selection of an appropriate modeling environment or language. It includes proper verification of the transformation of the conceptual model into software. As discussed in [24] the design objectives, functional requirements, data structures, and progress must be tracked. Such tracking is aided by current software engineering tools. Design principles such as top-down design and structured programming should be applied to assist in the systematic development of the simulation model. Incorporation of appropriate documentation is essential to the evolution of the simulation model.

The above elements are needed to ensure the success of a communications network modeling effort. They do not guarantee success, however, as simulation has several limitations: simulation models can be expensive to construct and execute, simulation does not give the insights obtainable from closed form analytical results, and, thus, system optimization is difficult using simulation. As an aid in selecting a simulation tool, in the next section we will focus on how state-of-the-art communications network modeling environments represent target systems.

6.3 PARADIGMS FOR COMMUNICATIONS
NETWORK REPRESENTATION

A variety of tools are available to provide an environment for the DES of communications networks. Each of these tools contains a system configurator, simulation exercisor, post-processor, and model library [51]. A system configurator is used to describe the network operations. Control of the execution of the simulation model is handled by the simulation exercisor. This may be a complex task involving the simultaneous specification of multiple cases and/or the distribution to the simulation computation over a network of processors. As previously discussed, the post-processor aids the user in analyzing the results; these results usually contain extensive statistical and graphics capabilities. Simulation studies tend to evolve over time. Furthermore, there are many common communications network components, e.g., window flow control procedures. This leads to a desire to reuse elements of simulation models. To assist in this reuse, model libraries are used. Existing tools approach these four tasks from a different perspective. However, the main distinction between the tools is their approach to representing the network processing in the system configurator. In this section, some of the desired features of an environment for the DES of communications networks will be described, mainly focusing on the framework of the system configurator. This will be followed by an introduction to some of the common paradigms for communications network representation.

6.3.1 Desired Features of Environments for
Computer-Aided Modeling and Design
of Communication Networks

The purpose of this section is to present the attributes needed in a modeling and design environment for communication networks. The environment defines the structure the network manager will use to construct, execute, and analyze system models for target networks.

Graphic Modeling. A major attribute of many modern approaches to network modeling is their graphical orientation. Graphical workstations provide the necessary computing power to allow the user to describe the network using icons that are typically interconnected to show the flow of information through a network. An important goal is to make the details of the DES implementation (i.e., the specific software) invisible to users. The primary advantage of the graphical orientation is that it allows for the construction of models without the need for the user to write any software. The graphical interface provides a mechanism for describing systems that are more closely matched to the system operation as compared with textual programming languages. Icons can be used to easily represent complex network subsystems. Typically the user does not need to be an expert in a specific programming or simulation language to efficiently use graphical-based tools. A goal of the graphical orientation is to minimize debugging through the use of a rich model library and then, when new models are needed, focus the detailed debugging stage on logical and modeling errors instead of faults in language syntax. Furthermore, the graphical orientation lends itself to animation or the visualization of the flow of information through the network. Animation further eases the debugging process and it also allows the analyst to obtain a qualitative feel for the system under study by observing its evolution in time.

Event and Protocol Processing. The modeling environment must be able to capture the asynchronous nature of communications networks. Random information (e.g., interarrival times and message lengths) and queueing are some of the elements that lead to the asynchronous nature of communications networks. The extended queueing network paradigm is specifically aimed at capturing this phenomenon, as will be described later.

The performance of a communications network is a function of protocols, that is, the algorithms that define how each network element is to respond to an event. A typical event is the arrival of an information packet. Graph-based models such as Petri nets and state transition diagrams are often used to model the protocol aspects of communication networks [43,44,52,53].

Note that to provide an appropriate modeling environment, both the extended queueing network and graph-based paradigm had to be modified to accommodate the full range of functions needed to model both asynchronous events and protocols.

Hierarchy. Modern communication networks are both extremely complex and geographically distributed. A major problem in network management, as well as modeling, is handling this complexity. A hierarchy is needed to manage the complexity of the network topology, protocols, and data structures. A hierarchical framework allows system elements to be constructed from either other elements or primitives. At each layer in the hierarchy, the user deals with a level of detail that is appropriate for that layer. Details of the layer below and the one above are hidden. Communication networks are particularly well suited for hierarchical modeling. A hierarchy also provides a convenient way to describe the geographical extent of communications networks. Switching offices in telephone networks are often described by a hierarchy. Hierarchy is an important element of the telecommunications management network (TMN) as discussed in Chapter 3.

Any modeling and design environment for communications networks should take advantage of the concepts of the open system interconnection (OSI) layered architecture [3]. The layered architecture is also aimed at minimizing design complexity. Each layer in the architecture performs specific functions and provides services to the adjacent layers. The details of how the services are performed are hidden from the adjacent layers. Both the interfaces between layers and the data structures used to communicate between layers must be well defined. The implementation of each layer can vary, however. As a consequence of this layered architecture, data structures used at high layers are encapsulated into the data structures of lower layers (see Fig. 6-6). Note that the header information at each layer provides that layer information on how to process the packet. A modeling environment should accommodate a layered architecture through the use of a hierarchy of data structures that allow encapsulation and the construction of network layers for elements that provide specific functions via a fixed interface. Such an environment would allow for the direct substitution of layers to study specific implementations. The layered architecture also encourages software reuse. Once models of

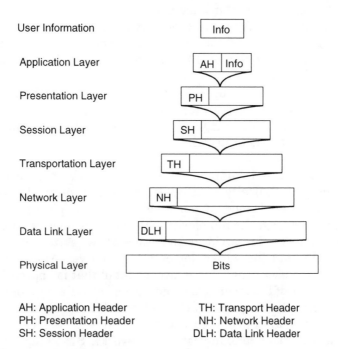

AH: Application Header   TH: Transport Header
PH: Presentation Header   NH: Network Header
SH: Session Header   DLH: Data Link Header

Figure 6-6 Data Structure Encapsulation in the OSI Layered Architecture

specific layers are constructed, they can be reused in many studies, even if the implementation of other layers change.

User Interface. The next set of attributes relate to an intelligent user interface and the performance analysis mechanism. A graphical orientation has been assumed, thus the user will construct models by interconnecting icons or blocks. The modeling environment should use expert system techniques to perform extensive consistency checking. The goal of the consistency checking is to provide the user an immediate or early indication of an error. This will reduce the usual cycle of building a model, having the software crash, and then debugging. Consistency checks protect the user whenever possible. It is hoped that a model that passes the consistency checking should execute. Additionally, the environment should facilitate teamwork. Consistency checking should alert the user to changes in models of network elements introduced by other team members. The intelligent user interface should also provide on-line documentation and help. The interface should provide a mechanism for the inclusion of documentation for new models. The modeling environment is intended to provide performance estimates for the system under test. A flexible mechanism for the instrumentation, that is, collection of performance data from the models, is essential. Typically a simulation model is not executed once. The model is executed many times with parameter variations or even with system modifications during the evolution of a study or system design. The modeling environment must allow for the convenient analysis and comparison of results from several simulation studies and system design alternatives. That is, it should provide a convenient mechanism for performing design iterations.

It is important that all aspects of modeling, analysis, and design mentioned above be handled within a unified environment. The user interface of a framework should have a modular structure containing a model library, graphical editors, simulation and database managers, as well as a post-processor.

6.3.2 Common Paradigms for Communications Network Modeling

Construction of simulation models of communications networks is a time-consuming and error-prone process when general purpose languages, such as C or FORTRAN, are used. Even when specific simulation languages like SIMSCRIPT [23] are employed, the development of simulation models is difficult. Also, reuse and extendibility of such models is problematic. To overcome these difficulties, new graphical approaches to the simulation of communication networks have been developed. These graphical approaches provide the modeler with a visual environment for constructing models of communications networks. These approaches are based on the following different ways of viewing the operation on the network.

The Extended Queueing Network Paradigm. Several graphical approaches have been based on the extended queueing network paradigm, e.g., Q+ (formerly PAW) [41], GENESIS [36], SES/Workbench [42], and RESQ [38]. In this paradigm, communications links, buffers, disks, and other elements are modeled as resources. A set of entities, i.e., packets, flow among the resources contending for their use [25]. Recently BONeS [54] has added a resource modeling capability. Extended queueing network (EQN) models contain a broader set of network elements compared with traditional analytic queueing models. These additional elements allow for the characterization of a wide class of systems. Specifically, standard queueing networks do not model important system characteristics, i.e. simultaneous resource possession, parallelism, and synchronization. Standard queueing networks that are composed of sources, queues, and sinks must be augmented. In the extended queueing networks, system modeling is done with queues, packets (or jobs), nodes, and rules to control the flow of packets. The extended queueing network is particularly well suited for capturing the asynchronous and queueing aspects of systems. Further, it is often possible to obtain analytic approximations directly from extended queueing network models (see [41]). However, it is sometimes awkward to model the detailed protocol functions using this paradigm.

The Petri Network Paradigm. Petri nets were originally developed to study the interconnection properties of concurrent and parallel activities [55]. They have found extensive applications in modeling communications network protocols. The elements of Petri nets include tokens, places, triggers (transitions), and arcs. Places contain tokens whereas arcs connect places to triggers. A trigger is enabled if each of its input places contain one or more tokens. Upon firing, a token is removed from each of its input places and a token is added to each of its output places. Petri nets are particularly well suited for the study of protocol correctness and lack of deadlock. For example, Petri nets are used to specify and prove progress without deadlock and work without delay properties of protocols [56], and for protocol correctness [54]. This paradigm has been modified to capture the time aspects of communication networks, queueing models. Data structures have been associated with tokens for network modeling. These are sometimes known as *times Petri nets*. Stochastic activity networks are such a combination of Petri nets and queueing models [57–59]. A modification of Petri nets for network simulation is given in [53]. Another modified Petri net modeling paradigm is called *PROT net* [52], and a simulation tool called TOPNET based on PROT nets has been developed [44].

The Finite State Machine Paradigm. Finite state machines (FSMs) provide another mechanism to represent protocols [55]. Upon receiving an input while in a given state, the FSM will change state (based on a state transition function) and produce an output. FSMs are typically represented with state transition diagrams, which is an advantage of FSM-based tools. Such diagrams are ubiquitous in network standard documents. FSMs must be augmented as was done in [43] to capture the queueing and timing nature of networks. A tool called OPNET [43] is based on the FSM paradigm.

The Data Flow Block Diagram Paradigm. The data flow block diagram paradigm is based on the observation that descriptions of protocols and networks require a description of the packet structure, i.e., the data structure, and the operations that will be performed on these structures by the network and its protocol. The network and its protocol are described here by an arrangement of functional blocks that transform and, possibly, queue their input data structures to produce output data structures. A tool based on the data flow block diagram paradigm is called *BONeS* [54]. Blocks are constructed in a hierarchy where at the bottom of the hierarchy are basic operations called *primitives*. This paradigm is composed of data structures, arcs, ports, and block. Data structures contain the information that is passed from one block to another. Typically data structures will contain the information that is encapsulated within a packet with the possible addition of modeling specific information, e.g., packet length or creation time. It should be emphasized that these data structures can closely follow the standard protocol packet formats and that they are organized in a hierarchy. Arcs are the lines that connect modules. Arcs represent the path for the flow of data structures within the protocol. Data structures will flow along an arc from one block to another. Arcs are connected to modules via ports. A network is specified in the data flow block diagram paradigm via a set of nested, hierarchical block diagrams. That is, new blocks can be constructed from other preexisting blocks via block diagrams, which can then later be used in the construction of still higher level blocks, and so on. These elements provide a mechanism to model both the synchronous (protocol processing) and asynchronous (event driven) nature of networks.

All of these network representations have their domain of applicability, i.e., systems where one method is well matched. For example, extended queueing networks are excellent for modeling store and forward systems. Petri nets and FSMs are well suited for studying protocol correctness. The block diagram approach can be tailored to have the characteristics of FSMs, Petri nets, and EQNs [46]. Applications of each representation outside its domain may be awkward. From the network management perspective, the paradigm that best deals with the specific performance issues of interest should be selected. Given this background, the next section will present some criteria for selecting a modeling environment for applications related to network management.

6.4 CRITERIA FOR SELECTING A MODELING ENVIRONMENT FOR NETWORK MANAGEMENT APPLICATIONS

The selection of a performance prediction tool for network management applications must consider many factors, including technical, economic, and organizational. The relative importance of these factors depends on the user. Below is a list of such factors.

Cost. Performance prediction tools for communication networks range in cost from several hundred to tens of thousands of dollars. This represents a wide range of capabilities.

Availability. Some tools are commercial products while others are proprietary.

Training required. The level of specialized training required to master these tools varies greatly. The training required depends on how easy it is to conceptualize the modeling paradigm. The modeling paradigms described above require different levels of modeling abstraction, e.g., the transformation of the operation of the target network into a Petri net model may be a nontrivial task requiring a high level of network and modeling expertise. The degree to which the tools provide online help should also be considered. Some tool vendors provide tutorials and direct training classes.

Documentation. Related to training is the level and quality of the documentation provided by the vendor of the basic framework. The extent of online documentation and its ease of use should be examined. The ease of documenting the developed models should also be considered.

Flexibility. As pointed out, a major motivation for using simulation in network management is its flexibility to change with the target network and the rapidly evolving technology. Thus, the tool must not be limited in its ability to support different network functions.

Model Reuse. Tools should also be evaluated on the ease of reusing previously developed network models. The ability to exchange simulation models between modeling groups is a factor to consider.

User Interface. The ease of use is an important consideration. The interface should aid the user in constructing the model and obvious inconsistencies and semantics errors should be automatically identified.

Model Debugging. The modeling environment should contain debugging tools (e.g., animation) to assist the user in correcting logical errors.

Portability. With the rapid increase in computational capacity coming with new workstations, it is important that the tools be portable. This portability could come from using standard window systems.

Computational Efficiency. Simulation is a computationally extensive task. The implementation of the simulation engine within the tool effects its execution speed. Efficiency is an important consideration.

Harmonious Operation in a Network Environment. The computing environment for these tools often consists of a network of workstations, possibly from different vendors. A desirable attribute of a tool is its ability to take advantage of a networked computing environment. This may simply take the form of being easily able to specify the workstation where the simulation code will be executed.

Powerful Post-Processor. A powerful post-processor is required to ease the user's task of interpreting the results from the simulation. The ease of forming confidence intervals using any of the schemes described above is desirable. Furthermore, the tools should aid in the iteration over system parameters to easily form performance curves.

Animation. Animation is a powerful tool for visualizing the operation of the systems as well as debugging simulation models. The animation capabilities of the tools should be considered.

Usage. Consideration should be given to the nature of modeling effort. That is, a tool to provide life cycle support of a network will most likely be different from the tool that would be selected for a one-time modeling effort.

Interface with the External World. Simulation tools will become more valuable to network managers as they continue to include interfaces to the external world, e.g., the target network. The extent the tool provides such interfaces should be a major factor in the evaluation process for network managers.

Given these factors, the relative merits of specific modeling environments can be evaluated. In addition, consideration should be given to general-purpose and special-purpose languages as compared with the newer graphical-based modeling environments. Such a comparison is given in Tables 6-1–6-3.

TABLE 6-1 Relative Merits of General Purpose Languages

Advantages	Disadvantages
Wide availability of the languages.	Longer programming time.
Few restrictions imposed on the model.	Difficult verification.
May be knowledgeable in the language.	Generally limited ability to reuse models.
Generally more computationally efficient.	Model enhancement difficult.

TABLE 6-2 Relative Merits of Special Purpose Languages

Advantages	Disadvantages
Requires less programming effort.	Must adhere to the format requirements of the language.
Provides error-checking techniques superior to those provided in general-purpose languages.	Reduced flexibility in models.
Provides a brief, direct vehicle for expressing the concepts arising in a simulation study.	Availability.
Possesses ability to construct user subroutines required as a part of any simulation routine.	Cost.
Contains set of subroutines for common random numbers.	Increased computer running time.
Facilitates collection and display of data produced.	Training required to learn the language.
Facilitates model reuse.	

TABLE 6-3 Relative Merits of Computer-Aided Analysis and Design Environments

Advantages	Disadvantages
Provides complete integrated performance analysis environment.	Must learn framework.
Graphical based.	May be tied to specific hardware platform.
Typically integrates language, database, prior knowledge, and statistical analysis packages.	Flexibility may be reduced,
Management of models and input/output data.	Execution time increased.
Facilitates model reuse and group model development.	Cost.

Procuring a performance prediction tool can represent a significant investment. Also, the characteristics of the tool will have to be dealt with long after the purchase. Care and thought should go into the selection process.

6.5 CASE STUDY: DETERMINATION OF THE SENSITIVITY OF SYSTEM PERFORMANCE AS A FUNCTION OF LOAD FOR A CAMPUS INTERNETWORK USING SIMULATION

As the communications needs continue to grow on both academic and business campuses, their network managers are faced with dealing with a rapidly evolving system that must satisfy the growing needs of their user communities. Simulation provides a mechanism to address the performance issues associated with such a campus network. In this example, a large network is considered. At the level of interest of the network manager, the details of the protocols are not considered directly, rather the architecture and topology of system is constructed by reusing existing component (subsystem) models and the performance issues addressed. In this example [54] the BONeS tool were used.

The purpose of this section is to show how network managers can use simulation in the study of a substantial network. Here the network manager is interested in addressing the performance issues for a large campus network. The main concern is the performance of alternative network designs given common network elements. For network managers it is important that models of the network components exist and are reused and specifically configured for the specific study. (See [60] for the details of the individual component models.) The topology of the campus network under consideration is given in Fig. 6-7. This system consists of a fiber distributed data interface (FDDI) backbone, which serves as a high-speed data trunk among the smaller LANs.

Connected to the network are a number of workstations capable of generating and receiving data packets. The packet arrival process at each workstation is Poisson, and the destination of each packet will be one of the other workstations in the network. Each workstation, with the exception of the workstation creating the packet, is equally likely to be chosen as the destination for each new packet. Without knowledge of the specific traffic patterns, these are idealistic assumptions made for this example; in realistic networks the arrival process is often not Poisson and the traffic is not uniformly distributed across all the possible destinations. Again this points to the need for better empirical data on the demands for network resources.

The bridges in this model connect the various subnetworks to the backbone network. In this study, bridges perform basic operations—including buffering, filtering, and forwarding—and they perform protocol conversion of frames if they bridge two dissimilar protocols. Bridges operate by looking at all data frames and copying them if the destination address of a frame is on the opposite side of the bridge. The bridge will store the frame until it can be transmitted on the destination subnet

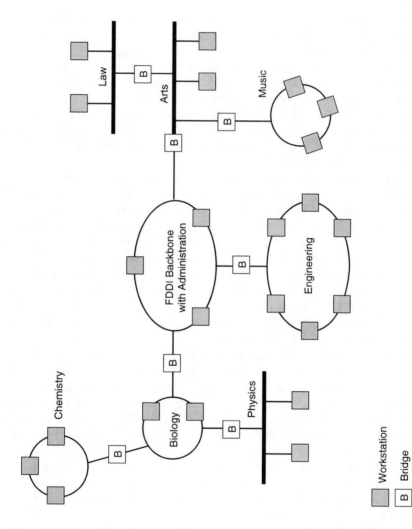

Figure 6-7 Topology of the Campus Network

according to the network protocol. Bridges may have finite buffer sizes, which causes packets to be rejected when the buffer is full.

An important aspect of this model is its use of the hierarchical approach to represent the campus network. At the top level, each subnet is represented with a single module. The subnet modules then contain a number of elements that represent nodes on the subnet. Each node contains the specific traffic source and media access control (MAC) model for the particular type of subnet. Two subnets, the Chemistry Token-Ring subnet and the Arts Carrier Sense Multiple Access with Collision Detection (CSMA/CD) subnet are shown in Fig. 6-8 to illustrate the hierarchy. A similar hierarchy is used for all subnets.

The focus of this example is on the evaluation of alternative network designs from a network management perspective. Thus the details of the protocols are not a concern here. Of critical importance for network managers is the performance as the system parameters are varied. In the specific scenario considered here, the network manager is concerned with the performance of the campus network as a function of user traffic. A simulation study over a set of eight traffic intensity values (see Table 6-4) was conducted by changing the interarrival mean of requests in each node.

TABLE 6-4 Traffic Intensity Values for Simulation Study

Iteration	Traffic Intensity (Mbps)
1	0.400
2	0.468
3	0.548
4	0.641
5	0.749
6	0.877
7	1.025
8	1.199

The nonlinear variation in traffic intensity gives a smaller separation between lower traffic intensities, providing more information in the "knee" of the curves over the range of loads of interest.

Values for the fixed parameters used in this study are:

Packet Length   The length of the data in bits. Here we set it to 6000 bits. All packets will have a data length of 6000 bits. All three MAC protocols will add

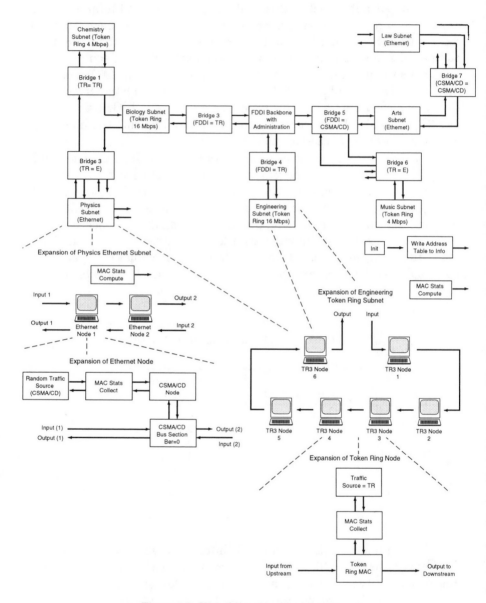

Figure 6-8   Hierarchy of the Campus Network

190

	framing bits to this number to determine the length of the frame they must transmit.
Interarrival mean	This is an exported parameter from the Poisson pulse generator inside the traffic source in each workstation. It defines the average spacing between generated pulses in seconds. Since we specified the average number of bits per second data generation rate per node with the parameter traffic intensity, the interarrival mean then equals (packet length)/(traffic intensity).
Propagation Delay	This parameter specifies the time it will take one bit to propagate between two adjacent CSMA/CD nodes. It was set to 1 microsecond.
Propagation Delay	This parameter specifies the time it will take one bit to propagate between two adjacent FDDI nodes. It was set to 10 microseconds.
Propagation Delay	This parameter specifies the time it will take one bit to propagate on the token-ring networks. It was set to 0.25 microsecond.
Maximum Queue Size	This is a parameter from the queue within each workstation. It specifies the maximum number of packets that can be stored in each workstation. It was set to 2000 packets.
Ring Latency (FDDI)	Set to 60.06 microseconds. It was calculated with the equation: (total propagation time around the FDDI ring + (no. of nodes) * (one bit time for hardware processing delay).
Token Holding (4 Mbps)	Set to 2.5 microseconds. It specifies the period of time that a workstation in a 4 Mbps Token Ring network can hold a token.
Token Holding (16 Mbps)	The maximum time a node on either 16 Mbps Token-Ring period network can hold the token. It was set to 10 microseconds.
T_Pr(0)	The threshold time for asynchronous traffic with priority 0 in an FDDI network. It was set to 10 msec, which is also the target token rotation time for FDDI.
TTRT	The target token rotation time for FDDI network. The average rotation of the Token on the FDDI backbone will be no greater than the value of this parameter. The maximum time

| | an FDDI node will wait before receiving a token is 2 * TTRT. TTRT is set to 10 msec. |
| TSTOP | Specifies the length of simulation real time for each run. It was set to 10 sec. |

The goal of this system study was to determine the throughput, delay, queue size, and bridge delay for the given design as the traffic intensity per node varied.

The throughput of each subnet plotted against the traffic intensity per node is shown in Fig. 6-9. The throughput of subnets with smaller channel capacity (Chemistry and Music subnets) saturate first as traffic intensity increases. These two networks saturate when the traffic intensity per node is around 0.64 Mbps.

The traffic entering the Biology and Chemistry subnets through the bridge connecting both networks reaches a maximum due to the saturation of the Chemistry subnet. Further increase in traffic intensity per node will thus prohibit any more traffic to flow from the Chemistry subnet to the Biology subnet. The rate of change of the throughput curve of the Biology subnet thus becomes smaller.

The Chemistry subnet is transmitting more traffic than the network can handle, so outstanding packets waiting to be transmitted are queued within the bridge. Figure 6-10 shows how the queue size grows as a function of load. As expected, at high loads the queue size will grow steadily until the buffer is filled to its capacity and any further incoming packets to the queue will be rejected. A rough estimate for the maximum queueing delay that a packet will experience when the queue is full equals the average time to transmit 2000 packets on the 16 Mbps Biology subnet (i.e., 3.75 seconds). Figure 6-11 shows that the mean queueing delay for the bridge between Biology and Chemistry is close to 3.5 seconds, providing a sanity check.

The throughput results (Fig. 6-9) also show that the Arts subnet saturates when the traffic intensity at each node is about 0.76 Mbps. In the case when a number of subnets saturate, the actual amount of traffic injected into the network is limited, therefore the throughput of larger networks increases at a slower rate at higher traffic intensity per node. This effect can be seen in the throughput curves for the FDDI backbone and Biology subnet. Both become flatter at a higher traffic intensity than they were at low traffic intensity.

Growing queue size causes queuing delay to become a dominant factor in end-to-end transmission delay when the traffic intensity is large. This effect is reflected through the dramatic increase in delay measurement in both end-to-end subnet transmission delay and queueing delay through bridges shown in Fig. 6-11 and 6-12.

These results give the network manager an indication of the

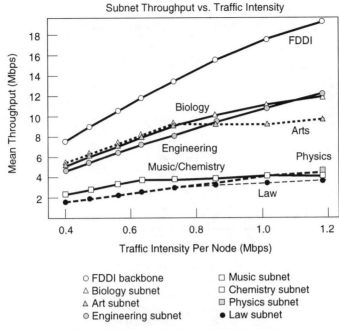

Figure 6-9 Throughput vs. Traffic Intensity

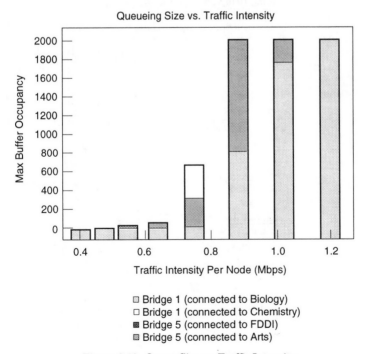

Figure 6-10 Queue Size vs. Traffic Intensity

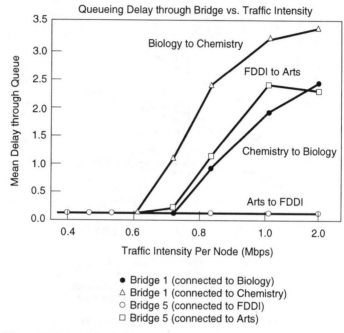

Figure 6-11 Queuing Delay Through the Bridges vs. Traffic Intensity

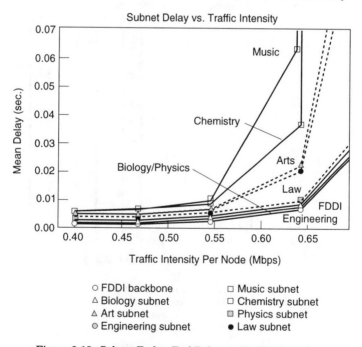

Figure 6-12 Subnet End-to-End Delay vs. Traffic Intensity

suitability of the given system design for satisfying user requirements. If the user requirements are not satisfied using this design, then the topology of the system can be easily changed and another system study performed.

6.6 CONCLUSIONS

Communications networks are rapidly changing. Network managers must have a set of tools to control their systems in both the long and short term. Performance prediction has traditionally been used for long-term network planning and design [2]. Computer simulation has been an important element of those processes. Performance tools are becoming tightly coupled with network measurement and management systems. In this chapter, simulation was introduced in a network management context. The representations used in state-of-the-art simulation environments were presented. The criteria for selecting a simulation paradigm and tool for network management applications were discussed. A representative case study was used to illustrate the application of simulation in determining the performance of a campus network. In the future, performance predictions environments may be available that will collect network topology information from the target system and monitor the traffic on the system to automatically generate models of demands for network resources. This information would then be easily used by network managers to quickly identify performance bottlenecks as well as evaluate scenarios for network growth using simulation.

REFERENCES

[1]. K. Terplan, *Communications Network Management*, Englewood Cliffs, NJ: Prentice-Hall, 1987.

[2]. R. C. Dohrmann, et al., "Minimizing communications network deployment risk through simulation," Technical Report, Hughes Aircraft Co. Space and Communications Group, Colorado Engineering Laboratories, February, 1991.

[3]. M. Schwartz, *Telecommunication Networks Protocols Modeling and Analysis*, Reading, MA: Addison Wesley, 1987.

[4]. B. W. Stack and E. Authurs, *A Computer & Communications Network Performance Analysis Primer*, Englewood Cliffs, NJ: Prentice-Hall, 1985.

[5]. C. H. Sauer and K. M. Chandy, *Computer Systems Performance Modeling*, Englewood Cliffs, NJ: Prentice-Hall, 1981.

[6]. M. Reiser, "Performance evaluation of data communications systems," *Proc. IEEE*, vol. 70, no. 2, pp. 121–195, February 1982.

[7]. F. A. Tobagi, M. Gerla, R. W. Peebles, and E. G. Manning, "Modeling and measurement techniques in packet communications networks," *Proc. IEEE*, vol. 66, no. 11, pp. 1423–1447, November 1978.

[8]. P. Heidelberger and S. S. Lavenberg, "Computer Performance Evaluation Methodology," *IEEE Trans Computers,* vol. C-30, no. 12, pp. 1195–1220, December 1984.

[9]. L. Kleinrock, *Queueing Systems*, vols. I and II, New York: John Wiley, 1976.

[10]. J. H. Hayes, *Modeling and Analysis of Computer Communications Networks*, New York: Plenum Press, 1984.

[11]. V. Ahuja, *Design and Analysis of Computer Communications Networks*, New York: McGraw-Hill, 1982.

[12]. D. Bertsekas and R. Gallagher, *Data Networks*, Englewood Cliffs, NJ: Prentice-Hall, 1987.

[13]. D. Gross and C. M. Harris, "Simulation," in *Fundamentals of Queueing Theory*, 2nd ed., chapter 8, New York: John Wiley and Sons, 1985.

[14]. T. G. Robertazzi, *Computer Networks and Systems: Queueing Theory and Performance Evaluation,* Springer-Verlag, 1990.

[15]. J. N. Daigle, *Queueing Theory for Telecommunications,* Reading, MA: Addison Wesley, 1992.

[16]. A. M. Law and W. David Kelton, *Simulation Modeling and Analysis*, ed. 2, New York: McGraw-Hill, 1991.

[17]. C. M. Sauer and E. A. MacNair, *Simulation of Computer Communication Systems,* Englewood Cliffs, NJ: Prentice-Hall, 1983.

[18]. M. Pidd, *Computer Simulation in Management Science,* New York: John Wiley, 1984.

[19]. W. G. Bulgren, *Discrete System Simulation,* Englewood Cliffs, NJ: Prentice-Hall, 1982.

[20]. A. Pritsker, *Introduction to Simulation and SLAM II*, New York: John Wiley, 1984.

[21]. R. E. Shannon, *Systems Simulation,* Englewood Cliffs, NJ: Prentice-Hall, 1975.

[22]. C. H. Saver and E. A. MacNair, *Simulation of Computer Communications Systems*, Englewood Cliffs, NJ: Prentice-Hall, 1983.

[23]. A. M. Law and C. S. Larmey, *An Introduction to Simulation Using SIMSCRIPT II.5,* Los Angeles: C.A.C.I., 1984.

[24]. Raj Jain, "The Art of Computer Systems Performance Analysis," New York: John Wiley, 1991.

[25]. J. K. Kurose and H. T. Mouftah, "Computer-aided modeling, analysis and design of communication systems," *IEEE Journal on Communications,* vol. 6, No. 1, pp. 130–145, January 1988.

[26]. Special Issue, "Computer Aided Modeling, Analysis, and Design of Communications Systems II," *IEEE Journal on Selected Areas of Communications,* vol. 6, no. 1, January 1988.

[27]. L. F. Pollacia, "A survey of discrete event simulation and state-of-the-art discrete event languages," *Simulation Digest* (ACM Press), vol. 20, no. 3, pp. 8–24, Fall 1989.

[28]. P. Cope, "Building a better network," *Network World,* pp. 34–40, December 28, 1992/January 4, 1993, pp. 34–40.

[29]. J. Y. Hui, *Switching and Traffic Theory for Integrated Broadband Networks,* Boston, MA: Kluwer Academic Publishers, 1990.

[30]. S. Francis, V. Frost, and D. Soldan, "Measured performance for multiple large file transfers," in *14th Conference on Local Computer Networks,* Minneapolis, MN, 1989.

[31]. W. T. Marshall and S. P. Morgan, "*Statistics of mixed data traffic on a local area network,*" Computer Networks and ISDN Systems, vol. 10, no. 3–4, October–November, 1985.

[32]. R. Gusella, "A measurement study of diskless workstation traffic on an Ethernet," *IEEE Transaction on Communications,* vol. 38, no. 9, pp. 1557–1568, September 1990.

[33]. R. Jain and S. A. Routhier, "Packet trains-measurements and a new model of computer traffic," *IEEE Journal on Selected Areas in Communications,* vol. SAC-4, no. 6, pp. 986–995, September 1986.

[34]. G. S. Fishman, *Principles of Discrete Event Simulation,* New York: John Wiley and Sons, 1978.

[35]. A. Alan and B. Prisker, *Introduction to Simulation and SLAM II,* New York: Halsted Press, 1984.

[36]. J. F. Kurose and C. Shen, "GENESIS: A graphical environment for the modeling and performance analysis of protocols in multiple access networks" in ICC'86, pp. 8.5.1–8.5.6, June 1986.

[37]. T. J. Shriber, *Simulation Using GPSS,* New York: John Wiley and Sons, 1974.

[38]. C. H. Sauer, M. Reiser, and E. A. MacNair, "RESQ—A Package for Solution of Generalized Queueing Networks," *Proceeding of the 1977 National Computer Conference.*

[39]. C. H. Sauer, E. A. MacNair and J. F. Kurose, "Queueing network simulation of computer communications," *IEEE Selected Areas in Communications,* vol. SAC-2, no. 1, pp. 203–220, January 1984.

[40]. Information Research Associates, PAWS/A User Guide, Austin, TX, 1983.

[41]. B. Melamed and R. J. T. Morris, "Visual simulations: The performance analysis workstation," *IEEE Computer,* vol. 18, no. 8, pp. 87–94, August 1985.

[42]. Scientific and Engineering Software, Inc., *SES / Workbench*™ *Introductory Overview,* Austin, TX, 1989.

[43]. A. D. Jerome, S. P. Baraniak, and M. A. Cohen, "Communication network and protocol design automation," *1987 Military Communications Conference,* Washington, DC, pp. 14.2.1–14.2.6, October 1987.

[44]. M. Ajmone Marsan, G. Balbo, G. Bruno, and F. Neti, "TOPNET: A Tool for the Visual Simulations of Communications Networks," *IEEE Selected Areas in Communications,* vol. 8, no. 9, pp. 1735–1747, December 1990.

[45]. A. Dupuy et al. "NEST: A network simulation testbed," *Communications of the ACM,* vol. 33, no. 10., pp. 64–74, October 1990.

[46]. W. Larue, E. Komp, K. Shanmugan, S. Schaffer, D. Reznik, and V. S. Frost, "A Block Oriented Paradigm for Modeling Communications Networks," *IEEE Military Communications Conference (Milcom '90),* Monterey, CA, October 1990.

[47]. H. V. Norman, *LAN / WAN Optimization Techniques,* Norwood, MA: Artech House, 1992.

[48]. J. L. Hess, "PlanNet Models Network Traffic," Corporate Computing, vol. 1, no. 1, pp. 45–46, June/July 1992.

[49]. M. A. Crane, *An Introduction to the Regenerative Method for Simulation Analysis,* Berlin: Springer-Verlag, 1977.

[50]. R. H. Myers, *Response Surface Methodology,* Boston: Allyn & Bacon, 1977.

[51]. P. Balaban and K. S. Shanmugan, "Computer-aided modeling analysis and design of communications systems: An introduction and issue overview," *IEEE Journal on Selected Areas in Communications,* vol. SAC-2, no. 1, pp. 1–8, January 1984.

[52]. J. Billington, G. R. Wheller, and M. C. Wilbur-Ham, "PROTEAN: A high-level Petri net tool for the specifications and verifications of communications protocols," *IEEE Transactions on Software Engineering,* vol. SE-14, no. 3, pp. 301–316, March 1988.

[53]. J. Jackman and D. J. Medevias, "A graphical methodology for simulating communications networks," *IEEE Transactions on Communications,* vol. 36, no. 4, pp. 459–464, April 1988.

[54]. K. S. Shanmugan, W. W. LaRue, and V. S. Frost, "A block-oriented network simulator (BONeS)," *Journal of Simulation,* vol. 58, no. 2, pp. 83–94, February 1992.

[55]. P. E. Green, Ed., *Computer Network Architectures*, New York: Plenum Press, 1982.

[56]. G. Kunkum, "An approach to performance specification of communication protocols using timed Petri nets," *IEEE Transactions on Software Engineering*, vol. SE-11, no. 10, pp. 1216–1225, October 1985.

[57]. W. H. Sanders and J. F. Meyer, "Reduced base model construction methods for stochastic activity networks," *IEEE Journal on Selected Areas in Communications*, vol. SAC-9, no. 1, pp. 29–36, January 1991.

[58]. A. Movaghar and J. F. Meyer, "Performability modeling with stochastic activity networks," *1984 Real-Time Systems Symp.*, Austin, TX, December 1984.

[59]. J. F. Meyer, A. Movaghar, and W. H. Sanders, "Stochastic activity networks: Structure, behavior, and applications," *Proceedings International Workshop Timed Petri Nets*, Torino, Italy, July 1985.

[60]. COMDISCO Systems, Inc., "Block Oriented Network Simulator (BONeS)," *Modeling Reference Guide, Version 2.0*, April 1992.

CHAPTER 7

Knowledge-based Network Management

Kurudi H. Muralidhar
Industrial Technology Institute
Ann Arbor, MI 48106

7.1 INTRODUCTION

7.1.1 Network Management Overview

Integrated communications networks are fast becoming the backbone of modern enterprises. Integrated communications networks tie together different parts of the enterprise to speed information flow, reduce information redundancy, and eliminate information handling errors. The trend away from centralized computing facilities and toward distributed systems continues.

Integrated networks can also create new problems, however. Once networks are in place, the enterprise must operate and maintain them—in other words, manage them. Managing these networks can be a complex problem—enterprise networks can easily contain hundreds or thousands of devices from a wide mix of vendors owing to the widely varying communications requirements in different parts of the organization. Chapter 1 provides a detailed definition and overview of network management. Problems in network management and associated technologies are described in Chapter 2. Chapter 4 describes the protocols and services required for network management. These are useful for the discussions in this chapter.

To define the context for this chapter, the basic elements needed for network management of a layered communications architecture, as in the seven-layer open systems interconnection (OSI) model, are as follows:

Network Administrator. The person or persons who use the network manager application to perform network management functions.

Network Manager Application. An automated tool with a special human–machine interface that the network administrator (the person) uses to monitor and control network activities. The network may have more than one network manager application. The application implements the configuration, fault, performance, security, and accounting management functions.

Network Management "Agents" and "Managers." The network management agent resides in individual network components that are to be managed. The manager resides only in the manager station in a centralized scheme; in a distributed scheme, the manager will reside in multiple stations. The manager provides communication services to the network management application. For example, it passes requests from the application to the agents, receives responses as well as other information from the agents, then gives these to the network management application. The agent performs two main functions: managing the resources within its local network component via the layer management entities (LMEs) and communicating with the manager.

LMEs. The LMEs interact with each protocol to maintain basic information about the configuration and status of the layer.

Network Management Service and Protocol. The network management service provides a set of communications functions to the network manager application. The network management protocol defines how the manager communicates with the agents in individual network components. ISO's common management information service (CMIS) and common management information protocol (CMIP) are examples [1–4].

Management Information Base. The management information base (MIB) is the concatenation of the information that each LME maintains on its protocol layer resources. The MIB is included in the LMEs in each network interface. In addition, the network manager application will maintain an information base for the domain for which it is responsible [5]. Figure 7-1 shows a typical network management environment [6]. Figure 7-2 shows a typical manufacturing local area network with management elements [7].

7.1.2   What Is Meant by Knowledge-based Network Management

Knowledge-based (KB) network management is capable of capturing a network management expert's views, synthesize the knowledge and ex-

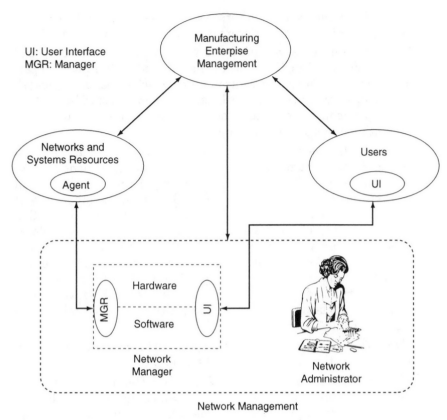

Figure 7-1 Manufacturing Local Area Network with Management Elements

isting conditions, and analyze the current situations based on the captured knowledge [8]. It provides decision making support based on the
management elements, constraints on those elements, their behavior, and
the relationships between the elements. Such support mechanisms belong
to construction of specific structures of elements (synthesis), often seen in
configuration management, and analysis of specific structures of elements
(analysis), often used in performance and fault management [9].

The basic *synthesis* problem seeks, given a set of elements and a set
of constraints among those elements, to assemble from the elements a
structure that satisfies the constraints. Such types of problems are typical
in configuration management, accounting management, and security
management. There are three types of synthesis problems, each based on
how time in constraints is represented: *nominal, ordinal,* and *interval.* In
the nominal type, the structure (configuration, accounting, or security)

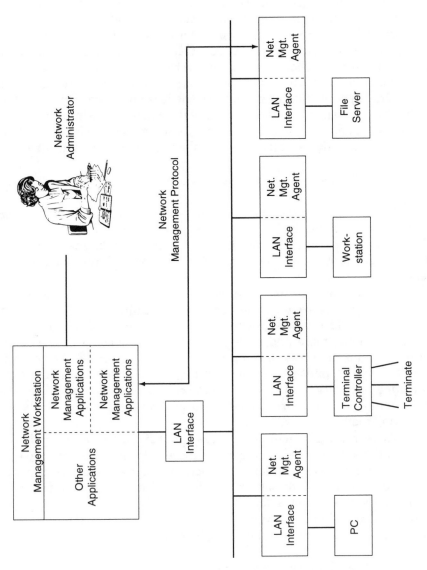

Figure 7-2 Manufacturing Local Area Network Elements

203

being assembled must satisfy constraints concurrently, but there are no constraints ordering changes in its state over time. In the ordinal type, commonly referred to as *planning*, some elements of the structure participate in *before/after* constraints. The interval type, sometimes described as *planning in time*, leads to scheduling problems.

Analysis begins with a known structure (from the synthetic class) and reasons about the relation between its behavior and the elements that make the structure. There are two major forms: *prediction* and *interpretation*. This is normally the case in the fault and performance management functions. In prediction, the reasons are made from the structure and the behavior of its elements to the behavior of the whole. In the interpretation, reasons are made from the structure and its observed behavior to the state of its elements. Interpretation often involves monitoring and diagnosis.

7.1.3 Why Use Knowledge-based Systems?

There are many reasons for using knowledge-based systems for managing networks, including dealing with increased network complexity, ensuring real-time response for managing, dealing with network management problems consistently, and training and retaining skilled personnel.

Over the years, the use of networks has grown steadily and more and more organizations are using networks strategically. The proliferation of PCs has also added to the growth of network usage. With the increased use of networks, more and more applications are placing stringent requirements on networks, leading to increased complexity. Today's networks come with multiple media; multiple services; high-speed, multiple switching technology; and different protocols. These have added to network complexity, requiring sophisticated knowledge-based tools for managing them.

With the advancement in networks and their strategic importance to organizations, there is a need to make these networks operate very efficiently and to operate in a fault-free mode with high efficiency response to the needs of the applications. In the case of control and manufacturing applications, the networks have to respond to the real-time needs of the applications. These requirements also necessitate sophisticated knowledge-based tools for managing the networks catering to real-time needs.

Knowledge-based systems can be used for training personnel by embedding expert knowledge in the system. It will help them to be trained quickly and will serve as a backup when they want to learn more. It can also help reduce cost of maintaining expensive skilled or expert personnel. It can replace the expert in most routine cases, thereby allowing experts to concentrate on nonroutine, exceptional cases.

7.1.4 Some Existing Knowledge-based Systems

Knowledge-based systems can exist in two distinct types: assistant and autonomous. In the assistant mode, the KB systems work offline and help administrators to manage. In the autonomous mode, the systems work online and make management decisions. Historically, most knowledge-based systems have been developed for the telecommunications industry. This appears to be changing because of the proliferation of local area networks within most enterprises. Some examples of knowledge-based management systems are listed below. Some are prototypes, others are commercial products.

Some assistant mode tools are:

- Blondie-III, a prototype network configuration expert system developed by Dutch PTT to assist human planners in making decisions about installing new lines or using existing lines [10].

- STEM (Strategic Telecommunications Evaluation Model) from Analysis Ltd.[1], analyzes multiservice networks and identifies preferred choices between various technical solutions.

- Planicom, developed by Sofrecom[1] (a subsidiary of France Telecom), assists in network planning, routing, and circuit grouping.

- Seteli, developed jointly by the Finnish Post and Telecommunications Agency and the Helsinki University of Technology, combines engineering and financial knowledge domains [11].

Some autonomous mode tools are:

- Automated cable analysis (ACE) for telecommunications, diagnoses cable faults [12].

- Troubleshooter, an AT&T system, diagnoses trouble in Datakit networks [12].

- Nexpert, from Neuron Data (Palo Alto, California), a rule-based, object-based tool (shell) that assists in building systems that monitor events, isolates faults, and attempts repair.

- LAN Command Advanced, from Dolphin Networks, an expert system for Novell LANs that makes intelligent inferences based upon approximately 1,000 conditions on the LAN.

---

[1]Contact the company for more information.

7.1.5 Organization

In Section 7.2 of this chapter, approaches and methodologies that are applicable to knowledge-based network management systems are described. Development schemes, including requirements for knowledge-based systems, knowledge-based solutions, network management functional areas, and services required are described in Section 7.3. In Section 7.4, an implementation example is given, describing a single application of knowledge-based fault management system for a manufacturing network. Applicability of knowledge-based systems to local area networks (LANs), wide area networks (WANs), and Internets is described in Section 7.5.

7.2 APPROACHES AND METHODOLOGIES

7.2.1 Methods for Knowledge Engineering

To build large knowledge bases for complex real-time applications such as network management, we have found several techniques—such as reason maintenance, temporal reasoning, production systems, and planners—as most applicable. In the past, each of these was an area of separate research, but demands put upon real-time systems requires that several of these methods be integrated to address the network management problem.

Reason Maintenance. Reason maintenance is a function that ensures that a database of assertions is kept in a consistent manner. It prevents the knowledge-based system from believing assertions P and NOT(P) to be true at the same time. There are two basic types of reason maintenance systems: truth maintenance systems (TMS) [14] and assumption-based truth maintenance systems (ATMS).

TMS maintains one view of the world. Every assertion in the database can be viewed as *in* (believed) or *out* (not believed) at any time. Each assertion A has associated with it a set of in-justifiers (assertions that support A) and out-justifiers (assertions that contradict A). An assertion is *in* if all of its in-justifiers are *in* (believed) and none of its out-justifiers are *in*. Thus a TMS forms a network of assertions where each assertion A is connected to other assertions A^+, which form A's in- and out-justifiers. Assertion A in turn can also be an in- or out-justifier for other assertions in the database. When assertion A (or any assertion) is modified, the TMS traverses the network of assertions, ensuring that the effects of the change are reflected in the other database assertions.

ATMS differs from a TMS in that it allows contradictory assertions in the database. The ATMS, however, ensures that no conclusion can be reached using contradictory assertions. An ATMS can be viewed as multiple alternative worlds. Each world is consistent, although assertions in one world may contradict assertions in some other world. This allows the reasoning system to try alternatives without affecting the current state of the world.

Temporal Reasoning. Knowledge-based systems working in real-time need the ability to reason about time. Temporal reasoning systems provide this capability by providing functions including [15]:

1. Temporal database
2. Database access language
3. Projection capability
4. Temporal reason maintenance capability

The database consists of a set of assertions in addition to information that specifies when the assertions will be true or false. This information can specify absolute times when the assertion will be true or they may be specified in relation to some other assertion. For example, proposition P could be true at 0300 GMT 4 July 1990 and remain true for two hours. Proposition P could also be represented as being true one hour after proposition Q becomes true. The temporal reasoning system also allows the user to specify constraints that should not be violated. The system monitors the constraints and alerts the user if a constraint is violated. This can be used by a planner to recognize an unworkable plan or by a monitor to recognize an abnormal condition.

In addition, a temporal reasoning system can simulate actions to a limited extent. This allows it to predict intervals that will be true (or false) at a later time because of current conditions. Thus a consistent database is always maintained.

Production Systems. The knowledge base consists of a series of rules and a control strategy for evaluating the ruling. Each rule consists of a left hand side (LHS), which is a set of conditions and a right hand side (RHS), which is an assertion that is true if the conditions hold.

A sample rule might read:

(if (connect error & D)
  (elevated packet count &N)
   (on &D &N)
  then (network congested &N))

This rule would read: If a device D is getting connect errors, and the number of packets going over subnet N is elevated, and device D is on subnet N, then conclude that the connect errors are the result of network congestion on network N.

Two basic control strategies, called *forward chaining* and *backward chaining*, are used to evaluate production rules. A forward chaining system examines all the rules of the LHS at the start of a cycle. Rules that have all their preconditions met are collected. One of the rules is selected and its RHS is asserted into the database. With backward chaining, the system acts like a theorem prover. It begins with the conclusions and attempts to prove them by proving all the LHS assertions.

Planners. Planners represent one of the most complex parts of any real-time knowledge-based system. The planner starts with a set of goals and a library of actions available to it. Goals consist of system states that are desired for some reason. Actions can be thought of as parts of programs that will affect the world. The planner selects a series of actions that, when executed, will take the planner from the current state to the desired goal state.

In real-time systems, this becomes a very difficult problem. Planners must deal with constantly changing situations. Plans may become invalid because of changing conditions and even goals can become obsolete. Worse, often a plan must be generated and executed in a certain span of time, which can require that a plan begin execution even before it is completed.

7.2.2 Assistant Mode Systems

Assistant mode systems are the simplest type of intelligent network manager. They function as assistants to users and operate offline. A user determines that a network problem has occurred and uses the assistant knowledge base to help solve the problem. The assistant knowledge base asks the user questions about the network and observable conditions. It then diagnoses the problem and provides the user with both the diagnosis as well as possible solutions. An advantage of this approach is that the knowledge base can be used for training in addition to diagnosis. This is done by adding a program that uses the knowledge base to postulate problems and provide symptoms to trainees who diagnose problems themselves. These systems are generally implemented as production systems using any of a number of off-the-shelf tools.

These systems can be very powerful and can become cost-effective. However, they suffer from several drawbacks. They do not provide real-time response since they operate offline. Lack of real-time response limits (but does not eliminate) their usefulness since they are used to diagnose problems only and not to determine that a problem exists. Users must make the initial determination that a problem has occurred. These two factors limit their use in many applications.

7.2.3 Autonomous Mode Systems

Unlike assistant mode systems, autonomous mode systems are an integral part of the network management system. Autonomous mode systems run constantly throughout the network and generally decide what to work on and where problems may reside. The ability to do this requires a far greater degree of performance. An autonomous mode system must monitor the constantly changing situation on the network and decide for itself if anything needs to be done. For faults that can take long periods of time to unfold, this can become a complex task.

Autonomous systems do have advantages over assistant mode systems, however. Since it is constantly running, an autonomous manager can isolate problems faster than a human can. This allows greater network performance and resiliency.

Characteristics

Distributed vs. Central Autonomous Network Managers. Autonomous network managers can be centralized or distributed. A centralized manager resides on a single node and monitors the entire network from that location. This has the advantage of providing network managers with a single location to interface with the system. Also, the manager need not worry about how to interface with other managers to accomplish tasks.

This approach also has several disadvantages, however. A centralized manager is an obvious single point of failure. In addition, with modern interoperable networks, other managers may be unwilling to surrender control of their networks to another system.

A distributed manager solves many of these problems. A distributed manager places agents on several terminals and each has responsibility for its local area. This also improves performance since local problems are resolved locally, which reduces network traffic. Finally, with a distributed system, more processing power can be brought to bear on problems as the need arises.

Categorize and Filter Events. Most conventional network management systems report errors as they arrive and write them to some sort of log file. This pattern makes diagnosis more complex since network technicians must wade through hundreds of messages to get at the few needed to isolate the problem.

With an autonomous manager, events can be filtered and categorized so that operator workload is reduced. When an event comes in, the autonomous manager can categorize the message. It could be a symptom of a new fault not previously known, a symptom of a known fault, or a new fault related to a known one.

For example, suppose a series of faults causes the knowledge-based network manager to conclude that a device on a network has failed. The system can then open a special file for associated error messages and flag the device as down. The human manager would then be made quickly aware of the problem and would have all the evidence needed to confirm one easily accessible location.

Later, the manager may receive a connect fail message from an entity attempting to connect with the failed device. The network manager would recognize that this failure relates to a known failure (assuming the device was still down) and would store the message in the same previously created file.

Sense, Interpret, Plan, and Act. The control for an intelligent autonomous network manager (ANM) is based on control theory. At the top level, control consists of the following four steps:

1. Sense
2. Interpret
3. Plan
4. Act

Sense. First the ANM must have access to all information available about the network and update internal data structures to reflect this information. In general, this information will consist of the status messages normally sent to the manager by each device. In a distributed environment, they will also include status reports made by other ANMs.

Interpret. After a current view is gathered, the ANM must assess that view to identify problems. This assessment can include statistical analysis of counter values returned by devices and heuristics run to identify problems.

The result of this assessment gives updated status information to the users. In addition, the assessment will produce a series of goals that are to be worked on. These goals can range from identifying a piece of information needed to complete a diagnosis to a major restructuring of the network to deal with a major failure.

Plan. The goals generated in the previous step are analyzed here. For each goal, a plan must be developed to meet the goal. This plan must also be integrated with existing plans so they do not interfere with each other.

Note that sometimes a goal can be achieved by doing nothing. Suppose, for example, a goal is to find out if a station is still active. Generally this would result in a plan that polls the station and asks for its status. However, since all we need to know is if the station is still working, it may be better and faster to do nothing and wait for its normal status message to arrive.

Act. The act step takes the plans generated above and monitors their execution. It notes the preconditions that must be met for each step in the plan and places the plan into execution when the proper preconditions have been met.

The act step also looks for failures of plans. For example, if an attempt to read a counter value from a device fails, then the plan associated with it is aborted. When this happens, the fact that the plan failed and the reasons for failure are recorded so that the next pass through the control loop can assess the reasons for failure and the planner can generate a new plan, if needed.

7.3 DEVELOPMENT SCHEMES

7.3.1 Requirements

There are several requirements that should be considered when developing a knowledge-based network management system. Some of the requirements are related to the response time required for managing networks and some are related to the architecture of the knowledge-based system. Other important requirements are related to the user interface and the packaging of the overall system for making it more affordable.

Real-time Issues. More and more network applications need fast response times and require a response in a bounded time. This poses a real-time constraint on the network management system that is capable of fast and bounded response time. This requirement has an impact on the way the knowledge is represented and analyzed using reasoning algorithms.

Architectural Issues. From an architecture point of view, knowledge-based network management systems must support all the functional areas required for managing small or large networks. It must support management functions, services, and the manufacturing information base.

The architecture of knowledge-based network management systems must also guarantee response in a bounded time to meet the real-time needs. It must grow in a linear fashion, i.e., directly proportional (as opposed to exponential) with network size and must be capable of handling large, complex networks.

The architecture must be able to sense and act on the information to both detect and correct faults. It must also be able to gather more information to further analyze network faults or performance tuning. Another issue is the ability to follow situations over time so that the network administrators are not flooded with an overload of information. The architecture must support the ability to filter out both extraneous and unwanted information. In addition, the architecture must support explanation of any actions the knowledge-based system takes.

User Interface Issues. The user interface for a knowledge-based system must be easy and psychologically compatible. It must support varying degrees of a user's expertise. No assumptions about the skill set of users can be made by the system.

The knowledge-based system must also support the expert's view of the network and provide easy-to-follow instructions. It must support consolidation of relevant information and filter out any unwanted information.

Packaging. The knowledge-based system must be packaged to be affordable; i.e., it must support varying degrees of functionality to accommodate a lower price for managing small networks. It must be modular to facilitate addition and deletion of components based on the network needs.

The system must also be packaged to support distribution over the network to support decentralized management. It must end the *avalanche effect* where network failures cause other failures to flood the operator screen.

The system must be packaged to provide a separate window for each situation with a global view of the network. It must assist operators and administrators in focusing more on problems by eliminating unwanted information.

7.3.2 Possible Solutions

Distributed Artificial Intelligence. Distributed artificial intelligence (AI) deals with how intelligent agents[2] coordinate their activities, and practical mechanisms for achieving coordination among AI systems [16].

---

[2]AI agents differ from the network management agents previously discussed. The network management agents serve only to interact with LMEs and communicate with the network management station. The AI agents would be equivalent to the network management agents with the addition of intelligent network management application functions.

Theoretical analyses of intelligent coordination have involved concepts from game theory, organization theory, economics, team decision theory, and speech act theory. These analyses have helped formalize aspects of intelligent coordination, although these formalisms to date fail to capture all of the complexities that arise in nontrivial domains.

Simultaneously, several researchers have studied coordination issues that arise in particular domains and have developed relevant coordination techniques. For example, when large tasks need to be broken down and allocated among agents, a form of contracting must take place to pair tasks with agents. The Contract-Net protocol [17] provides a structured approach to forming contracts, and this protocol has been extended to work in multiple stages for distributed constraint satisfaction problems [18].

Alternatively, when data are inherently distributed among agents (as in distributed network management), the agents must exchange relevant information in a structured way to eventually converge on more global views. Techniques employing communication and organizational structuring have been developed for such problems [19,20]. More recent work has tried to build a common framework to support these and other coordination techniques [21].

Some amount of distributed AI work has concentrated on managing communication networks [18]. The work has focused on protocols for exchanging constraints on resource usage; their goal is to exchange the minimal amount of information necessary in order to restore as many high-priority connections as possible. This work needs significant extension to handle the variety of tasks involved in network management, such as fault diagnosis, performance management, and configuration.

Another segment of research [8,22] has focused on using AI techniques to diagnose errors in a network of cooperating problem solvers. This work uses a model of constraints on interactions and depends on a central diagnostic module to collect and analyze the information from the network.

Knowledge-based Planning.   Recent work in planning uses AI techniques to find sequences of actions that achieve a combination of goals. Early research emphasized strategic planning, where effective action sequences were developed before any of the actions were taken. Planners like STRIPS [23], NOAH [24], and MOLGEN [25] take a strategic view, building an entire plan under the assumption that nothing unexpected will occur in the environment. When these planners were applied to real world problems, they suffered serious problems since the real world tended to be far more variable than the planners could predict. When things didn't work exactly as planned, these planners needed to stop and replan.

As a result of the limitations found in strategic planners, other work addresses the other extreme with situated activity (sometimes called *reactive planning*). Situated activity planners plan only a very short time into the future. They use the present situation to index into a library of actions, select the appropriate one, and implement it.

In dynamic environments, it might make more sense to plan immediate actions rather than entire action sequences. Such situated activity planners strive to achieve overall goals by pursuing specific relevant actions and do not waste time planning for eventualities that never arise. Other types of planners attempt to find a compromise between the strategic and reactive extremes, such as planning for contingencies or incrementally adding details to abstract plans. In addition, planning and acting in the real world means reasoning about time constraints, so planners have had to incorporate temporal reasoning.

Reasoning Architectures. AI systems reason by applying diverse kinds of knowledge to a problem. To encode this knowledge and control its application, an AI system needs to have a reasoning architecture. One traditional reasoning architecture is the production system, where knowledge is represented as an if-then rule and where the control mechanism matches rules to data and fires the applicable rules [26]. The results of these rules in turn trigger other rules, and the system continues until it solves the problem or runs out of rules.

An alternative architecture is a blackboard system [27], which treats different types of knowledge as separate entities that interact by reading from and writing to a shared blackboard. The modular decomposition of the knowledge into fairly large, independent knowledge sources allows considerable flexibility in how systems are built and extended. The blackboard concept can be extended so that the blackboard is a sophisticated database that is capable of representing temporal and default relationships between entries, allowing blackboard systems to incorporate temporal reasoning and truth maintenance capabilities. However, there exist many other reasoning architectures that possess different strengths and limitations, such as SOAR [28], which incorporates learning and automated subgoaling into a rule-based system.

7.3.3 Underlying Management System Requirements

Underlying the knowledge-based network management solution is a set of functional areas and supporting services. The functional areas define the objectives for the management activities and the services provide the tools for carrying out management actions.

System Management Functional Areas. The ISO standards effort has defined a basic set of facilities that network management should provide. Each is described below in the context of a knowledge-based system. For more information, see Chapters 1, 3, and 4.

Configuration management comprises mechanisms to determine and set the characteristics of managed resources. There are two facets of configuration management: static and dynamic. In a static mode, the knowledge-based system must examine system and network parameters for consistency and resolve discrepancies between dependent parameters. In a dynamic mode, the system may be required to initialize and terminate resources, redefine the relationships between certain resources, or execute system reconfiguration.

Fault management comprises mechanisms to detect, isolate, and recover from or bypass faults in the network. Faults are abnormal conditions that require intervention by the network administrator. These should not be confused with errors, which may not require a network management scheme for correction. A knowledge-based network management system should be able to recommend or initiate corrective action and learn from the results. In situations where down time is unacceptable, fault management should also include the ability to anticipate faults.

Performance management comprises mechanisms to monitor and tune the network's performance, where performance is defined by user-set criteria. It requires learning about the network's performance through the gathering of statistics with respect to throughput, congestion, error conditions, delays, etc. It should undertake fine tuning activities, compare results, and learn from the results. It must also determine what are the network's normal and abnormal operating conditions.

Accounting management comprises mechanisms for controlling and monitoring charges for the use of communications resources. It includes the setting of limits and tariffs, informing users of the costs incurred, and combining costs where multiple resources are involved. The expert system should be able to compute the most effective use of resources and report where expenses are out of bounds.

Security management comprises mechanisms for ensuring network management operates properly and for protecting network resources. Security management tries to ensure the integrity of the data and to protect resources from unauthorized use. When security has breached, the management system should immediately take "police" action to protect the resources and identify the offender.

Management Services. The tools provided by the OSI specifications include:

- The ability to read attributes of remote resources.
- The ability to write or replace the values of attributes on a remote system.
- The ability to notify the manager application when an abnormality has been detected and to provide the manager with enough information to take corrective action.
- The ability to affect a resource as a whole, such as initializing, terminating, or resetting a resource.

7.4 IMPLEMENTATION AND EXAMPLE

In this section, a single application of a knowledge-based system called *OSI fault manager* (OSIFaM) for diagnosing faults in a manufacturing automation protocol (MAP) local area is discussed.

7.4.1 MAP Fault Management

The MAP 3.0 specification (manufacturing automation protocol version 3.0) includes definitions that directly impact fault management implementations. For example, fault management must be cognizant of the architectural components that may be used in MAP LANs, each component's expected behavior, the network objects that network management may manage, and their associated characteristics.

The architectural components that can be used in MAP networks are full MAP systems, which use a 7-layer protocol stack; miniMAP systems, which use a 3-layer protocol stack; MAP/EPA systems, which use both a 7-layer and a 3-layer protocol stack, bridges, routers, and gateways.

The manageable objects in MAP are called *resource classes*, which are similar to object classes in the ISO network management model and roughly equate to protocol layers. For each resource class, MAP defines a set of characteristics, actions that can be performed on the class, and event reports that the class may generate. The characteristics may include counters of internal protocol events (e.g., number of octets sent by the network layer, number of token pass failures in the MAC sublayer), configuration parameters (e.g., MAC slot-time, transport retransmission time), and thresholds associated with specific counters (e.g., transport timeout threshold). Actions are directives that the manager can request a particular resource class to perform (e.g., adjust the physical layer transmit power). Event reports are indications of noteworthy occurrences sent from the managed system back to the manager (e.g., the transport timeout counter in system X exceeded its threshold).

Thus, the information that will be available to the fault management system is well-defined. However, these definitions also limit the addition of new information types that would aid fault management, as they would require changing the MAP specification. Thus, the manager has to derive all that it needs from what is available. For example, a system's failure to respond to the manager is an indicator of a fault in the network.

7.4.2 OSIFaM Objectives and Requirements

A primary objective in building an OSI fault manager (OSIFaM) is to create a tool that reduces the expertise needed to diagnose and correct faults in OSI networks [29,30]. In addition to this objective, several others can also be fulfilled:

- Develop a tool for fault management of OSI networks, which will be extensible to other open system architectures (for example, TCP/IP), using MAP 3.0 as a first case of an OSI network. Focus on the object-oriented techniques wherever applicable.
- Implement ISO fault management and follow other pertinent developing ISO network management standards.
- Reduce the skill needed for fault management by filtering out redundant and extraneous information or by capturing expert diagnostician knowledge in the knowledge base.
- Improve the speed and accuracy of fault functions over manual methods and over nonexpert fault-diagnosis systems. Our goal is to make real-time diagnoses that can distinguish between multiple faults that occur over time.
- Justify or explain all knowledge-based system actions and decisions, when requested, so that the human user can understand the system's actions and develop trust in the system's correctness.

7.4.3 Network Management System Architecture

The design approach for OSIFaM was predicated on adding fault management knowledge-based capabilities to an existing network management platform. Figure 7-3 depicts an integrated network management architecture that includes a knowledge-based fault management system. Figure 7-4 shows the general goal architecture for knowledge-based network management systems. The knowledge-based systems box represents one or more knowledge-based systems such as OSIFaM. The user interface box represents a common interface for knowledge-based systems in this architecture. The network image (information base) is the

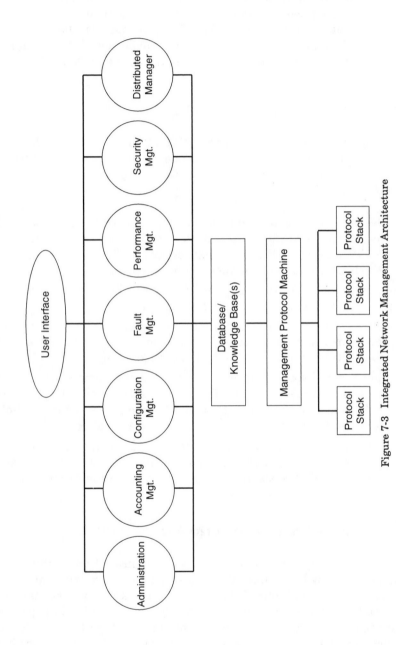

Figure 7-3 Integrated Network Management Architecture

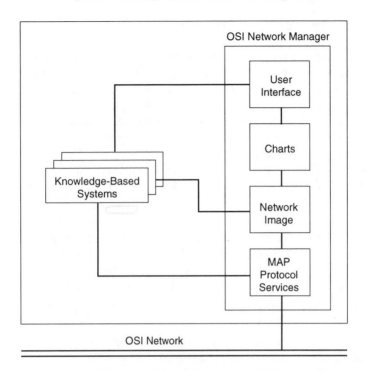

Figure 7-4   Knowledge-based Network Management System Architecture

repository of information on the network resources being managed; it includes basic status, topology, and configuration information on the devices in the management domain and is used by all of the knowledge-based systems. The MAP protocol services represents the interface to a MAP network.

The general architecture also shows the interconnection between the knowledge-based systems and other elements in the network manager platform. All of the knowledge-based systems can communicate with each other to request information or functions particular to each. Each knowledge-based system can also access the network image to obtain basic network information and can request and receive protocol services directly from the protocol services function.

A key feature of this architecture is its modularity. The knowledge-based systems are depicted as separate modules for several reasons. They can be developed independently and each contains some knowledge unique to it. Furthermore, this gives an option, if required, to place the knowledge-based systems in physically separate systems for performance or reliability reasons. This also permits addition or deletion of modules to tailor the network management station to an organization's specific management needs.

7.4.4 Knowledge-based System Architecture

Figure 7-5 shows the components in OSIFaM's architecture. The basic information elements common to these components include counters, events, diagnosis, temporal intervals, and the network image.

Counters. A counter indicates the number of occurrences of errors or other relevant network events. Counters typically will indicate the condition of a specific protocol layer. Examples of counters are the number of transport protocol data units retransmitted, the number of data link modem errors, and the number of network protocol data units sent.

Events. Events describe significant occurrences in the network as they happen. These events may be potential problems, such as too many transport layer retransmissions, but can also indicate nonerror conditions. Events can be either externally generated by managed system agents or be internally generated by OSIFaM.

Diagnosis. A diagnosis is a schema that describes a failure in the network.

Temporal Interval. A temporal interval is a structure added onto any assertion, e.g., events and diagnoses that can be stored in the temporal database. It gives information on when an assertion can be believed.

Network Image. The network image is a set of schemas that describes the network itself and its constituent elements. OSIFaM uses several schema types to describe the network elements, includ-

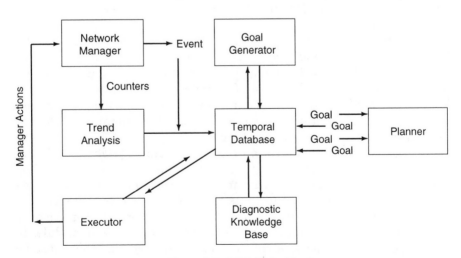

Figure 7-5 OSIFaM Architecture

ing systems, subnetworks, and interconnections (bridges, routers, and gateways).

The major components in OSIFaM are the trend analysis processor, the temporal database, the diagnostic knowledge base, the goal generator, the planner, and the plan executor.

Trend Analysis Processor. The trend analysis processor identifies anomalous trends in network activity, as indicated by excessive changes in the counter values. The trend analysis algorithms are based on statistical analysis methods for determining when a change in a counter value differs from a baseline period with statistical significance. For example, sudden or long-term increases in a counter, such as number of retransmissions, may indicate a network problem.

The trend analysis processor receives messages from the network manager, that give the name of the counter being received and the amount of change in the counter (delta) since the last update. It receives messages as they arrive and converts their contents to a suitable internal format. In general, this means adding the counter value to a list holding a moving history for the specified counter (and removing the oldest entry). It then identifies an appropriate trend analysis algorithm to process the new counter value. If the analysis shows that the delta is within the range, no other action is needed. Otherwise, a schema is created that describes the occurrences of an event. The event is then posted to the temporal database.

Temporal Database. The temporal database (TDB) is the central database of the OSIFaM system. It contains assertions as well as information on the time intervals over which the assertions are true. In addition, the TDB has a simulation capability that is used to predict events which are expected to happen.

The TDB also processes queries from other OSIFaM components. Information on assertions and temporal relationships between intervals may be queried.

After a database update, the TDB projects the effects of the actions into the future as far as it can. This is done by simulating the fault and taking note of future events and their durations. The end result is a time map that details the current situation in the network and gives the user information on what can be expected in the future.

Diagnostic Knowledge Base. The diagnostic knowledge base consists of rules that can explain events. These explanations can be either a diagnosis of a new problem or an explanation of an event in terms of previous diagnosis. In the case of a malfunction diagnosis, a schema is created to describe the problem.

The only input to the diagnostic knowledge base are events from the TDB. When a new event is entered into the TDB, the diagnostic knowledge base attempts to explain the event. Events can be explained in one of three ways:

1. The first way involves diagnosing a problem. If an event, perhaps along with others, can identify a specific malfunction, a diagnosis schema is created to explain the event and the diagnosis is added to the TDB. This diagnosis can be complete or partial and may be updated later. In addition, the name of the diagnosis schema is added to the "causes" slot in the relevant event schemas.

2. In the second case, the event is a side effect of a known malfunction. For example, consider a connection failure to a system that is known to be down. In this case, the event (connection failure) is not a cause for a problem, rather an effect of whatever fault caused the system to fail. If an event is found to be the side effect of some failure, the name of the diagnosis schema is added to the "caused by" slot in the event schema.

3. Finally, the diagnostic knowledge base may be unable to classify the fault as either the cause of a specific failure or as a side effect of a failure. In this case, the "causes" and "caused by" slots are left free.

Goal Generator. The goal generator has the responsibility of deciding what active measures the OSIFaM should take to identify malfunctions. The goal generator, in conjunction with the planner and the plan executor, allows OSIFaM to work in an autonomous manner. The goal generator monitors the TDB, looking for problems that need to be solved, and generates goals to solve the problem. Some of these goals may be meta-goals, which act to focus the attention of the Planner so that it can work on whatever goal is best, given network conditions.

The goal generator monitors the TDB and looks for situations where work is needed. This work may include the following:

- Completing a partial diagnosis
- Determining how to reconfigure the network after a failure
- Fixing a malfunction
- Generating a status report for the network administrator

The goals are identified, prioritized, and posted to the TDB. In addition, the goal generator may give deadlines to the planner stating when the plan must be finished generating a plan, or how long the plan may take to execute.

Planner. The planner creates plans for any actions that OSIFaM must take. These actions include requesting reconfiguration of the network after a fault, correcting malfunctions, or localizing malfunctions.

The planner looks for goals from the goal generator, which are posted on the TDB. When a goal is posted, the planner generates a plan to accomplish the goal within a time frame specified by the goal. After the planner generates the plan, it posts the plan to the TDB where it is executed by the plan executor.

The planner may use several planning technologies, according to the time constraints present. For near-term goals that pop up, a reactive planner may be used to generate quick plans. Some of these plans are standards procedures to deal with problems and some may be only holding actions designed to hold things together while a better plan is generated. As the time constraints for the completion of goals moves farther out, the planner may use other methods. In this way, the OSIFaM can put more "thought" into generating better plans when time is less of an issue.

Plan Executor. The plan executor executes the plans generated by the planner. This requires not only the generation of management actions to be sent to the network manager, but the overall monitoring of the plan to ensure that it works.

The plan executor deals with the moment-by-moment execution of plans by the OSIFaM. It monitors the TDB, waiting until a plan is ready for execution. This is determined by looking at the plan preconditions, which may include the time the plan is to start.

When the plan preconditions have been met, the plan is read from the TDB and each action is sent to the network manager to go over the network. While this is going on, the plan executor monitors the progress of the plan. By looking at the plan, the executor can determine what effects the plan should bring about by looking at the effects slot in the plan schema. If, after the action has executed, the desired effects have not been achieved, then there is reason to think that this plan will not work. In this case, a problem report is created and posted to the TDB and execution of the plan is aborted. The goal generator uses the problem report to direct the planner to generate an alternative plan to accomplish the goal or to abandon the goal altogether.

7.4.5 Example

Following is an example of how OSIFaM can monitor a network and identify and isolate a fault. The network in this scenario is a simulated MAP network consisting of two LANs connected by a bridge. Figure 7-6

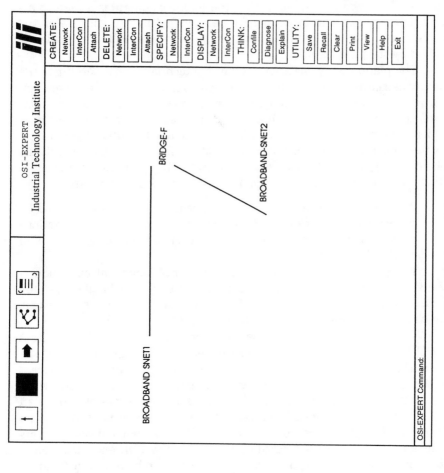

Figure 7-6 Example Networks

shows how the networks are involved in this demo scenario. At the top of the display are general controls for the KB shell tool and OSIFaM. At the right are a series of controls that allow users to build and configure MAP networks as well as turn on and off the OSIFaM diagnostic and monitoring capability. In the center of the display is a map of the network. The network consists of two subnets connected by a bridge.

At scenario time 4, the network manager receives several events that were posted to the TDB. The diagnostic knowledge base then attempted to explain the events and concluded that a cable had broken somewhere on BROADBAND-SNET2.

With the failure cause identified but not localized, the goal generator created a goal to localize the failure. The planner took the goal and decided to poll the devices on the affected subnetwork, with the logic that there should be a device that is working, but with downstream neighbors that cannot be reached. The cable break should be between these two systems. Finally, the plan executor executes the plan to localize the failure.

Figure 7-7 shows the progress of the plan at time 5. The display of the whole network has been replaced with a display of the affected subnetwork. The window at the top provides the user with error messages received by the system. Connections to devices ROBOT-11 and POPUP-34 have failed and are highlighted in black. We now know that the break is between ROBOT-7 (who reported the failure) and INSPECTION-1. The user interface has recognized that the two devices are down and highlights this fact to the user.

Figure 7-8 shows the results at the end of the scenario. Connection attempts initiated by OSIFaM to INSPECTION-1 and ROBOT-13 have failed, and this is reported by the plan executor as the result of its actions. This gives the diagnostic knowledge base the information it needs to localize the fault and it updates its diagnosis to reflect the location of the cable break. The user interface then changes the user display to show the additional down devices and places an X through the location of the cable break.

7.4.6 Lessons Learned

The current version of the OSIFaM runs either on a TI Explorer or a Sun-3 workstation using the Carnegie Group's KnowledgeCraft™ shell. So far, it is a demo capability that can diagnose three faults: a modem failure, a break in the network cable, and system congestion. It has diagnostic reasoning and trend analysis capabilities, with a discrete event model of managed system behavior embedded in the diagnostic rules.

The observed results of this effort are the following:

- The number of messages sent to the network administrator for any given fault is reduced drastically (at least in order of magnitude).

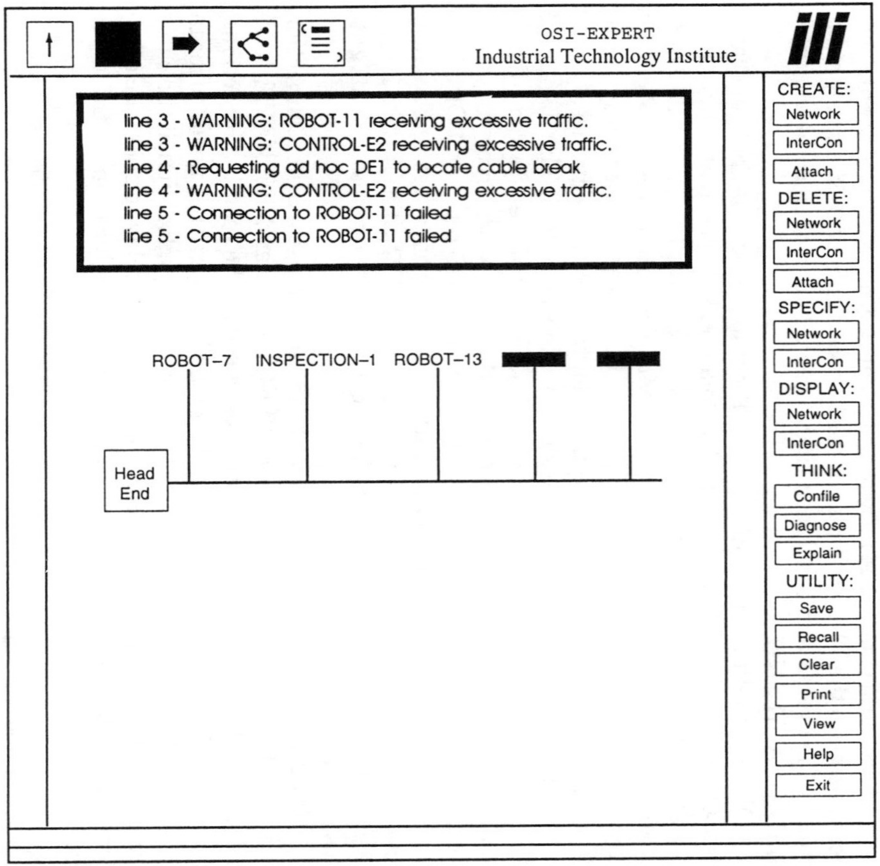

Figure 7-7 Progress of the Plan at Time 5

226

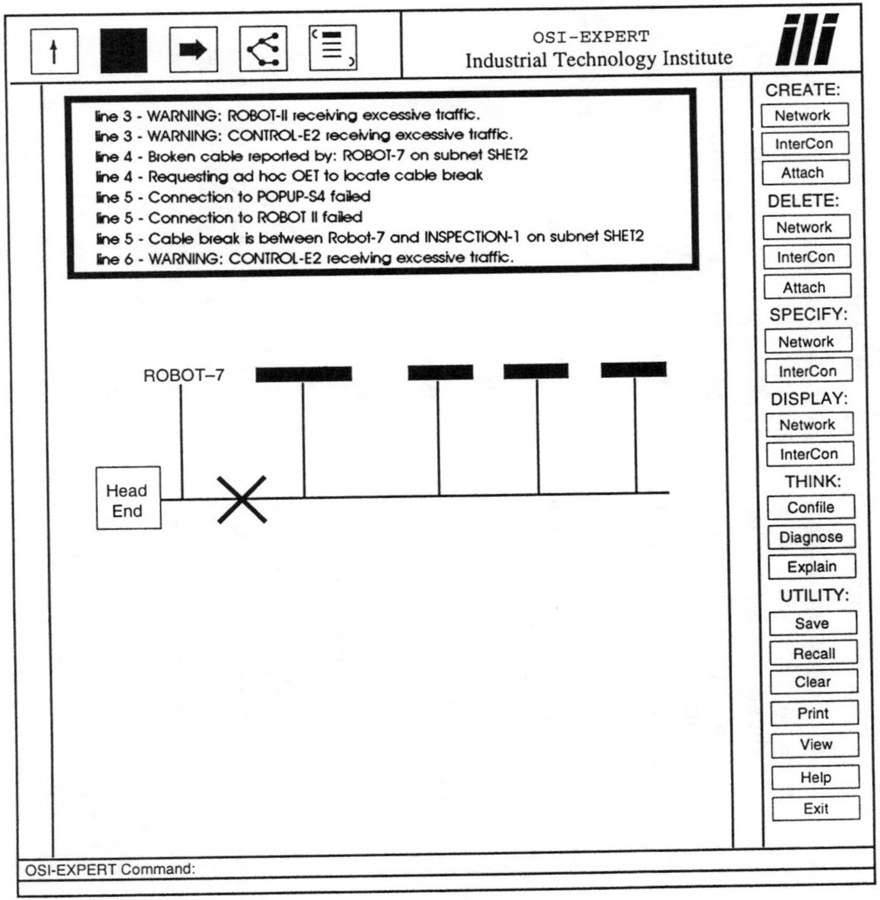

OSI – EXPERT
Industrial Technology Institute

CREATE:
Network
InterCon
Attach

DELETE:
Network
InterCon
Attach

SPECIFY:
Network
InterCon

DISPLAY:
Network
InterCon

THINK:
Confile
Diagnose
Explain

UTILITY:
Save
Recall
Clear
Print
View
Help
Exit

line 3 - WARNING: ROBOT-II receiving excessive traffic.
line 3 - WARNING: CONTROL-E2 receiving excessive traffic.
line 4 - Broken cable reported by: ROBOT-7 on subnet SHET2
line 4 - Requesting ad hoc OET to locate cable break
line 5 - Connection to POPUP-S4 failed
line 5 - Connection to ROBOT II failed
line 5 - Cable break is between Robot-7 and INSPECTION-1 on subnet SHET2
line 6 - WARNING: CONTROL-E2 receiving excessive traffic.

ROBOT-7

Head
End

OSI-EXPERT Command:

Figure 7-8 End of the Scenario

227

Because the system keeps track of which notices are due to specific faults, it can filter the information sent to the network administrator to only a few items of real importance.

- The trend analysis mechanisms are not appropriate for all network counters. For example, some counters (e.g., number of octets sent and received) can normally be highly variable and generate a large number of "false" events. Thus, other mechanisms, such as utilization measures, may be more informative in these areas.

- A general-purpose shell may be needed to make the system flexible even in operations, but the general purpose shell adds enough overhead to slow the operation of the system considerably. This degrades performance and makes real-time performance difficult to achieve. Thus trade-offs in overhead and flexibility must be evaluated before the final design of the knowledge-based network management product.

- Knowledge-based system development is a complex software development effort and needs additional care in managing cost and schedule overruns.

7.5 APPLICABILITY TO VARIOUS NETWORK ARCHITECTURES

7.5.1 Local Area Networks

Clearly, the state of practice in LAN management is to move away from system-based management and toward a distributed management scheme. This includes open systems networks as well as LAN operating systems. As the "network becomes the computer," the user needs to manage not only the communication aspects of the network but the distributed applications running on a particular node.

The OSI model and the structure of management information supports this extensibility requirement. One need only expand the management information base to satisfy the new requirements. The network administrator or operator of a management system tends to be a network user rather than the commercial provider. The knowledge-based system must take into account the relative skill level of the user and provide assistance accordingly. It must not hinder the expert user from taking timely corrective actions in the event of a fault, yet some situations may require a step-by-step, interactive approach to problem solving, and, in still other cases it may require the system making an immediate best guess.

When designing or selecting a knowledge-based network management solution for a LAN, a number of considerations need to be weighed.

Factors such as LAN size, criticality of downtime, less-than-optimal performance, and time spent configuring and reconfiguring systems need to be examined.

7.5.2 Wide Area Networks

This chapter has focused on LAN management using MAP as a case study. When managing WANs with knowledge-based systems, special considerations must be taken in the design of the knowledge base and the inference engines. These issues include knowledge domain, timing issues, number and type of intermediate systems (routers, gateways, etc.), number of nodes, geographical dispersion of nodes, and isolation of the management system(s).

Management of WANs tends to focus on switching, link control, routing and congestion avoidance, access control, and billing for usage. These focuses are managed in the context of configuration, fault, performance, security, and accounting management. The design of the knowledge domain and inference engines, then, should be concentrated in these areas.

Timing will affect both the integrity of some management information and the "repair" activities of fault management. Counters and remotely set time-stamps may not be trusted if the delay is too long. If rules have time factored in their fault correction schemes, then those rules and inferences may have to be adjusted for possible delays over a WAN. WAN management systems must focus on problem avoidance.

Since WAN management concentrates on a somewhat different set of requirements, there is a need for a hierarchical system of managers. In a knowledge-based system management environment, it requires specialized knowledge domains. The agents on a WAN must contain the knowledge base and communicate with the manager on a higher level.

7.5.3 Internetworks

The management of interconnected networks that have different communication protocols and management schemes is becoming a reality. The standardization of management information structure and protocols by ISO expedites the melding of differing management schemes. There are a number of different products that will manage both OSI and TCP/IP networks, for example. This is accomplished either through protocol gateways or proxies. In a knowledge-based system, this requires extra work in the area of the knowledge base development.

Security may effect the accessibility of management information and the ability to take corrective action for repair or fine tuning. It is conceivable that given a set of interconnected networks, there will be various security policies that come into play. This may hinder or prohibit correction or repair activities the manager may need to take.

7.6 CONCLUSIONS

Management of networks is becoming extremely important to ensure fault-free and efficient operation of networks. The cost of management by supporting experts will increase considerably, thereby necessitating sophisticated tools for managing networks. It requires a careful analysis of network management needs to select a system for managing. Knowledge-based systems promise cost-effective and efficient solutions for managing networks of any size.

In designing and developing knowledge-based systems for managing networks, one should keep in mind the trade-off issues associated with size/complexity of networks vs. cost of the management system. The management system must be very affordable. In developing knowledge-based systems, there are several risks associated with cost and schedule overruns. The development efforts must be carefully managed to ensure minimization of cost and schedule overruns. The real-time requirements and affordability requirements pose several constraints on the architecture, development environment, and target platform for the network management product.

Knowledge-based network management systems are becoming more distributed using concepts of distributed artificial intelligence to support decentralized and autonomous management of networks. They are becoming available on PCs with stripped knowledge-based shells to improve response time and reduce cost. They are supporting multiple networks and various management functions. In addition, knowledge-based systems are made available using standards and standard platforms for portability ease.

Continued research and development efforts are required in developing efficient methods and techniques for network management using knowledge-based systems to manage complex networks in the twenty-first century. Additional work is also required in distributing network management to autonomous entities (individual nodes on the network) to improve the response time and reduce the overall cost of network management systems.

7.7 ACKNOWLEDGMENTS

The author expresses deep sense of appreciation to Dr. Edmund Durfee of the University of Michigan and Michael McFarland and Allen W. Sherzer of the Industrial Technology Institute for their invaluable contributions to this chapter. In particular, the author thanks Michael and Allen for writing portions of this chapter and sharing their experience with network management system development. The author also thanks Sandy Halasz for her great help in preparing this manuscript.

REFERENCES

[1]. International Organization for Standardization, ISO 9595/1, *Management Information Service*. Part 1, Overview, 1990.

[2]. International Organization for Standardization, ISO 9595/2, *Management Information Service*. Part 2, Common Management Information Service Definition, 1990.

[3]. International Organization for Standardization, ISO 9596/1, *Management Information Protocol Specification*. Part 1, Overview, 1990.

[4]. International Organization for Standardization, ISO 9596/2, *Management Information Protocol Specification*. Part 2, Common Management Information Protocol, 1990.

[5]. International Organization for Standardization, *Draft International Standard ISO 10165*. Structure of Management Information Parts 1, 2, and 4, June 1990.

[6]. C. A. Joseph and K. H. Muralidhar, "Network management: A manager's perspective," in *Communications Standards Management*, Boston: Auerbach Publishers, 1989.

[7]. C. A. Joseph and K. H. Muralidhar, "Integrated network management in an enterprise environment," *IEEE Network Magazine*, July 1990.

[8]. Shri K. Goyal, "Knowledge technologies for evolving networks," *Integrated Network Management II*, IFIP, Amsterdam, The Netherlands, pp. 439–461, 1991.

[9]. H. V. D. Parunak, J. D. Kindrick, and K. H. Muralidhar, "MAPCon: A case study in a configuration expert system," AI EDAM, vol. 2, no. 2, pp. 71–88, 1988.

[10]. Gidi van Liempd, Hugo Velthuijsen, and Adriana Florescu, "Blondie-III," *IEEE Expert*, vol. 5, no. 4, pp. 48–55, August 1990.

[11]. Ahti Salo and Raimo P. Hamalainen, "Seteli: The strategy expert for telecommunication investments," *IEEE Expert*, pp. 14–22, vol. 5, no. 5, October 1990.

[12]. Rodney Goodman, "Real time autonomous expert systems in net-

work management," in *Integrated Network Management*, New York, NY: North-Holland, pp. 599–624, 1989.

[13]. Jim Cavanaugh, "The AI Zone," *LAN Magazine*, pp. 63–70, vol. 7, no. 3, March 1992.

[14]. J. Doyle, "A truth maintenance system," *Artificial Intelligence*, vol. 12, no. 3, pp. 231–72, 1979.

[15]. T. Dean and D. K. McDermott, "Temporal database management," *Artificial Intelligence*, vol. 32, no. 1, pp. 1–55, April 1987.

[16]. A. Bond and L. Gasser, *Readings in Distributed Artificial Intelligence*, San Mateo, CA: Morgan Kaufmann Publishers, 1988.

[17]. R. Davis and R. Smith, "Negotiation as a metaphor for distributed problem solving," *Artificial Intelligence*, vol. 20, no. 1, pp. 63–109, 1983.

[18]. S. Conry, R. Meyer, and V. Lesser, "Multistage negotiation in distributed planning," in *Readings in Distributed Artificial Intelligence*, San Mateo, CA: Morgan Kaufmann Publishers, 1988.

[19]. V. Lesser and D. Corkill, "Functionally accurate, cooperative distributed systems," *IEEE Trans. on Systems, Man, and Cybernetics*, vol. SMC-11, no. 1, pp. 81–96, January 1981.

[20]. D. Corkill and V. Lesser, "The use of meta-level control for coordination in a distributed problem solving network," in *Proceedings of the Eighth International Joint Conference on Artificial Intelligence*, Karlsruhe, West Germany, pp. 748–756, August 8–12, 1983.

[21]. E. Durfee, *Coordination of Distributed Problem Solvers*, Norwell, MA: Kluwer Academic Publishers, 1988.

[22]. E. Hudlicka and V. Lesser, "Modeling and diagnosing problem solving system behavior," *IEEE Trans. on Systems, Man, and Cybernetics*, vol. SMC-17, no. 3, pp. 407–19, May–June 1987.

[23]. R. Fikes and N. Nilsson, "STRIPS: A new approach to the application of Theorem proving to problem solving," *Artificial Intelligence*, vol. 2, no. 3/4, pp. 189–208, Winter 1971.

[24]. E. Sacerdoti, *A Structure for Plans and Behavior*, New York, NY: Elsevier, 1977.

[25]. M. Stefik, "Planning with constraints (MOLGEN)," *Artificial Intelligence*, vol. 16, no. 2, pp. 111–139, May 1981.

[26]. R. Davis and B. Buchanan, "Production rules as a representation for a knowledge-based consultation program," *Artificial Intelligence*, vol. 8, no. 1, pp. 15–45, 1977.

[27]. P. Nii, Blackboard Systems: the blackboard model of problem solving, part one," *AI Magazine*, vol. 7, no. 2, pp. 38–53, Summer, 1986.

[27a].P. Nii, "Blackboard Systems: blackboard application systems, black-

board systems from a knowledge engineering perspective, part two," *AI Magazine*, vol. 7, no. 3, pp. 82–106, 1983.

[28]. J. Laird, A. Newell, and P. Rosenbloom, "SOAR: An architecture for general intelligence," *Artificial Intelligence*, vol. 33, no. 1, pp. 1–64, September 1987.

[29]. C. A. Joseph, M. McFarland, and K. H. Muralidhar, "Integrated management for OSI networks," *IEEE Globecom Proceedings*, vol. 1, San Diego, CA, pp. 0565–0571, December 2–5, 1990.

[30]. C. A. Joseph, A. W. Sherzer, and K. H. Muralidhar, "Knowledge-based fault management for OSI networks," in *Proceedings of the Third International Conference on Industrial and Engineering Applications of AI and Expert Systems,* pp. 61–68, July 1990.

CHAPTER 8

Configuration Management

Gordon Rainey
Bell-Northern Research
Ottawa, Ontario
Canada K1Y 4H7

8.1 INTRODUCTION

The network management (NM) category *configuration management* is considered herein to include the processes of network planning, resource provisioning, and service provisioning. Note that this is a broad interpretation of configuration management but it is useful to consider these three processes together. Collectively they represent more than 50% of typical network operating expenses and leverage both the capital budget and revenue potential of the operating company. The perspective adopted here addresses operations as they might apply to a common carrier, private network operator, or an enhanced service provider. To simplify ongoing references to these network operators, they will be collectively referred to generically as *operating company* or *service provider*.

8.1.1 Scope of Chapter

This chapter defines and describes configuration management as a set of processes with specific triggers, inputs, internal functions, and outputs. These processes are undergoing rapid evolution and, in some cases, reengineering to keep pace with the challenge of configuring more complex networks for more complex services and to take advantage of emerging computing technology and techniques both in the operating system (OS) environment and in the network element (NE) environment.

Present modes of operation are described as well as strategic objectives driving evolution toward the twenty-first century. Target modes of operation are discussed as well as transition aspects constraining or enabling their attainment.

The processes described typically allow control information to flow from management users to OSs and NEs and status information to flow in the reverse direction. Management users are primarily operating company staff accessing the network through OSs. However, there is incentive to extend to customers and end users (as management agents) direct access to OSs and NEs to more quickly achieve control or retrieval of status information.

8.1.2 Network Planning

The network planning process translates the enterprise, corporate, or operating company's business strategies into a set of cost-effective plans to configure and evolve the managed network. Network planning, therefore, establishes the architecture or configuration of the network.

8.1.3. Resource Provisioning

The resource provisioning process ensures procurement, deployment, and configuration of the necessary resources in time to meet anticipated customer or end-user service demand for an economic engineering interval. The network resources in question include both hardware and software. Network planning, particularly in the form of the current plan, provides a key input to the resource provisioner.

8.1.4 Service Provisioning

The service provisioning process, initiated by the customer or end-user service request, further configures selected resources, including NE datafill, to respond to specific customer or end-user service profiles and aggregated traffic demand across the network.

8.1.5 Relation to Other Management Functions

Configuration management is a primary network management function [1] because it is responsible for putting the network in place. It also fine tunes the configuration for delivery of services to specific customers and databases. Finally, it establishes resource database and maintains resource status for use by other network management functions, such as fault management and performance management.

Fault management further ensures that the network resources operate in a trouble-free and error-free manner. This may be achieved proactively by anticipating problems and preventing their occurrence or by quick repair when a failure occurs. Fault management references the configuration management resource inventory for identification of network elements in network surveillance and testing. It interworks with the resource management database if the maintenance status changes and when faulty equipment is replaced or when a fault is circumvented by network reconfiguration.

Performance management ensures that the network is operating efficiently and delivers the prescribed grades of service. Performance management operational measurements are key inputs to both the planning process to track growth of demand and to the resource provisioning process in order to flag actual or impending resource shortages. As with fault management, performance management depends on the configuration management resource inventory in referencing its monitoring measurements and directing its network control messages.

Accounting management ensures that the customer is accurately billed for network services rendered. This will include access to auxiliary network management services such as direct exercise of management control and status information retrieval. Service provisioning typically triggers the appropriate mechanisms for billing the services being provided.

Security management is a utility function supporting other prime network management functions (by allowing or restricting management user access), including network planning, resource provisioning, and service provisioning. Security management prevents unauthorized intrusion, and allows flexible privilege control for operating company personnel. For effective self-service, security management must be extended to provide sufficient authorization and privilege control granularity for customers and end-users.

8.1.6 Goals of Configuration Management

At the highest level, the goals of the operating company can be expressed in terms of quality of service and growth objectives, as well as profitability of the particular communication segment and its contribution to the overall corporate bottom line. Profitability goals are typically broken down in terms of revenue generation, cost containment, and expense reduction. Configuration management processes have a large impact in all of these areas.

Network planning plots a course among several potential network evolution alternatives in order to guide and leverage the expenditure of large amounts of capital for network resources in anticipation of even larger service revenues to pay for the operation of the network and generate a profit. Anticipating the service needs of the target market and the technologies that can best meet these needs is most critical to the success of the enterprise. Marketing and business development groups typically deal with the former and network planners with the latter.

Network planners normally use a three-pronged approach for choosing between network evolution alternatives. The service-rendering capability of the network resources is the key consideration. The capital cost of the technology is the second critical factor. The OS costs and operating expenses are also factored in. Because network resources have a long service life cycle, life cycle cost optimization—considering both first cost and operating cost—is used in prescribing network resource selection and deployment guidelines.

The expense for staff incurred in planning is typically small relative to resource provisioning or service provisioning and relative to the capital expenditures being leveraged. There is a continuing emphasis on more powerful planning tools to process the large amount of information required to compare alternatives and produce sound plans for networks of increasing service and technical complexity.

Resource provisioning has its largest direct impact on capital containment not only because it controls actual resource procurement, but also because engineering and installation effort for network growth and modernization is typically capitalized [2]. In addition there are secondary impacts on revenue generation because service provisioning is dependent on the right network resources being in place in time to meet service demand. Following the plans generated in network planning, specific jobs are engineered and capital expenditure is incurred in the procurement and installation of network resources in the resource provisioning process. Minimizing the capital cost of resources can be achieved by competitive tendering, negotiating bulk discounts, and reducing resource deployment lead times.

Service provisioning goals [3] typically address the strategic objectives of expense reduction, revenue protection, and service improvements. Because service provisioning must respond quickly to customer requests, and because the current mode of operation involves a human interface as well as semiautomated processes, there is still much room for expense reduction and service response time reduction. With increasing competition, service responsiveness is a more critical consideration for service providers. Finding ways and means of quickly delivering services are key goals of service provisioning. Therefore, service provisioning velocity is also becoming a grade of service measure with increasingly stringent parameters.

8.2 NETWORK PLANNING

8.2.1 Network Planning Process

The network planning process translates the operating company's corporate strategic objectives and business plans into a set of long- and short-term network evolution plans. Because planning deals with the future, there are elements of uncertainty about future service demand, availability, cost of technology, etc. There is also a need to compare evolution alternatives. Planning is the process of comparing alternatives and recommending the best balanced course of action.

Figure 8-1 shows a generic network planning process starting with interpretation of the operating company's strategic plan and ending with the delivery plans to the resource provisioners. Note that the steps in the network planning process allow for iteration as changing circumstances and new information dictate. This points to a need for planning models and tools that allow for rapid recycling.

8.2.2 Goals of Network Planning

The key goal of network planning is to produce a set of plans that meet the operating company's strategic objectives and give the resource provisioners clear direction. The planning horizon should include both short-term and long-term objectives. These plans should accommodate new service introduction, growth, and modernization as well as provide guidelines for equipment selection and deployment.

8.2.3 Network Planning Functions

Figure 8-1 is a general planning sequence. The following sections discuss the functional blocks of the planning process in more detail. Given the enterprise strategic planning objectives as input, the network planner must translate these into a set of workable plans for the resource provisioners.

Interpret the Strategic Plan. The operating company's strategic plan contains the high-level objectives and business rationale for evolving the enterprise and the philosophy for evolving the network. It establishes which businesses the operating company is in and which it is not. The strategic plan is the total declaration of the operating company initiatives to achieve their profitability requirements. Because the network is one of the main vehicles for achieving these objectives, provisions of the strategic plan must therefore be interpreted as the first step in network planning.

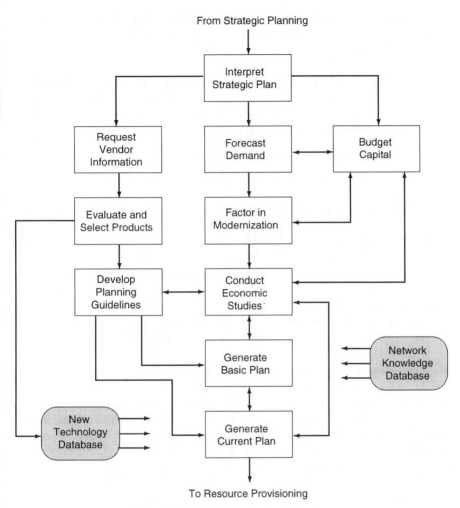

Figure 8-1   Network Planning Process

A key challenge of network planning is to ensure that sales plans and network capabilities mesh with profitable network evolution scenarios.

Forecast Demand. Demand forecasting is a key function because it largely determines the size, topology, and service capabilities of the network as well as its revenue potential. Forecast parameters typically include the number of network access lines and services, traffic, and community of interest for each service in the portfolio. Planning forecasts project historical data for existing services, blend in economic projections, and take advantage of market research, particularly for the launch of new services. Increasingly, the service revenue projections are being used to condition the allocation of capital for growth and modernization. The technology capabilities of network resources must match the requirements of the market demand.

Factor in Modernization. Because of equipment obsolescence, network planning must also include provisions for modernizing older components. Modernization may be precipitated by the escalating operating costs being incurred to deliver an adequate grade of service from the older equipment or by the fact that the older equipment cannot deliver the desired services. Analog to digital modernization has been underway for some time, and, more recently, transformation from copper- to fiber-based access and transmission.

Typically capital expenditures for growth and modernization are of the same order of magnitude in long-established common carrier networks. Modernizing one part of the network may make reusable equipment available for another part. Operating companies typically do not buy new equipment of older design beyond a certain point in the design life of equipment—they become self-sufficient in the older equipment type by recycling it from one part of the network to another. Recycling allows the operating company to modernize the more critical parts of the network with state of the art equipment while lowering the capital cost of continuing to grow the less critical part.

The depreciation rate of equipment, established by regulating bodies, is a key factor in establishing the service life of equipment and the rate of modernization. The increasing pace of technology and service evolution are dictating faster obsolescence and shorter depreciation intervals. There is still a marked difference between the planning life of public utilities (20 to 25 years) and, say, LANs or electronic data processing (EDP) equipment (5 to 10 years).

Budget Capital. Available capital places a practical constraint on what can be achieved. Operating companies usually split the capital budget into growth and modernization components, on a geographic basis, to ensure equilibrium of network evoluation. Job sizing and timing depend on available capital. Mechanisms must also be in place to accommodate the impact on projects of inevitable budget cuts. These impacts include the recycling effects of project cancellation, deferral, downsizing, etc., or, alternately, project acceleration as a result of bulk discount procurement contracts that could speed up deployment for modernization.

Conduct Economic Studies. Economic studies [4,5] establish the overall viability of a proposal or the best among a set of alternatives for evolving the network. Network planners may also determine costs of network alternatives to support assessment of new service thrusts and the definition of marketing strategies. Capital, expense, and revenue dollars cannot be compared directly because factors such as income tax and depreciation allowance affect them differently. The time value of money must also be considered because financial outlays are typically distributed through time.

Economic parameters such as net present value (NPV) and pay-back period are used to determine the viability of launching a new service, for example, taking into account revenues, capital costs, and operating expenses. Other economic parameters such as present worth of capital expenditures (PWCE) and present worth of annual charges (PWAC) help determine the best technical alternative, assuming revenues are constant.

Economic studies also establish economic engineering intervals for job sizing and timing, which the provisioners can apply to their particular circumstances. Because projects have a "get started" overhead regardless of size, there is a natural tradeoff necessary between job size and interval.

Operating companies typically have engineering economic models and systems in-hand, tailored to their own economic environment, to facilitate planning studies. Recently these have become more accessible to planners, having been converted from the mainframe environment to the planners' workstation spreadsheet environment.

Evaluate and Select Products. Technology selection is a complex process because networks can be large, they may support multiple services, they usually contain complex resources whose designs are undergoing rapid technological evolution, and they are typically supplied by multiple vendors. In fulfilling this role, planners typically coordinate with technology groups who specify the required capabilities of new products, conduct trials on new equipment, etc.

Technology identification is through contact with traditional suppliers, vendor marketing bulletins, trade shows, industry periodicals, technical conferences, lab visits, requests for information (RFIs) to key vendors, etc. Selection is typically based on a comparison of cost and capabilities and advertised capabilities require confirmation via joint technology and field trials. Successful systems are then standardized for procurement and deployment. The result is a new technology portfolio of equipment that is capable of meeting the needs of the service portfolio cost effectively.

Communication networks are rarely supplied by a single vendor and typically two or three vendors are chosen for each key technology area to ensure competitive procurement pricing, a second source of supply, and so on. Having an excessive mix of technology from different vendors, however, tends to complicate subsystem interworking and increases training costs.

Most large organizations develop corporate standards and specifications for equipment comparison and selection. These corporate standards are typically a subset of applicable industry standards but may also include in-house arrangements developed for internal use. These procurement specifications are appended to RFIs and requests for quotation (RFQs) and used as references for equipment comparison and selection. The selected products are standardized for internal procurement and deployment and their characteristics are input to new technology databases used as references for planning guidelines, basic plans, and current plan generation.

Generate Planning Guidelines. Once the preferred system or systems have been selected for a particular application, network planners issue planning guidelines so that basic (long term) and current (short term) planners and resource provisioners do not have to go through an extensive selection process, but can focus on detailed analysis for their particular applications based on a narrower range of recommended alternatives. Planning guidelines must combine network, market, technology, and operations knowledge into a set of recommended network evolution solutions that include incremental requirements on the associated OSs. Planning guidelines must be comprehensive to ensure the most cost-effective and orderly evolution and management of the network with less proliferation of technology, spare parts, test tools, and training expense, while retaining flexibility for future contingencies.

Generate Basic Plan. The long-term network evolution view for specific applications or parts of the network are generated by fundamental planners. The fundamental planner typically projects demand and modernization requirements for a predefined application or segment of the network, up to 20 years, based on the imbedded equipment, the anticipated long-range service demand, modernization considerations, and referencing planning guidelines relevant to the application. The fundamental planner creates basic plans that outline the way in which the network is to evolve to meet market demands.

Generate Current Plan. Based on approval of the basic plan and associated capital allocation, the current planner will produce and maintain more detailed plans involving specifics on service forecasts, modernization, resource quantification, configuration, and cost—including, typically, up to a 5-year view. These are periodically updated to maintain currency. Field work may be necessary to ensure that field conditions are accurately reflected within the data available to the planners. The current plan references the target architecturé in the basic plan but applies it to smaller geographic areas of the network. Current plans must also be aligned with network evolution scenarios based on the latest marketing strategies. They are also an essential input to budgeting, to advise the financial departments on what levels of capital to raise for network expansion, as well as to advise resource provisioners of any impending resource shortage. The flagging of impending resource shortages by current planners is one of the main triggers for initiating a resource provisioning job. The current plan allows the provisioner to operate independently of the planner for normal demand requirements.

Historically planning was paper based and crude by today's standards. With the ready availability of distributed processing power, network knowledge databases, networking, etc., all orchestrated by planning application software, the planning process is producing better plans faster, and these can be recycled more frequently as changing conditions dictate.

8.2.4 Evolution Drivers

The only justification for a communication network is to provide communication services. Network provider incentives to improve network planning, therefore, relate to delivering services more cost effectively in an increasingly competitive environment. For many traditional service providers this means reengineering the planning process to become more responsive to customer needs. Anticipating these needs ahead of actual demand and ensuring the network evolution plan is adaptable to a growing array of elaborate communication services is a key challenge. Selecting between a broad array of technical alternatives is part of this challenge.

Current Limitations. Current network planning processes, developed in an era of monopoly services and, in some cases, captive suppliers, are not responsive to emerging needs in a competitive environment. Current guidelines for NE deployment typically do not generally optimize profitability based on new marketing opportunities.

Opportunities for Improvement. Network planning opportunities, therefore, include better alignment of the network planning process with strategic initiatives in a competitive environment. Targeting the right market niches and accurately forecasting service demand is also critical. The increasing diversity of services, especially data services, is making decisions on what services to offer more difficult.

Capitalizing on opportunities afforded by emerging technology, particularly in the area of service adaptiveness and reliability, will decrease risk. The increasing diversity of services leads service provisioners to demand more flexible equipment. Partnering with selected vendors in planning for and bringing new network capabilities into trial will also speed up service introduction. This may include custom development of customer-specific features that may not be part of a vendor's general plans.

Creating easy access to network knowledge will facilitate revenue opportunity or deployment cost assessment. Creation of multi-application corporate databases is a step towards this end. Planning tools, to quickly contrast alternatives, process "what-if" scenarios, and arrive at recommendations based on bottom line parameters, are essential. Emerging OS-based tools in the planning process are becoming more integrated. Open architecture principles are becoming more evident in accessing multi-application corporate databases and other supporting OSs.

8.2.5 Target Process

The future network planning process should allow a more comprehensive (end-to-end) view of service-driven resource topology. This view must include consideration of both the managed network and the managing network, i.e, both NEs and OSs. Introduction of complex new services and technology should be simplified by the planners so that it is business as usual for resource provisioners. Service- and technology-specific procedures, although sometimes necessary for a new service launch, should be merged into an adaptive common planning stream. To reduce planning cycle times, the typically lengthy and iterative estimate approval and budgeting processes must be streamlined.

The target process would also focus on market segment macro-forecasting for planning and budgeting as opposed to detailed microforecasts. Microforecasting will, however, be a necessary part of resource provisioning, particularly for the per-service stratification for service-sensitive resources and for just-in-time provisioning of deferrable plug-ins, connectivity jumpers, etc.

8.2.6 Transition Considerations

Making the transition from the present mode of operation (PMO) to the target or future mode of operation (FMO) is usually more complex if there is an extensive imbedded base of processes and procedures to evolve.

Because the network planning phase is at the leading edge of network management activities, it is constantly undergoing change to accommodate new services and technology and does not currently interwork with other processes on an on-line, real-time basis. It is therefore the process that is easiest to adapt to emerging requirements and opportunities. It is also the one that should be reengineered first because of the leverage it has in directing the substantial configuration management process and the implied capital expenditures. How to best achieve open architecture, integrated systems, and corporate databases are becoming major issues of transition.

8.3 RESOURCE PROVISIONING

In the context of this chapter, network resources refer to NEs delivering access, switching, transport, and network database functionality. These NEs are comprised of hardware and software resources. Hardware categories include infrastructure, common equipment, plug-ins, and connectivity elements such as jumpers. Software categories include generic programs and software patches.

Additional "soft" network resources include directory numbers (DNs) for end-users, office codes (NNXs) and private network location codes, network planning areas (NPAs), as well as translations (datafill) required to tell the software how the office or NE is configured.

8.3.1 Resource Provisioning Process

The resource provisioning process, triggered by the need to grow, modernize, upgrade and/or rearrange the network, ensures that adequate resources are deployed and configured to meet service needs for an economic engineering interval. Major resource provisioning activity, involving infrastructure and/or common equipment, is typically managed via formal project management techniques, while short-term activity such as deployment of plug-ins, cable splicing, and jumpering is handled by work orders. Software, although deployed coincident with initial jobs or major extensions, is often upgraded periodically, independent of equipment additions. The same is true for software patches, introduced between scheduled software releases as temporary fixes to correct major bugs in the main software programs.

Figure 8-2 gives an overview of a generic resource provisioning process for major resource provisioning projects. The major functional blocks in the process are further described in Section 8.3.3.

8.3.2 Goals of Resource Provisioning

The goals of resource provisioning are to meet the resource needs of the communication network in the most cost-effective way. As indicated above, the primary purpose of the communication network is to meet the service needs of its customers. Some means to achieve these goals follow.

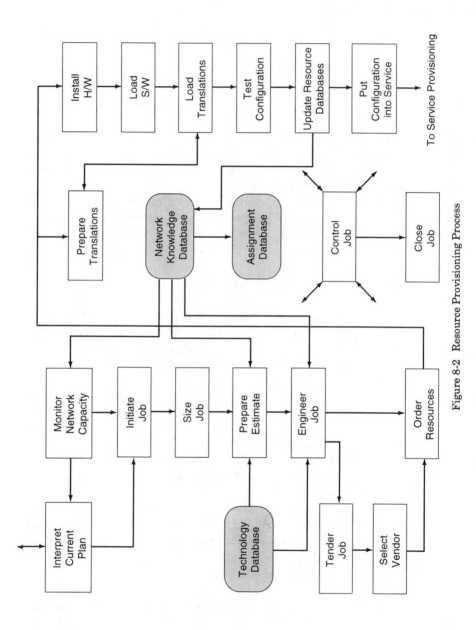

Figure 8-2   Resource Provisioning Process

To Service Provisioning

Just-In-Time Provisioning. A key objective of resource provisioning is to ensure that network resources are deployed just ahead of actual demand, but are not deployed needlessly, excessively, or prematurely. Just-in-time provisioning [3] is a goal aimed at reducing both lead time for deployment and the engineering interval. It also addresses the need to ensure that equipment is deployed not only where it is needed but where it will be needed. Accurate demand forecasting and trade-offs between installation visits and capital carrying charges are required to achieve just-in-time provisioning.

Dedicated Resources. Dedicating resources (or dedicating plant) is a thrust that prescribes leaving resources, particularly access resources, that have already provided service, in place even if a customer or end-user has recently discontinued service [3]. This is done under the assumption that another customer or end-user may subsequently soon require the same or similar service at that location, and it is less costly and more service responsive to leave the resource in place rather than reassign it elsewhere. The concept of soft dial tone goes a step further by extending the dedication to include end-user access to service negotiation resources in order to establish services.

Self-Inventorying. NEs are becoming more intelligent and they typically require configuration translations datafill for real time control. Self-inventorying NEs would have the awareness of their own resources but would also be capable of uploading OSs upon query. With self-identification features, self inventorying could be extended to automatically update the NE database when plug-ins, software programs, or patches were added to the NE. Self-identification would also include the capability to ensure that the wrong plug-in is not put in a slot.

Configuration data would include hardware and software quantities, configurations, and states. Resource status for dedicated plant, DNs and line equipment numbers (LENs), etc., including aging and reservations for large planned jobs, would be kept for service assignment purposes. NEs are viewed as the appropriate stewards of their configuration database. This provides them with quick access to their configuration data for real time operation and eliminates the problems of database alignment with duplicate OS versions.

Self-Engineering. NEs capable of self-inventorying and performance monitoring could also do internal load balancing automatically and remove the need (in routine cases) for external operating company intervention. Extensions of this capability also include flagging resource shortages and prescribing resource additions.

8.3.3 Resource Provisioning Functions

The following functions apply largely to medium and large projects other than jobs that can be handled by routine orders or work orders. Smaller resource provisioning jobs are typically handled through less formal processes.

Interpret Current Plan. The need for major resource additions is typically identified in the current plan for evolving a particular part of the network. Other triggers may include unexpected large customer requests, changes in strategic plans or budgets, etc. Regardless of the specific trigger, the first step in resource provisioning is to interpret and rationalize the situation and provisions of the current plan. This will typically involve consulting with the current planners to consider the latest available demand, modernization, and budget information to confirm the need to proceed with resource provisioning activity immediately or, alternately, defer the project.

Monitor Network Capacity. In addition to timing triggers from the current plan, there is increasing emphasis being placed on comparing remaining capacity in the network to projected demand on an ongoing basis, and flagging the need to add more resources. This allows for shortening lead times to provision deferrable plug-ins, or even common equipment, when coupled with an automated ordering system. The way resource provisioning monitoring works is as follows. The historical demand on each resource pool is projected and compared with remaining capacity. This implies the capability to forecast growth on a network element (or resource pool) basis. If the projected demand exceeds the capacity within the resource-ordering lead time, a flag is raised to notify the resource provisioner to review the situation. The provisioner may proceed with an order directly if it involves plug-ins, etc., or initiate a project if it involves common equipment or network infrastructure. The provisioner may also apply additional "street smarts," such as sales information (not available to the monitoring system and not reflected in the current plan), and respond accordingly.

Initiate Job. Job initiation results when the time for provisioning major new resources is imminent and the need is confirmed. The scope of the job and lead time varies depending on whether it involves infrastructure, common equipment, or software upgrade. It involves identifying the project, providing advanced notification to coordinating groups and suppliers, and setting up a job control structure to track progress. Related jobs must be similarly initiated and scheduled.

Size Job. In the current plan, the incremental capacity required, the major equipment elements involved, building and network implications, etc. will typically be covered, in broad terms, in the current plan. All of these aspects require detailed verification at the front end of the resource provisioning phase. Utilization information for interconnected network elements must be correlated to assess whether they also need relief. How much equipment to install will depend on economic intervals, available funding, strategic plans, building capacities, provisioning guidelines, local rules, etc. Modernization requirements will be combined with growth requirements. The fixed start-up cost for a job means that jobs should not be too frequent but sized per planning guidelines for an economic engineering interval.

Prepare Estimate. Job estimates form the basis for approval of capital allocation to a project, necessary to proceed with resource procurement. Resource pricing is obtained from the equipment vendors. Sources include price lists and recent quotes. In addition to the vendor's software, engineered, furnished, and installed (SEF&I) components, operating company costs also typically include additional components such as federal, state, and municipal taxes; engineering, installation, and acceptance testing labor; interest during construction; the cost of refurbishing reused equipment; removal of old equipment; etc. Many operating companies and vendors have developed pricing tools to facilitate the costing of equipment additions and generation of job estimates.

Engineer Job. Once the proposed job is sized and the estimate approved by operating company management, a detailed specification of the equipment and how it is to be configured is prepared. This will also entail equipment layout and determination of building facilities impacts such as power, lighting, air conditioning, etc. Reserving required space and making provisions for building facilities is part of engineering the job. The equipment specification will be used as the basis for job tendering, if competitive bidding is involved, and for the equipment order.

Tender Job. If the job calls for multiple competitive bids, RFQs are tendered to prospective suppliers. The supplier list will typically be generated from suppliers of equipment approved as standard for deployment by network planning. If the job is an extension to an existing system, the choice may be limited to the original supplier.

Select Vendor. If competitive bidding was involved, selection criteria is then applied to the responses and a final vendor is selected. Selection may be based on the lowest first cost, how closely the bid met the original specification in terms of system features, the past performance or operation cost of similar equipment from the supplier, some combination of the above, and so on.

Order Resources. Once the equipment vendor is selected, an order is placed for the resources specified in the operating company job specification. The job specification then forms the basis for further refinements in terms of changes to equipment quantities, timing and pricing, and other contractual terms.

Control Job. Once the job is initiated, all aspects including scheduling, monitoring progress, overcoming impediments, minimizing service disruptions, and seeing the job through to completion on budget must be monitored. Summarizing the impact of job construction expenditures on the capital budget bottom line is one of the inputs that allows senior management to manage allocations to upcoming projects.

Prepare Translations. In this context, the term *translations* means the configuration data that the stored program controlled network element requires to perform call processing and network management functions. These typically represent both resource and service configurations. The resource data is sometimes referred to as *office parameters*. The service data is sometimes referred to as *service profiles* and may include line translations, routing translations, etc. Initial installations and major extensions of stored program controlled equipment usually imply a substantial effort to prepare accurate translations. Translations for major jobs are typically prepared and verified on offline systems. They may also be refined during the installation and testing interval to ensure that they represent the latest view of the equipment configuration and services as the project moves toward cut-over and the in-service data. On an ongoing basis, the service profile translations are updated as part of service provisioning (see Section 8.4.3).

Install Hardware. Equipment for major jobs is typically delivered directly to the installation site from the vendor's plant. Depending on the particular situation and type of resources, installation may be done by the vendor, a contractor, or operating company personnel. In any case, operating company personnel are always involved to ensure the installation work is properly executed with minimal or no service disruptions.

Load Software. Software loading is the process of entering software into NEs. This may occur coincident with an initial installation or major hardware extension. It may also occur independently of a hardware addition, as part of a software upgrade.

Load Translations. Translations include NE resource datafill representing both resource configurations and service profiles. Normally office parameters specifying equipment quantities and configurations are datafilled during resource provisioning. Service translations are entered incrementally during the course of service provisioning. For major cutovers associated with modernization or customer transfers, however, larger blocks of service translations must be generated and batch-entered during the course of resource provisioning. These are used to test the configuration. As the cut-over date approaches, recent changes to these translations must be entered to ensure that at cut-over, the service profiles match the customer needs.

Test Configuration. Acceptance testing is carried out by the operating company or jointly with the vendor in order to ensure that the resources added operate properly and to the operating company's satisfaction. Testing will verify hardware, circuit operation, interconnecting facilities to other network locations or to operation support systems, software and datafill integrity, etc. Because there are always new technologies, processes, systems, and services that the operating company provisioners are not familiar with, there is a need for technical support during provisioning and, particularly, in the testing phase.

Update Resource Databases. Once the resources are deployed and tested, the resource databases to be referenced for subsequent resource provisioning, service provisioning, fault management, and performance management are updated.

Put Configuration into Service. Subsequent to job acceptance and update of resource databases, the added resources are put into service at cut-over (for modernization or customer transfers) or made available for new service provisioning (in the case of a project performed purely for growth).

Close Job. Upon completion of all work, the job is officially closed and manpower is redirected to other projects.

8.3.4 Data Management

Resource provisioning requires substantial amounts of data inventorying and transfer. For resource provisioning, data transfer is typically infrequent, of high volume and high speed, and usually coincident with major hardware and/or software additions. Configuration data includes hardware resources and their states; software program code as well as vintage information for main loads and patches; connectivity information including circuits, paths, and trails; resource assignment data in support of service provisioning; etc. For initial jobs and major extensions, the vendor and/or the operating company must generate the configuration data appropriate for each NE and load it into the NE, if it is stored program controlled. This data must be loaded into the NE to make it aware of its own configuration for call processing and network management functions.

Subsequent ongoing resource provisioning, service provisioning, and fault management activity may alter the quantity of resources, their configuration, or their state. As resources are added they become available for service assignment. When service is discontinued the resources may become available for reassignment or may be left dedicated for some time.

Data Duplication and Synchronization. Operating companies typically maintain configuration data in operating system databases to enable and facilitate fault management, service provisioning (assignment), performance (network) management, etc.

Given the dynamic evolution and reconfiguration of the network, it is difficult to keep accurate data. Keeping OS resource provisioning databases in sync with the actual deployed resource configuration currently is a major problem. Having intelligent NEs steward their own resource data and make it available as part of a uniform distributed database and processing environment is seen as a solution to this problem. Thus intelligent NEs offer the possibility of eliminating duplication of data and associated data entry error by maintaining the "official" copy of its own data, available upon query. If data must be duplicated for security reasons, then a means must be provided for frequent updates to ensure accuracy.

Resource Reservation, Dedication, and Aging. The aging of out-order DNs or dedicated plant must be factored into the resource data management process. Previously assigned DN resources are typically not assigned immediately on discontinuance of service but are aged for some time to avoid wrong number calls. Calls to recently changed numbers are intercepted and if a move was involved, the new number is given in an announcement. After sufficient time the aged DN is reassigned to service. Similarly dedicated plant must be aged before reassigning it. Resource provisioning data management, particularly for assignment purposes, must apply the aging functions to these resources. Both types of resources can be reassigned earlier under resource shortage circumstances if held-orders would otherwise occur. A held-order is a request for service that cannot be met due to a lack of resources.

Similarly resources may be reserved for specific later use and are not made generally available for assignment. Resource provisioning data management must therefore also keep track of reservation status.

Data Management Interfaces. Resource provisioning interfaces for configuration datafill or software loading are currently vendor specific. Some involve loading data from physical tapes plugged into NE load modules, but the trend is toward remote downloading from the vendor or an operating company OS. Open system interconnection (OSI) standards are being formulated to ensure future multivendor consistency. Translation from generic resource abstract representation to vendor-specific NE datafill would then be a vendor design responsibility.

8.3.5 Process Variations

As mentioned above, resource provisioning for different types of resources having different procurement triggers, lead times, and/or engineering intervals are typically managed differently. These resources can be categorized as follows.

Resource Type	Lead Time	Engineering Interval
Infrastructure	long	very long
Common equipment	long	long
Software upgrades	scheduled	short
Patches	short	short
Plug-ins	short	short

Software Upgrades. Software for periodic feature upgrades is not handled in the same way as common equipment or plug-ins. The availability of new software for upgrades, applicable to the installed base of equipment, is typically driven by the vendor's software development schedule. Once the operating company has successfully tested the new software in selected sites, it may then be deployed more generally in the network.

Plug-In Administration. Plug-ins for lines, trunks, channel units, etc. need not be provisioned with the common equipment that houses them and are typically deferred until short-term demand dictates. They are then deployed just in time. Procurement for deferrable plug-ins does not follow the formal job engineering process. They are usually ordered in bulk and, particularly for large networks, are distributed as required from central matériel distribution centers. Plug-in needs for both resource provisioning and fault management are typically aggregated.

Patch Administration. When software bugs are detected in new software issues already released, temporary software fixes, developed by the vendor, are distributed and loaded into affected network elements. Permanent solutions are then developed for the next major program release.

8.3.6 Evolution Drivers

Key evolution drivers for improvement in the resource provisioning process include:

- Lower capital carrying cost via:
  —shorter lead times
  —shorter engineering intervals
  —self inventorying resources
- Lower expense via:
  —self-aware resources
  —self-engineering resources
- Better service via:
  —fewer held orders
  —faster service response

Current Limitations. Currently too much resource provisioning is done after the customer service request. This results in delays annoying to customers as well as last minute operating company rush and expense. The brute force remedy, overengineering, is expensive in terms of capital carrying charges.

The intervals to install jobs or major extensions to working equipment must be reduced. Service disruptions during extensions due to resource provisioning activity must also be eliminated.

Software upgrades, which are quite frequent, typically require several hours for a major switching system. Service and resource translations must often be extracted from working systems, updated, repackaged, and reentered. Rekeying of data is also prone to error. As more of the network becomes software driven and programs and datafill more voluminous, effective ways to cut these unload/reload intervals and reduce configuration datafill errors must be devised.

Strategic Objectives. Because of increasing customer expectations and operating company competition, resource provisioning must be better tuned to the service needs of the market.

Just-in-time provisioning is another key objective for resource deployment. Just-in-time minimizes the capital carrying charges and the possibility of stranded resources, but the deployment must precede the actual service demand, otherwise there is risk of a held order. Also, frequent site rush visits are expensive so an economic trade-off between the size and frequency of resource provisioning must be struck. Because labor costs typically increase faster than capital expenditures, the trend is towards overprovisioning and dedicating resources. However, better planning and provisioning tools will allow better tracking and forecasting of demand for just-in-time provisioning.

Separating the resource provisioning process from service provisioning and getting resource provisioning out of the service provisioning critical path is also required in support of service velocity objectives.

Opportunities for Improvement. Aids to major resource provisioning projects include:

- Firm price quote tools
- Job engineering tools
- Translations data handling tools
- Just-in-time provisioning tools

For ongoing short-term resource provisioning, the key to improvement is empowering the resource provisioner to make better short-term decisions faster by easing access to network information and preprocessing of routine calculations. An example of this is automating the resource provisioning process for connectivity elements (such as jumpers) and deferrable plug-ins (such as line cards) by projecting service demand, monitoring and comparing demand to remaining capacity, and flagging impending resource shortages. Algorithms can be then applied to calculate the optimum dispatch order size and timing. This could be followed by automating the order process to dispatch or configure more resources, combining the dispatch request with other needs to visit particular sites through workforce management optimization, and downstream tracking of the effectiveness of the process to allow feedback for further improvements.

Centralization of software administration is another possibility for major improvements. Vendor software could be delivered to a central point for downloading to a cluster of NEs instead of manual software loads at each site.

8.3.7 Target Process

Function Partitioning. As mentioned above, some resource inventorying, internal load balancing, and configuration engineering functions should be moved into intelligent NEs [6]. Other management functions, such as service provisioning, fault management, and performance functions could then be done at the NE level through direct interaction with the resource database. NE software loading, which is usually handled manually by loading physical tapes onsite, should be done remotely from a central location in the future. With the rapid increase in software-controlled NEs, this will allow qualified specialists to do the job more efficiently. This will require high-speed facilities with file transfer protocols to carry out the task.

Data Partitioning. The NE-based resource inventory is seen as a key enabler. It removes the need for detailed vendor specificity at the OS level, the cost of data redundancy, and the problems of data synchronization. Making the resource inventory data abstractions quickly available to OSs for service assignment and other resource provisioning functions is a key challenge requiring a standard resource configuration model. This is currently being addressed in the standards forums.

OSI Interfaces. Resource provisioning interfaces of the future should be based on high level abstractions which mask out unnecessary vendor specific detail. Two types of interfaces are required. A high speed file transfer protocol [file transfer access and management (FTAM)] for software download and resource translation entry and a transaction protocol [common management information service element (CMISE)] for incidental query, deletion, or changes to individual data elements.

8.3.8 Transition Considerations

Reengineering the current process and repartitioning functions and data may appear to be desirable but usually there is resistance to change because of the cost associated with change and some skepticism about promised benefits [7]. Standards to interwork OSs and NEs, according to the new scheme, also take time to develop and achieve consensus. On the other hand, there are customer and competitive pressures that demand faster service delivery and efficiency improvements.

Because resource provisioning is the process that deals with the physical deployment and configuration of NEs, there will always be an element of material handling involved. However, the amount of physical work associated with NE deployment is decreasing as more of the functionality becomes software configurable. Also, jumpers are being replaced with software-controlled, data-configurable cross-connects.

8.4 SERVICE PROVISIONING

This section deals with the area of service provisioning, a major process within configuration management. Service provisioning begins with a customer request for services and ends when the network is actually delivering that service. Thus service provisioning covers the processes necessary for the installation and management of services within the communications network. Service provisioning involves assigning physical and logical resources to individual customers or groups of customers. This includes everything from telephone numbers and dial plans to calling features and voice mail, as well as all logical connections required to carry traffic from an originating point to a termination point.

8.4.1 Service Provisioning Process

For any service provider the service provisioning process must be optimized in order to realize the maximum earning potential of the services within the network. If services can be put into the network cost effectively and quickly, they are more likely to be successful. The following example will illustrate this point.

Many service providers offer new services on a trial basis to all of their customers in a localized area. In this way customers get to evaluate the features without paying anything, while the services provider can demonstrate the usefulness of the new capabilities. In order to provide these new services in this manner, the service provider requires automated processes for the installation (and deinstallation) of the services.

Figure 8-3 shows a typical service provisioning process. Although all

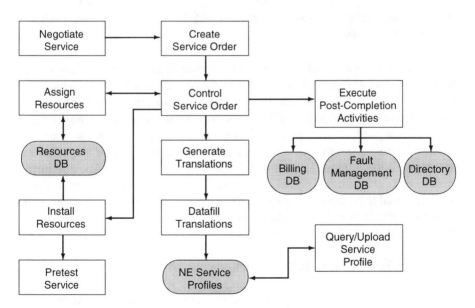

Figure 8-3   Service Provisioning Process

of these functions will likely exist in a service providers process, functions can be combined together or split apart, depending on the particular business needs of the service provider. Each piece of the process shown in Fig. 8-3 will be explained in Section 8.4.3.

8.4.2 Goals of Service Provisioning

Overall Goals. The overall goals [3] of service provisioning are simple—provide the service to the customer, when they want it, with the least possible expense. Realizing these goals is somewhat more difficult and requires some complex mechanisms on the part of the service provider. Two such mechanisms are as follows.

Flowthrough. Flowthrough service orders are those that pass through the service provisioning process unimpeded. These service orders should pass from function to function within the process in an automated manner, such that the process is completed with no manual intervention. Flowthrough requires machine-to-machine interfaces in place between all systems involved in the process. Flowthrough is a key method of reducing the cost of provisioning services to the customer. It is also the chief mechanism used by service providers during the last decade to reduce costs and improve service delivery times. However, as the services become more complex, and as more services become fully automated in their processes [e.g., plain old telephone service (POTS)], reducing the cost of the process further will have to be achieved through alternative methods (as described below).

Service-on-Demand. Service-on-demand is the concept where end-users have the ability to determine and provision their own services. This concept has been around longer than service order flowthrough [e.g., direct distance dialing (DDD)], however there are many additional capabilities that can be provided. Two applications within service-on-demand that can save the service provider costs, yet provide faster response to the customer, are terminal activation service (TAS) and automatic set relocation (ASR). TAS allows end-users to add features to their service profiles from their station sets. ASR allows users to relocate their sets to other physical locations served by the same switch, also through the use of the user interface on the set.

8.4.3 Service Provisioning Functions

This section describes each of the functions within the service provisioning process. All service providers may not perform all of the items described in each function, as these are generic descriptions that would apply to a typical service provider.

Negotiate Service. Service negotiation is the function where the user and the service provider discuss the service to be provided. The needs of the user are matched with the tariffed services available from the service provider. Upon successful negotiation, the user is generally given a completion date for the service order.

Create Service Order. Service order creation is the point where the actual pieces of the service profile are assembled. Typically at this stage the exact pieces of the service are determined, the address would be verified, plans would be made for any physical work, and credit checks would be performed.

Control Service Order. Service order control is responsible for ensuring that all of the pieces of the process happen as they should. If a service order passes this point, yet fails later on, it will normally be passed back to the control point for correction and/or manual intervention. All coordinating activities of the process happen at this point, therefore this is where the ties between the physical work (installation) and the logical work (translation) happen.

Assign Resources. All physical and logical resources are assigned to the service at this point. Physical resources include:

- Loop
- Line card
- Facilities
- Network termination equipment
- Customer premise equipment

Logical resources include:

- Directory numbers (DNs)
- Line equipment number (LEN)
- X.25 packet addresses
- Cross connections

Not all of these resources are necessarily assigned for every service. Assignment requires interworking between service provisioning systems and resource provisioning databases. Once assigned by the service provisioning process, the status of the assigned resource in the resource database is changed from *available* to *in service*.

Generate Translation. Once the assignment of the logical and physical resources has taken place, the translation of the service provider service and resources descriptors can take place. This translation is from the service order format into the information format (currently vendor specific) needed by the NE, to provide the service.

Datafill Service Profiles. After the translation from the service order to the NE-specific information has taken place, this data can be filled into the appropriate NE. This process is generally the first to be automated because of the manual intensity of the work and the possibility of keying errors in transcribing service-provider specific service codes into vendor-specific feature codes.

Install Resources. Installation is simply the installation of the necessary physical resources for the customer. This could typically include the installation of the loop drops, loop jumpers, main distributing frame (MDF) jumpers, the network termination (if any), and the customer premises equipment (CPE) (if it is provided). Resource provisioning goals described in Section 8.3 are directed at minimizing the physical installation work required after the customer request for service. Dedicated plant guidelines prescribe leaving resources in place when service is discontinued in order to eliminate the need to do installation work when service is subsequently requested at the same location. Pre-provisioning approaches and left-in-place policies have similar installation reduction objectives.

Pretest Service. Preservice testing is a final check to confirm that all aspects of resource and service provisioning relating to a particular service order are in working order. Preservice testing may not be carried out for all services. It largely depends upon the policies of the services provider and the type of service being provided. Simple services, such as plain old telephone service (POTS), typically do not have any preservice testing performed by the service provider. On the other hand, complicated services, such as DS-1 based special services, do have preservice testing.

Execute Postcompletion Activities. Postcompletion activities include additional administrative activities that must be performed but are triggered by service order completion. These include updating databases for billing, maintenance, or directory entry that result from the new service being put in place. For some services and service providers, a follow-up may also be performed to ensure the service is working as expected.

Query/Upload Service Profile. This capability is relatively new and is only present for certain services [e.g., international standard digital network (ISDN)]. Service profile query/upload provides the user with the ability to query the service they have been provided and, in some cases, the ability to program the CPE (e.g., telephone set) for the service. In the case of a service such as ISDN, the user could activate this capability and download their CPE with the DN and the features that were supported by the service. The user could then query the features available on their CPE.

8.4.4 Data Management

As a result of looking at the typical process for service provisioning, it becomes clear that there is a large amount of data involved in the process. If the databases of the NEs are taken into consideration (since many NEs are now software controlled, with their own data to manage), the service provisioning process would have one of the largest amounts of data of any process of the service provider. For this reason, the databases the service provider uses to store this data are critical to the smooth operation of this process. Separate databases have often been used at the various stages of the process, which has resulted in the need for synchronization and audit capabilities. Because these capabilities are often difficult and costly to adequately provide, a mismatch in the data would result.

As a result of the mismatch in data, many service providers are rethinking how databases are put into the process. One of the ways to help alleviate this problem is by coordinating all database activities. This coordination could involve setting up a centralized database or a logically centralized database that may be physically distributed. This one database would then contain as much of the data the service provider would use on as regular a basis as possible. This does not entirely solve the problem of distributed databases. As noted previously, NEs also have their own databases. This results in a need to coordinate the databases involved in the process with those in the NEs.

8.4.5 Service Provisioning Tools

Because most of the service provisioning process is related to the management of data, the tools required to manage this process become simply data management tools. These tools include ones that allow the user to effectively query the data, enter new data, edit existing data, and delete old data. Tools are also required to audit the data in the network and synchronize data between processes.

These types of tools sound very much like the types of tools provided by common database management systems. They are not as easy to provide in this process, however, because of the large volume of data and the complexities of that data.

To fully understand how much data is involved, the simplest service can be studied: POTS. For POTS the following information must be kept to a bare minimum:

- Name of customer
- Billing address of customer
- Address where service is provided
- Directory number of customer
- Loop information (including makeup of loop)
- Credit history of customer
- Digitone or dial pulse dialing
- Information about CPE (if provided)
- Service providing equipment
- Office equipment number
- Billing information

This is by no means a comprehensive list of the information stored, although it does indicate the amount of information stored by the service provider.

8.4.6 Interfaces

The number of systems involved in the provisioning process may be greater than the number of boxes shown in Fig. 8-3. This results in the interfaces provided between these systems being crucial to the success of the process. In the past, without exception, these were proprietary (and sometimes closed) interfaces that evolved from simple beginnings. Only now are more and more of these interfaces being published and becoming standardized. This is a result of a number of factors, two of which are a drive for a multivendor network and more services being introduced into the network. One of the first interfaces to face opening up and standardization is the interface to the NEs. This interface is the first because of the multivendor nature of the network elements. The two types of interfaces that are being pursued in North America for service provisioning are TL1 (Transaction Language one) and OSI based interfaces. TL1 is a man machine language (MML) based on International Telecommunications Union—Telecommunications (ITU-T—formerly CCITT) recommendation Z.300, while OSI is a series of standards in development from ISO, CCITT, and ANSI. The standardization of these interfaces will be a long process because of the complexity of the interface and the large embedded base of systems.

8.4.7 Process Variations

The process described previously in this chapter was a typical process used by the service provider to provision a new service. In reality, there is no such thing as "typical."

Why Do Process Variations Exist? Every service provided has different requirements when being provisioned. These different requirements occur because of the difference in the service being provided. The following section highlights some of the differences between these services.

Types

POTS. POTS is the simplest of services, but it also has the largest volume. For reasons of volume and simplicity, this process is the most likely to be fully automated with flowthrough. POTS is also the service that is the most likely to follow the flow shown in Fig. 8-3.

Centrex. Centrex has the addition of the Centrex group, which must be provisioned. The Centrex group contains all of the information which relates to everyone in the customer group. This would include such things as dial plans and simulated facility groups and group features. The amount of provisioning needed for a Centrex group is considerably more than for a single end-user, and is considerably more complex. Most service providers still have many manual processes as a result of this complexity. Some of these processes are additional steps in the flow shown in Fig. 8-3. One such step is the development of the service parameters for the Centrex group as a whole.

Special Services. Special services typically have low volume and range from relatively simple to very complex. Service providers may also support the full range, which results in supporting dozens of services. As a result of these characteristics, special services have many additional steps in each part of the process to ensure that they are provisioned correctly.

AIN. New services such as advanced intelligent network (AIN) based services may incorporate new steps, particularly during introductory implementation.

AIN introduces two new components to the process of service provisioning:

- service creation environment (SCE)
- service management system (SMS)

The SMS is mainly a system that manages the service on a network-wide view because of the network-wide services provided by AIN.

The SMS is normally for the provisioning of the service control point (SCP) only.

The service creation environment (SCE) is an entirely new process within service provisioning. The SCE is the point where the service is created from a predefined set of building blocks or primitives. As part of this process the newly created service would be tested.

8.4.8 Evolution Drivers

There are two main drivers that will force the evolution of the service provisioning environment within the service providers:

- competition
- new services

Both of these drivers feed off of each other. As competition increases, service providers will introduce new services to keep abreast (they will also try to cut costs). As new services are introduced into the network, new competitors will enter the market to provide the new services. These drivers will result in some changes to the existing processes. As costs are cut, the existing processes will be further streamlined and automated. New processes will be added and integrated to support the new service.

Current Limitations. There are a number of limitations that will result in slowing down the rate of change within the service provisioning process. Some of these are specific to service provisioning while others are more generic. Standardized interfaces are not available for service provisioning, nor will standards be available for all services in the near term. What is required, but not yet available, is a framework for generic standardized services (POTS and national ISDN). Work is currently underway to develop a standardized OSI interface for provisioning. There are still limitations in the database technology available to service providers for supporting the volume and complexity of the data required.

Strategic Objectives. The many different service providers have many different strategic objectives, although there are common elements. One of these common elements (as it is in any process where there is a competitive environment) is to cut costs.

Within the service provisioning process there are two common ways [8] to cut costs:

- automation
- process improvements

Automation occurs, in part, through mechanized interfaces, which can be realized through the standardization of those interfaces. Interfaces are discussed above, process improvements are discussed in the next section.

Opportunities for Improvement.   In any complex process there are opportunities for improvement—only two of which are discussed here. The first is to maximize the use of the equipment that is already in the process. The main example of this is the use of the database within the NE. Typically most processes do not use the NE databases because they were not usable in the past. However, as NEs become more intelligent this will change [9,10].

The second is the use of the customer in the process. If customers are given the capability to run the process and manage it, they will likely feel the service provider is more responsive.

There are many examples of how this can be provided. The following are but a few:

- better help systems for features through automated voice response
- customer activated features and changes
- customer activation of new services [11,12]

8.4.9 Transition Considerations

There are many factors that must be taken into consideration when making changes to such a complicated process. The following are some of these considerations:

- speed of service to the customer
- support of new services
- quality of the service delivery
- cost

8.5 CONCLUSIONS

Configuration management is a major grouping of network management functions for:

- planning what resources should be deployed and how they should be designed (network planning),
- executing the deployment of these resources to meet service demand (resource provisioning), and

- assigning customers to these resources when they request service (service provisioning).

Configuration management functions interwork with other network management functions, which ensures that:

- the network operates trouble-free (fault management),
- the network operates efficiently (performance management),
- access to network management functions is on a privilege basis (security management), and
- the customer is billed for services rendered (accounting management).

1. Network planning translates the strategic business objectives of the operating company into plans that guide the deployment of network resources for growth and modernization.
2. Resource provisioning implements the network plans by procuring and installing resources required for an economic engineering interval.
3. Service provisioning responds to the customer service demands by allocating available resources for the requested service.

REFERENCES

[1]. ITU-T Recommendation M.3400, "TMN Management Functions," 1992.

[2]. G. E. Sharp and G. Tseung, "Steps toward generic OAM intelligence," in *IEEE Network Operations and Management Symposium*, Memphis, TN, April 1992.

[3]. R. Morin, "Bell Canada's direction in service provisioning," *IEEE Network Operations and Management Symposium*, Memphis, TN, April 1992.

[4]. L. G. Grant, W. G. Ireson, "Principles of Engineering Economy," The Ronald Press Company.

[5]. AT&T, "Engineering Economy—Third Edition," New York: McGraw Hill.

[6]. P. A. Birkwood, S. E. Aidarous, and R. M. K. Tam, "Implementation of a distributed architecture for intelligent network operations," *IEEE Journal on Selected Areas in Telecommunications,* vol. 6, no. 4, 1988.

[7]. P. Boissonneault, G. E. Sharp, and S. E. Aidarous, "An operations

architecture for the future," in *IFIP/IEEE International Workshop on Distributed Systems: Operations & Management,* Santa Barbara, CA, October 1991.

[8]. S. E. Aidarous, D. A. Proudfoot, and X. N. Dam, "Service management in intelligent networks," *IEEE Networks Magazine,* vol. 3, no. 1, 1990.

[9]. Bellcore FA-001134, "Information Networking Architecture (INA) Framework Overview," August 1990.

[10]. Bellcore FA-TSV-001294, "Framework Generic Requirements for Element Management Layer (EML) Functionality and Architecture," *Framework Technical Advisory,* issue 1, August 1992.

[11]. M. Ahrens, "Soft dial tone service activation," *IEEE Network Operations and Management Symposium,* Memphis, TN, April 1992.

[12]. G. Rainey and D. Coombs, "An Architecture for Service Management Evolution," *IEEE Network Operations and Management Symposium,* Memphis, TN, April 1992.

CHAPTER 9

Fault Management

Charles J. Byrne
Bellcore
Red Bank, NJ 07701

9.1 INTRODUCTION

Networks are vulnerable to faults in equipment and transport media. Faults include malfunctions of hardware and programs and data errors. The reliability of network components is determined by the quality of their design and manufacture and is in the domain of configuration management. Coping with the residual rate of faults is the responsibility of fault management.

The primary goal of fault management is the restoration of service in the presence of faults. The second goal is identification of the root cause of each fault: a root cause is the smallest repairable network component that contains the fault. The third goal is timely and efficient repair of faults, conserving the cost of labor, the carrying cost of spare parts, and the cost associated with a resource being unavailable for use until it is repaired. The fourth goal is to collect and analyze measurements of effectiveness of fault management in terms of service interruptions and the costs of repair as a guide to the allocation of resources to achieve the desired balance between service and costs. Part of the costs attributable to fault management are rebates to customers due to service outage and loss of future business to competitors.

9.1.1 Reliability: Relevance and Trends

The reliability of components are the responsibility of configuration management. However, reliability is an important factor in the perceived success of fault management. Reliability is characterized by the mean-time-to-failure of components, while fault management is characterized by the mean-time-to-repair or mean-time-to-restore-service. The customer perceives the product of the two as unavailability. Inevitably, the fault manager is held responsible for the bottom line. Consequently, a fault manager has a strong interest in reliability.

Certainly, advances in technology are contributing to increased reliability:

- Optical fibers are replacing copper wire, which is relatively vulnerable to lightning, corrosion, and moisture.
- Larger-scale integrated circuits are replacing smaller-scale integrated circuits.
- Digital transport and switching eliminates many fault modes associated with analog equipment.
- Controlled environmental vaults and other enclosures protect equipment outside of buildings from environmental hazards.

Nevertheless, a residual set of fault modes will persist, along with a set of human error fault modes such as:

- Cutting of buried cable by construction activity (the infamous backhoe syndrome).
- Errors during growth and rearrangement of equipment.
- Program bugs.
- Database errors.

Environmental hazards also induce faults:

- Hurricanes
- Floods
- Fires
- Lightning

These hazards often induce large numbers of faults over an extended area in a short time. Planning for such disasters is part of fault management.

9.1.2 Restoration of Service

Reactive strategies and proactive strategies are alternatives for the restoration of service. A reactive strategy emphasizes rapid repair. A proactive strategy emphasizes configuration management's choice of a network architecture that provides for deployment of redundant equipment and links so that a resource with a fault can be quickly isolated and replaced by spare resources. Another proactive strategy is to analyze trends in degradation and make repairs before the customers perceive a serious degradation of the quality of service. Proactive strategies are growing in favor, particularly among common carrier network operators.

9.1.3 Identification of Root Causes

Once a fault is detected, determination of the root cause must precede the process of repair. In some cases, particularly those involving new installations and environmental hazards, there may be multiple root causes. Rapid localization of one or more root causes depends on the prior deployment of detection checks, diagnostics, and test access points. For networks that cover extended areas, centralized control and reporting is essential for rapid and efficient identification of the root cause or root causes. Automation of the process of identification of the cause is needed to cope with the increasing complexity being designed into networks.

9.1.4 Repair of Faults

Once one or more causes of a fault have been determined, the next steps are:

- Assignment of labor, tools, and spare parts
- Dispatch of labor, tools, and spare parts to the location of the fault
- Repair of the fault
- Verification of proper operation of the repaired resource
- Return of the resource to a state of availability for service

Equipment design that specifies replaceable modules and connectorized cables promotes quick repair. Improved reliability has encouraged centralization of the labor force and spare parts, resulting in efficiencies of training, experience, and improved inventory control.

Rapid replenishment of the spare pack inventory through automation of reorders and rapid shipment is a promising area for future improvement. Examples of future improvements in tools are the growing use of hand-held test equipment and the deployment of repair manuals in portable computers. Further research may improve algorithms used to determine efficient routes for labor and vehicles so that travel time to accomplish multiple repairs is minimized.

Techniques for the detection and correction of software faults (program bugs and database errors) will grow in importance as software becomes an ever-larger part of networks.

9.1.5 Measurements of Effectiveness

What cannot be measured cannot be managed effectively. Fault management collects quality measures such as mean-time-before-failure, mean-time-to-repair, effectiveness of personnel (individually and by team), and service interruptions. Additional quality measures are the number of rebates issued due to service interruptions, customer perceptions (measured by surveys), and loss of market share to competition offering superior reliability or fault management. The cost measures specifically attributed to fault management are labor costs (overall and per fault), carrying costs of the spare parts inventory, and costs of redundant resources deployed for the purpose of restoration of service. Of these measures, reports on reliability, mean-time-to-repair, and service interruptions are based on data gathered in the course of detecting a fault and managing the repair process. There is an opportunity to design the appropriate measurement tools into the automated fault management function to increase the accuracy of these measures while controlling the cost of measurement. For example, the use of bar codes and scanners on spare circuit packs as they are removed from inventory and on bad circuit packs as they are returned for repair or salvaged promotes the correlation of the spare part movement with other fault records. Such automated measurement techniques will produce much better data than manual records.

Records of the history of fault management events, a measurement function, are very important for dealing with intermittent faults, that is, those that occur repeatedly and clear spontaneously. Although only a small percentage of faults are intermittent, they consume an inordinate amount of labor and spare parts.

9.2 OVERVIEW OF THE METHODS OF FAULT MANAGEMENT

The methods of fault management encompass the detection, reporting, and localization of faults, as well as aspects of restoration. The following aspects of fault management are discussed in further detail in Sections 9.3–9.5:

- Alarm surveillance
- Fault localization
- Testing

Section 9.6 discusses restoration of service and Section 9.7 discusses repair.

9.2.1 Definitions of Selected Terms

The vocabulary of fault management has evolved pragmatically, without careful attention to precision of definition. Even standards have not yet achieved full consistency. This chapter uses the following terms in the sense of Recommendation M.3400 [1] of the International Telecommunications Union—Telecommunications (ITU-T—formerly CCITT).

- An alarm is an unusual adverse event, usually an automatically detected failure, that potentially will be brought to the attention of the person or managing agent responsible for remedial action.
- A diagnostic finds faults in an internal feature of a unit of equipment or software that is part of infrastructure.
- A test finds faults in a transport resource that may be a subscriber access line, a leased circuit, or an infrastructure facility.
- Performance monitoring gathers statistical measures of degradation (e.g., transmission error rates) in circuits or related equipment.
- A trouble is a human perception of a problem with service, especially a customer's perception. The term is also used to describe a task assigned to a human or management agent to investigate and clear a trouble.

9.2.2 High-level Data Flow Diagram

Figure 9-1 shows how the individual parts of fault management relate to each other, to the transport performance monitoring part of performance management, and to other network management functions. This is a very comprehensive model, providing somewhat more automated functionality than is currently provided, even by large common carriers. However, the trend has been to automate more and more of this functionality and the trend is expected to continue.

Alarm surveillance is the part of fault management that is responsible for the collection of reports of failure events from many detection checks distributed throughout a network. It also receives information from the automatic restoration process of configuration management. The analyze alarms process of alarm surveillance enters the significant events into the surveillance condition database. Fault localization and transport performance monitoring (a part of performance management) also enter data in the surveillance condition database. This database also stores trouble reports from customers. Typically, several or even many manifes-

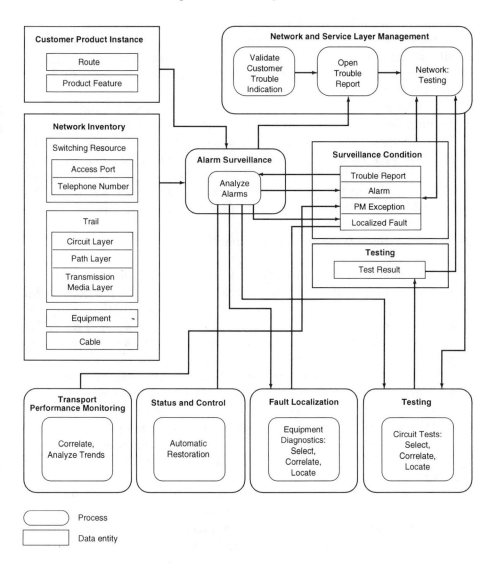

Figure 9-1 Overview of Fault Management

tations of a fault are reported and stored in the surveillance condition database. The analyze alarms process infers the root cause of the fault and its specific location from all of this stored data. The analyze alarms process can request equipment diagnostics and circuit tests to be run by fault localization and testing.

The network inventory database, administered by configuration management, is used by many of the functions within fault management. Particularly important is the dependency relationships that guide the analysis programs from superficial detection check information toward identification of the root cause of a fault. There are two types of dependency relationships: those within items of equipment and those that describe connections between units of equipment (i.e., trails).

9.2.3 Importance of Centralization

One important theme in Fig. 9-1 is the central collection of information concerning fault conditions. All relevant information proceeds toward the surveillance condition and testing databases, with appropriate analysis and correlation processes upgrading the data along the way. The centrally collected information is available in maintenance centers through appropriate human–machine interfaces.

Experience with the management of common carrier networks has dramatically demonstrated the advantages of locating experts in maintenance centers. First, there is an efficiency of scale in assigning problems. Second, there is an increase in learning rate and proficiency levels because the centralized experts become aware of relatively rare types of failure. Finally, the experts can easily share experiences and support each other.

Logistics theory and the application of the law of scale show that centralization of spare parts produces additional savings. Since the percentage variation in a combined central inventory is smaller, the average inventory can be reduced while maintaining a constant probability of being out-of-stock.

Successful centralization demands the ability to communicate with the distributed systems of the network, so that knowledge of far-flung events is efficiently available at the central sites. Also, there must be an efficient plan for the dispatch of staff and spare parts.

The degree of centralization depends on many factors, including distances between sites, transportation means, survivability requirements, and jurisdictional boundaries. An interesting methodology for placement of management centers is presented in a paper by Bouloutas and Gopal [2].

9.2.4 Role of Expert Systems and Intelligent Knowledge-based Systems

Many of the functions within fault management involve analysis, correlation, pattern recognition, and similar processes. They also use an extensive database containing descriptions of the structure of equipment and networks, which must be used in the process of inferring one or more root causes of a fault from many reports that may be related in subtle ways.

This functionality has provided, and continues to provide, an opportunity for application of the technologies of expert systems [3] and intelligent knowledge-based systems [4].

Today, the application of expert systems to mature network technologies is far advanced [5],[6]. But for new technologies, there is a problem: there are no experts in fault management until after extensive field deployment. Yet analysis capability is needed early in the deployment period. To quote from Paul E. Spencer et al. [7],

> How does one go about designing an expert system without an available expert?

The designers of new technology fill the gap. They must anticipate, through analysis, the more common failure modes and generate the rules needed by expert systems in early field deployment.

Expert systems must learn new rules as significant field experience is accumulated. This process need not be automated: in fact, learning may be more rapid if a designer uses the early field experience to revise the rules by analysis. This process can anticipate future failures not yet experienced by generalizing the information contained in the field reports. Furthermore, analysis of field reports by a designer can lead to the design of additional diagnostics, tests, and detection checks for alarm surveillance.

The reader is advised to keep the technologies of expert systems [8] and intelligent knowledge-based systems [9] in mind while considering the following discussions of the functionality of alarm surveillance, fault localization, and testing.

9.3 ALARM SURVEILLANCE

Alarm surveillance starts with the detection of a failure, where a failure is defined as the onset of a fault. A fault is a persistent condition in a physical unit that prevents it from performing its function. The value of the operational state attribute of an entity that models such a unit is "disabled" when a known fault is present. Failure events are detected by special circuitry or by program checks in software.

9.3.1 Types of Functional Failure Events

Some failure events are functional in nature: that is, the failure is specific to an externally visible feature. Examples of functional failure events are:

- Loss of incoming signal on a line
- Inability to access memory
- Inability to complete a protocol sequence with a remote unit

Hardware failure events are generated by specific circuit checks such as:

- Tripped circuit breaker
- Parity failure within a circuit board
- Mismatch between redundant circuits

Software failure events are generated by special procedures that monitor program and data integrity. Examples of such failure events are:

- Detection of illegal commands or responses
- Time out of an expected response from a remote unit
- Failure of an audit of database information
- Failure of defensive program checks

Environmental events are not directly associated with failures, but are associated with other types of events that require attention. Examples are:

- Intruder alarm
- Fire alarm
- Excess temperature

In some cases, one unit of equipment may be used to detect failures in another [10].

A data flow diagram illustrating alarm surveillance is shown in Fig. 9-2. Because alarm surveillance is often the first to detect the effects of a fault, it has a responsibility to notify fault localization and testing and combine their results with transport performance monitoring data in

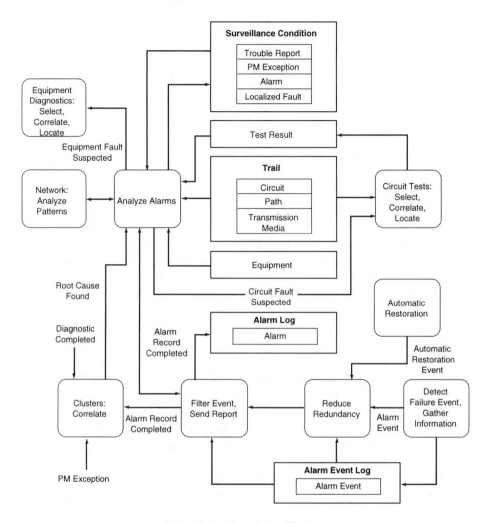

Figure 9-2 Alarm Surveillance

order to ensure that all maintenance resources contribute to the determination of the root cause. The analyze alarms process carries out this
responsibility, but before this process can act on a newly detected event,
other processes must prepare a data entity called an *alarm record*.

9.3.2 Detection

A comprehensive discussion of the means of detecting failures is
beyond the scope of this chapter. Some general observations may be useful, however.

Successful designers of detection tests, besides being ingenious, tend
toward the paranoid and have a working familiarity with Murphy's Law:
"Whatever can go wrong, will". The ensemble of detection tests often over-
laps, producing multiple check failures for single faults. Multiple check
failures are undesirable because they require analysis to determine the
root cause. Multiple failures are acceptable, however, because they minimize the probability of missing a fault altogether, a far more serious kind
of error.

False alarms are to be avoided because they lead to lack of confidence
on the part of the person responsible for remedial action. Fortunately,
false alarms are rare in digital optical networks. A troublesome residual
source of false alarms is electrical noise within buildings caused by lightning strikes. Such events should be considered in the design of check circuitry, especially fire and smoke detectors.

Failure events must be checked against previous state data (sometimes called *last look* data) to avoid reporting the same problem repeatedly. An alarm event is associated with a change from a normal state to an
unacceptable state and, in general, implies not only a check against an
acceptance threshold, but also a comparison with recent history. Effective
fault detection requires that the detection checks must be thoroughly
designed into the equipment [11], [12].

9.3.3 Gathering Information about Failure Events

When a check fails, a process gathers relevant information. For example, information may include the type and identity of the functional or
hardware unit, the type of check that has failed, and the time of day.
Certain attributes of the failed unit, such as whether the unit was enabled
and active (in service) at the time of failure, may also be included. An
important item of information is the level of importance. Alarms can be
designated as critical, major, or minor—depending on the nature of the
failure.

9.3.4 Other Types of Alarm Events

While alarm surveillance is primarily used to monitor the condition of equipment that is in service or on standby as a spare, it also monitors other kinds of events that are relevant to its mission. For example, alarm surveillance reports automatic actions that switch services to spare equipment, lines, or paths so that higher-level restoration functions can know the current configuration. Reporting of automatic actions involves state change events, as opposed to alarm events, but is often considered part of the alarm surveillance function. An *alarm clear* event is another example of a state change message that is considered part of this function. Such an event occurs when a unit that has been in the alarm state passes a diagnostic or other evaluation and is restored to the enabled state.

9.3.5 Alarm Log

An alarm log is a data entity that contains alarm records, entered in chronological sequence. Each alarm record contains information about an alarm event. Usually, it is identical to the information sent to a higher-level alarm surveillance process or to a user interface.

An alarm log serves several purposes. First, it preserves a sequential historic record of a series of events. Often the earliest events are detected by checks that are specific to the root cause of a fault. Second, if a number of alarm events are to be correlated, the log serves as a raw data storage buffer to allow events slightly separated in time to be collected for the correlation algorithm. Third, in cases where the fault is such that the failing device cannot communicate to centralized systems, a local log may capture the sequence of events leading to a failure.

9.3.6 Alarm Summary

An alarm summary is a report of the "alarm state" of a related set of resources (e.g., all the line terminations of a multiplexer). In the past, in very simple alarm surveillance systems, there was one alarm state for each type of alarm detection check and only a few alarm types for each resource. The trend for the future is to characterize the status of an object in terms of its operational states (enabled, disabled), administrative states (locked, unlocked, shutting down), and usage states (idle, active, busy), supplemented by a relatively rich set of qualifier values indicating, for example, why a resource is disabled. Reporting of these states is a matter for the status and control function of configuration management. The resolution of the potential overlap of this function with the alarm summary function is not clear.

9.3.7 Conditions for Alarm Reports

Each potential recipient of alarm reports does not usually want to receive alarms of all types from all managed objects at all times. For example, there may be a recipient that is only interested in alarms from transport managed objects, or only from failures associated with equipment that serves outgoing calls. Or an alarm recipient may know that a certain managed object is in the process of installation and wishes to suppress alarms while the work is in progress in order to avoid interference with other alarms. These rules, which are under control of the alarm recipients, are call *conditions* for sending the alarm. As each alarm is processed, a check is made of the condition algorithms for each potential recipient. Because there are a set of conditions for each potential recipient, this mechanism serves as a method of steering alarms to the proper destination, as well as selectively suppressing alarms. The mechanism also allows a single alarm to be passed to multiple recipients.

9.3.8 Reduction of Redundancy

Ideally, if all faults could be anticipated in the design, each would be associated with a specific detection test. But this is not practical. Therefore, detection tests are deployed strategically, each designed to detect a wide class of faults. To avoid missing some faults entirely, the classes of faults tend to overlap. Faults that fall into multiple classes cause multiple alarm events. A sound detection policy tends to produce more than one alarm event per fault because that situation is better than the possibility of having no alarm at all.

Analysis of the deployed detection tests, together with field experience of the most common faults, can identify relatively likely syndromes of multiple alarm events being associated with individual faults. Suitable algorithms can then identify such syndromes and eliminate the reporting of redundant alarms. This process requires examination of all alarms within a finite period, implying a mechanism to store a number of alarm events in a log. Therefore the process inherently involves a delay period, which is one of the costs of the process.

9.3.9 Correlation of Alarms

Just as a fault may generate multiple alarms within a single piece of equipment in a period of time, a fault may generate alarms in multiple pieces of equipment distributed in time and space. A very common event of this type is a cable cut: detection checks in all equipment receiving signals that route through such a cable will detect a loss of signal. Such systems may be widely distributed. A process that is aware of alarm records from many such units of equipment can correlate the alarms to infer the root cause. The process is easier if redundant alarms have already been eliminated within each unit of equipment.

Reduction of redundancy, correlation, and analysis can take place at multiple levels—within network elements, within subnetwork operations controllers serving clusters of network elements, and at the network level. That is why Fig. 9-2 shows separate processes at the different levels. At each level, the process has access to data appropriate to its domain. There is additional discussion of the role of subnetwork operations controllers in Section 9.10.1.

The analyze alarms process of Fig. 9-1 is particularly powerful because it has access to analyzed information from all maintenance functions as well as the power to call for additional diagnostics, tests, and analysis. As appropriate, it can request information from network-level processes that can work with maintenance processes in connecting domains. While other processes may be able to identify root causes, this process is the one that *must* identify the root cause. This process may be partly automated and partly supplemented by human analysis.

9.4 FAULT LOCALIZATION WITHIN EQUIPMENT

Assume that a specific fault,[1] perhaps one whose effects have been detected by alarm surveillance, is interfering with the proper operation of a piece of equipment. The purpose of fault localization, used in conjunction with other management applications, is to determine which piece of equipment contains the fault and exactly which replaceable module has failed in the faulty equipment.

Fault localization uses diagnostics as a means for finding the fault; it is closely related to testing and performance monitoring of transport circuits. It may be necessary to combine data from diagnostics, tests, and performance monitoring in order to determine the root cause of a fault.

Success in determining a root cause depends upon the quality of the design of the equipment, especially on whether sufficient design effort has been spent on the diagnostic tests, the hardware checks in the equipment itself, and the checks in the operational software.

A useful concept in the design of diagnostics is that of "visibility." From a remote workstation, through the operational software, can one "see" deeply enough into each replaceable module to determine whether it can perform its function? Such analysis, if done early in the design phase,

---

[1]Multiple faults may be present, particularly during installation of new equipment. The general principles of fault localization apply to multiple faults, but this discussion assumes a single fault in order to simplify the presentation.

can indicate areas where more check circuitry or programs must be added in order to achieve more resolution in discriminating among possible faults. Making diagnostics available early in the system test process is also important. First, they will be useful in identifying both design and implementation defects in the prototype. Second, undiagnosable faults that would be problems in the field have an uncanny tendency to show up in prototype testing. If the diagnostic programs are working, and if the laboratory experience is not dismissed as an artifact, design changes can eliminate a major source of user dissatisfaction.

One approach to design for maintainability is fault tree analysis [13, 14]. First, the probable fault modes of each component of a unit of equipment are identified—these are the potential root causes. The implications of each fault are followed, recursively, toward higher-level functional faults, creating a tree diagram. Each externally visible functional fault (symptom) is thus associated with a subtree that includes many root causes and many internal functional faults. These subtrees are then broken into smaller subtrees by designing specific failure detection checks or diagnostics to distinguish smaller groups of root causes. The process is complete when each externally (and centrally) distinguishable subtree contains root causes that are all within the same replaceable module.

Analysis of fault trees may consider single faults or two, three, or more simultaneous faults. An interesting case is the consideration of double faults, where one is an equipment fault and the other is an erroneous operator action [15].

Cable cuts tend to generate multiple faults, so the analysis algorithm for such faults is designed to identify the cable length associated with those groups of fibers or pairs that have the highest proportion of faults.

Note that the success of the fault location process depends on the design of the network and its components for maintainability. No expert system can compensate for the lack of information due to inattention to the quality of maintainability.

9.4.1 Diagnostics

Diagnostics of equipment, like tests or transport performance monitoring of trails, verify proper performance. Diagnostics, like tests, are often intrusive; that is, the equipment being diagnosed cannot perform its normal customer service function while the diagnostic is running. Before such diagnostics are run, the associated equipment should be locked out of service. Diagnostics are also useful tools for acceptance tests after equipment has been installed but before the equipment has been made available to customer service.

Usually a diagnostic stops on the first detected fault. However, a diagnostic may be designed to run in a mode that allows it to continue so that it can report multiple faults. This feature is particularly useful during installation of new equipment.

Fault localization for hardware is used to guide repair both during installation and after equipment has been placed in service. It is also used to guide service restoration.

9.4.2 Exercises

The term *exercise* describes a procedure in which a part of a network element is isolated and is systematically run through all its service features by test equipment, using the normal input and output ports of the equipment being exercised. Except for certain control features, an exercise is similar to a diagnostic.

9.4.3 Software Checks

Checks that run on software are *audits, checksums, operational tests,* or *program traps.* Such procedures are analogous to equipment diagnostics and are controlled in a similar way. An audit may compare redundant copies of data within a system, or it may compare copies of data in separate systems. Audits may run in response to a specific request or may run on schedule. An audit failure is usually sent to an exception manager for resolution, but in specific cases software can be automatically restored from a back-up copy or the transient memory used by a program may be initialized from a more stable source. The software checks of fault localization serve the needs of configuration management (particularly during the installation of software) and accounting management as well as fault management.

One special case of an audit is the comparison of data held in one system with similar data held in another. One of the systems may be designated as the steward of the data and the other system may be said to hold local redundant data. In such cases, should a discrepancy be found, the local redundant data is reset to the values of the stewarded data (assumed to be correct) and the audit result is reported. The choice between resetting data to a stewarded copy, rather than simply reporting a discrepancy to an exception manager, should be under user control.

Audits are also used to compare information held in software with values held directly in hardware. Such audits can sometimes detect malfunctions in either software or hardware.

A checksum is an integer whose value depends on the total contents of a physical unit of software. If a program is patched, or data is updated, the checksum changes. The checksum is used to verify that software has been properly transported or is of the correct version. Also, two copies of software can be compared by calculating and matching their checksums.

This procedure is much more efficient than a byte-by-byte comparison, especially if the two copies are physically separated. This kind of a check is especially useful in verifying software that is downloaded to a multi-function unit in support of service activation (see Chapter 8).

An operational test applies to programs. A program is executed and a controlled set of inputs are applied. Outputs are compared with expected values. If differences are observed, the differences are reported. A program trap helps to locate a fault in the design of a program by reporting the path taken through a branch point.

Figure 9-3 shows a data flow diagram for the fault localization function.

9.4.4 Selection of Diagnostics, Audits, Exercises, and Software Checks

A process that selects diagnostics, audits, etc. can be driven by a schedule or by a failure of detection checks. In the scheduled case, selection is relatively simple because modules are run individually. The only strategy involved is to decide how often to run each test. A refinement is to slow down the repetition rate or to skip modules when a unit of equipment is under heavy load.

The more interesting case is when the selection process is driven by a failure of a detection check and time is of the essence in determining the root cause. Then the selection process usually consists of choosing among a predesigned set of scripts, each being an ordered list of diagnostics. The nature of the detection tests guides the process of choosing the appropriate script. Two tactical conventions are in common use: all of the items in a script may be run and all results reported, or the script may terminate when a diagnostic reports a failure. The latter practice is more common because it is much simpler to verify proper design of the software and obviously gets quicker results. A much more complex strategy is to determine the choice of succeeding diagnostics, depending on the results of earlier diagnostics.

9.4.5 Running Diagnostics

The running of diagnostics and software checks is an art that is very much implementation-dependent. One design consideration is whether the entire resources of a computer system can be used to run the diagnostics or whether the system must continue its work while diagnostics run in the background.

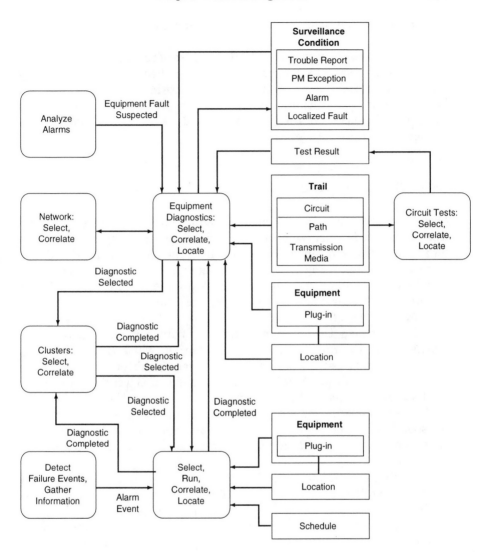

Figure 9-3   Fault Localization

9.4.6 Correlation of Results to Determine the Root Cause

This process is very much like the process of correlation of alarms. As an example, suppose that a receiver detection check reports a loss of signal. The fault could be within the receiver, within the far transmitter, or in the transmission medium. A correlation process could infer, from the results of diagnostics of the transmitter and receiver, which of the three elements contains the root cause.

9.5 TESTING OF CIRCUITS

Many types of tests are performed in communication networks, ranging from automated measurements (by means of operations systems, test system controllers, and remote test units or equivalent) to operational tests (e.g., a technician using hand-held test equipment or a customer with a telephone set or other customer premises equipment). This section focuses on functions to provide automated or (at least) remotely controlled mechanized testing.

By definition, the function of testing of transport resources is distinguished from the function of fault localization, which diagnoses equipment. While diagnostics are performed entirely within one system, testing often involves multiple systems, usually in separated locations. In testing, fault sectionalization is the procedure that determines the part of a circuit that contains a fault and fault isolation is the procedure that determines the specific fiber, pair, or replaceable module that contains the fault. In the context of testing, fault localization comprises both sectionalization and isolation.

Testing is used in service installation and maintenance. When services are installed, testing may be performed to determine if facilities and equipment have been properly installed and are operating at an acceptable level. Installation testing generally is used on complex services, where a reasonable probability of a poorly performing element or an error in installation may exist. Installation testing may be waived on simple services because the likelihood of a problem or the ability of testing to detect a problem without customer participation is small. To be most effective, installation testing should be end-to-end.

The performance criteria for installation testing may be more stringent than the criteria used for maintenance testing, to allow for degradation of components over their lifetime. For example, attenuation margins on a fiber may allow for future splicing losses. Especially stringent criteria may be placed on particular services. In some cases, the test analysis procedure may need access to the characteristics of the customer premises equipment.

Maintenance testing is performed to detect troubles, to verify troubles, to localize troubles, and to verify repair. Trouble detection, verification, and repair verification testing is most effective end-to-end. Sectionalization testing requires the ability to test parts of the service route.

9.5.1 Types of Tests

Equipment to provide test access is designed into the circuit, path, or transmission media trail[2] and placement is dictated by the maintenance policy of the network provider and the topology of the circuit. Generally, test access is required at interfaces between differing maintenance domains (e.g., loop and interoffice), or for testing specific line characteristics (e.g., telephone loop testing or T-1 local loop testing [16]).

Testing takes place at many levels; specifically, circuits, paths, and lines can be tested. Cost-effective design takes advantage of the interplay between different levels. For example, it is not necessary to provide both line and path testing on the same single-section facility because either test technique will unambiguously locate a fault within the facility.

Testing can be intrusive (incompatible with continuing service) or nonintrusive. If it is intrusive, a check should be made to determine whether an entity is in service before tests are performed. Alternately, the customer should be contacted to arrange for the test.

A data flow diagram for testing of circuits appears in Fig. 9-4.

9.5.2 Test Access

Within a subnetwork, there may be provision for the use of testing equipment that is outside of the subnetwork domain. If the testing function is internal, then an important consideration is cooperation with external testing functions to support end-to-end testing of transport services that cross domain boundaries and to provide fault sectionalization. It is particularly important to support sectionalization of a fault by providing test access at a subnetwork boundary.

---

[2]*Trail* is a new ITU-T term that comprises circuit, path, or transmission media connections and associated termination points. A trail of the circuit network layer has end-to-end functionality (for example, a telephone line). A trail of the path network layer is a transport medium (for example, a 64 kilobit DS0 path within a T1 carrier system). A trail of the transmission media network layer is associated with the physical medium and its terminations (for example, an OC3 optical line).

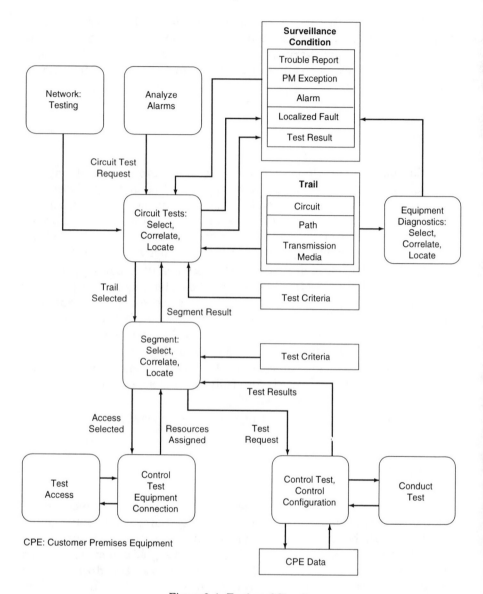

Figure 9-4 Testing of Circuits

9.5.3 Test Control

Testing is often provided by an overlay management network of test system controllers, remote test units, and test access. The simplest method of supporting testing for a subnetwork is the provision of testing functions compatible with the current testing systems into the service-carrying components of the subnetwork.

9.5.4 Trouble Reports

The testing function can be invoked by a craftsperson or by procedures of other functions of the maintenance group (see Fig. 9-1). Analog testing usually originates either from a customer trouble indication or from installation of new equipment or service. Digital testing may originate from a customer trouble indication, from installation, or from proactive surveillance alarms and performance management exceptions.

9.5.5 Test Database

There are several types of database modules that support testing:

- For each trail, an ordered list of test access points, supporting trails, supporting equipment units, and both software and jumper cross-connects
- Known fault conditions on trails, equipment units, and cross-connects
- Known test results

All test results are returned to the invoking procedure. Test results denoting failure are entered in the surveillance conditions data entities, perhaps after correlation with other information or other analysis.

9.5.6 Segment Selection

Once a trouble has been reported on a trail of the circuit network layer, the path network layer, or the transmission media network layer, the next step is to verify that one or more faults are present by making an end-to-end test,[3] if possible. If neither end of the circuit is accessible, then an access point near the center of the circuit is selected and tests are made on the segments that run to each end.

---

[3]This discussion assumes that the circuit is point-to-point rather than multipoint. The principles do apply to multipoint circuits, however.

The next step is to establish a strategy for breaking the trail into segments to establish the location of the fault to the best possible precision. The problem is to find the root cause with the fewest tests. Usually, it is assumed that only a single fault is present. However, the procedure should be designed to test that assumption. Alternative procedures should be provided for the multiple fault case, especially for support of installation.

The binary search strategy is a good one. An access point is chosen near the middle of the circuit and the segments to each end are tested (if neither end of the trail is accessible, then this step will have been done for verification of a fault). An access point near the middle of the failing segment is then chosen. This process is repeated until there are no available access points within the failing segment. At each step, "near the middle" does not mean actual distance, but is based on the number of available access points within the segment.

When the smallest testable segment that contains the fault is identified, the supporting trails are checked for known failures or, if none are found, they are tested. For example, if a fault is found in a network access circuit, the trail of the path layer network supporting the circuit would be checked and tested, and then the trail of the transmission media layer (often called a line) would be checked and tested.

An interesting variation of the testing strategy is to interweave tests of trails at the path layer network and trails of the transmission media layer network. Since provision of test access points is expensive, they can be assigned in staggered fashion; path layer access points at one network element, line layer access points at the next. This obviously complicates both the provisioning algorithm and the algorithm for selecting segments for testing, but it potentially saves capital investment.

When testing multipoint trails with a tree-like topology and using a single-fault assumption, an efficient strategy at each stage of testing is to test the longest branch that may contain the fault. As a general rule, the strategy of segment selection and test selection should be based on the type of symptoms and on the circuit topology.

9.5.7 Correlation of Results to Determine the Root Cause

A single root cause may cause many faults. The classic example is a cable cut that may affect scores of fibers or wire pairs and perhaps thousands of paths. When the provisioning rules provide test access points at every opportunity, the smallest testable segment is often specific to the fault, whether in equipment or transmission media. However, if access points are rare, then the smallest testable segment may contain a number of equipment items and cable splices with multiple branch points. In this case, it may be possible to locate the probable root cause by examining each of the potential root cause loci to determine whether all of its dependent trails are showing a fault. A variation if no locus shows 100% faults, is to determine the potential root cause locus that has the highest percentage of faults in dependent trails. There is obviously a trade-off between the costs of provisioning test access points and the difficulty of providing a correlation algorithm to infer root causes.

9.6 RESTORATION OF SERVICE

Continued service in the presence of a fault implies both provision of a prudent amount of spare resources and the flexibility to use those resources effectively.

Several restoration strategies are possible:

- Isolate the equipment found to contain the root cause. This allows the remaining good resources to successfully carry the service load, perhaps at a lower capacity.
- Switch service from bad equipment to good spare equipment. This may be done with one-to-one spare protection (e.g., using duplex processors [17]) or may be done with M spare units for a group of N working units. For example, one spare fiber might be provided in a cable for every 8 working fibers.
- Use the alternate routing capability inherent in loop or mesh networks to automatically maintain connections despite faults in individual branches or nodes.
- Rearrange services over complex routes to make room for a new route for a service. In the general case, this process would include interruption of low priority services (predetermined by service contract) to restore a priority service. This process is one of considerable challenge and is of growing interest.
- Restore software, either program or memory [18], from a backup copy. See Chapter 11 for an extended discussion of restoration strategies.

9.7 REPAIR

The repair process depends on the coordinated availability of three resources: labor, spare parts, and tools. All must be delivered to the site of the fault. The cost of labor and spare parts dominate the costs.

The keys to cutting labor costs are:

- A centralized labor pool
- An orderly training policy
- Selection of the appropriate worker to fit the root cause
- Means to schedule efficient dispatch routes

In very large networks, workers with relatively narrow training may be dispatched to replace spare modules or splice cables, under the guidance of experts at central locations. Mobile communication systems are of benefit in such an operation.

The keys to cutting the costs of spare packs are:

- Careful inventory control to ensure that a specific unit can be found when needed
- A prudent policy to determine whether to repair or scrap a bad module
- Careful handling of spares in the field
- Verification that insertion of a spare has really fixed the root cause, usually by reinsertion of the removed unit to establish that the fault returns

9.8 MEASUREMENTS OF EFFECTIVENESS

One important aspect of fault management is to collect data about how well the job is being done. This data should answer at least the following questions:

1. How often are faults occurring?
2. How many faults are affecting service?
3. How long are services interrupted when services are affected?
4. How often is the root cause correctly identified?
5. How many spares are being used?
6. How many hours are spent on repair?
7. Which craftspersons and teams are most efficient, in terms of mean-time-to-repair and the average number of spare circuit packs used per repair?

The purpose of making these measurements is to monitor the damage done by faults and to monitor the costs of repair. If the network is growing rapidly, it is desirable to normalize the data for questions 1 and 2 above, using some measure of network size such as capital cost or number of customers served. The other measurements should be normalized according to the number of relevant faults detected.

These measurements can be used to establish current and projected costs for the maintenance of the network. Comparisons of normalized measures from area to area establish the relative quality of fault management and identify the more successful management policies.

Finer-gauge measurements (breaking down the faults by type) focus analysis on the more frequent classes of faults to establish the relative reliability (mean time to failure) and repairability (mean-time-to-repair) of different types of equipment.

The objective measures discussed above should be supplemented by surveys of customer perceptions and estimates of gain or loss of market share due to the relative effectiveness of competition.

9.9 INTERFACE SPECIFICATIONS FOR FAULT MANAGEMENT

This section summarizes the special needs of fault management for interface specifications. It is based upon the more general discussions of interfaces in Chapters 2 through 5.

9.9.1 Protocols

Network and Transport Protocols. Connection-oriented network and transport protocols have an advantage over connectionless protocols for fault management, particularly for alarm detection. A connection-oriented network protocol often provides timely notification of failure of the operations communication link. This characteristic is similar to characteristics required of communication lines for fire and burglar alarms. On the other hand, it takes time to set up a connection, which could slow down restoration of service. Further, connectionless protocols are simpler to implement and require less processing power. Both the protocol and the underlying network should support detection of errors and loss of messages, with provision for error correction or retransmission.

Higher level protocols. In the networks of common carriers, the most common commercial interfaces for fault management use ASCII characters and a string-oriented message structure. Although several proprietary syntax structures exist, the man-machine language (MML) syntax of the ITU-T [19],[20] is in widest use. Based on this syntax, an extensive vocabulary and generic message set for the support of fault management are available [21]–[23].

Several protocols supported by individual computer manufacturers have been available to manage local area networks. This field is now dominated by the simple network management protocol (SNMP) [24]. This higher-level protocol supports the reporting of lists of data items, including alarm and alarm clear information [25], in a restricted subset of abstract syntax notation 1 (ASN.1) [26]. SNMP is suitable to report state attributes and state changes of small network elements and can be used by a central processor to poll for alarms although it would be awkward to use for fault localization and circuit testing in complex networks.

The standard for higher-level protocols for network management is OSI system management [27]. This set of higher-level protocols uses the full ASN.1 syntax at the presentation layer. In the application layer, OSI system management provides an object-oriented set of management services. A thorough discussion of OSI system management standards is presented in Chapters 4 and 5.

System management functions that are specific to fault management are defined in international standards (or draft standards) [28]–[30]. The alarm surveillance service is described further in ITU-T Recommendations M.3200, Annex C [31], and Q.821 [32]. ITU-T Recommendation M.3100 [33] defines the objects of the information model. Objects that closely relate to fault management are described in the following two sections.

9.9.2 Fault Management Behavior and Attributes of Managed Objects

Alarm event. A managed object that models something that contains a failure detection check has behavior that can generate an alarm event.

Alarm status. The alarm status attribute in a managed object indicates whether it is:

- In the cleared state, that is, no fault is detected.
- In the active-pending state, meaning that a detection check has failed but that a timing period is in progress to allow a transient fault to clear before an alarm event is emitted.
- In one of the active-reportable states, which indicate the severity of the alarm to be critical, major, minor, indeterminate, or warning.

Use of state attributes by fault management. Operational, usage, and administrative state attributes for managed objects are defined in international standards [34]. Individual managed objects may support all, some, or none of these attributes, depending on the nature of the modeled resource.

Managed objects that are subject to faults have an operational state attribute whose value can either be enabled or disabled. The alarm, procedural and availability status attributes have values that indicate the reason for a disabled state, if known. The operational state can also be checked by a discriminator's filter algorithm to determine whether events generated by low-level detection checks, such as performance monitoring threshold checks, should be suppressed, since there is little point in reporting degradation of a managed object that has already been reported as fully disabled.

The usage state attribute is sometimes checked by testing to see whether a circuit can be tested; if a circuit is busy or a group of circuits is active, testing will usually be delayed before intrusive tests are run.

Fault management uses the administrative state attribute to put a managed object in the locked state, immediately preventing it from being used while it is reserved for test, under test, or awaiting repair. Through use of this attribute, fault management can also put an object whose usage state is busy or active into the shutting down administrative state so that new users are prohibited and it will become locked when there are no longer any current users. The administrative state attribute also locks a managed object when it is in the process of installation. The control status attributes may record the reason or reasons for a managed object being in the locked state.

Response to requests for tests, diagnostics, etc. Some managed objects have a test or diagnostic behavior that can be invoked by an action. In this case the test result may be returned in a response to the action or a notification, or may be held for recovery by means of the get service.

9.9.3 Fault Management Support Objects

Event forwarding discriminator. An instance of an event forwarding discriminator is provided for each alarm destination. Each event forwarding discriminator reviews all alarm events and, subject to passing a filter algorithm, sends a notification to a remote management agent. Different types of alarms can be selectively routed to several destinations by providing one discriminator for each destination, with an appropriate filter algorithm.

Log. For the purpose of fault management, a log is an object that contains alarm record objects for future reference. Logs are useful when it is desirable to recover the time sequence of events and for buffering information to support procedures searching for redundancy, correlations, and patterns. Like an event forwarding discriminator, a log supports a filter algorithm that determines which events are to be logged.

Test object. A test object, when activated by the invoking of an appropriate action service, causes equipment diagnostics or trail tests to be run on one or more managed objects. There are two types of test actions:

1. *Uncontrolled:* The action directly invokes a test and the test results are received as a response to the action.

2. *Controlled:* The action causes creation of one or more test objects that, through their behavior, then invoke tests. Test results are sent in notifications from the test objects or are available by means of get services operating on attributes of the test objects.

The behavior of each test object determines the nature of the test. As an alternative, test behavior can be defined within the managed object.

The properties of several kinds of test objects have been defined [30]:

- Objects that model fault localization:
  - —Internal resource test object (models a diagnostic)
  - —Connectivity test object (models a diagnostic of a connection within a network element)
  - —Data integrity test object (models a software check)
- Objects that model testing of circuits:
  - —Connectivity test object (models a test of a connection between network elements)
  - —Loopback test object (models a test of a specified bidirectional connection that connects the receiver to the transmitter at the far end, so that transmissions from the near end are returned and can be compared to the signal sent)
  - —Echo test object (similar to a loop-back test object, but the communication path between near and far ends is not specified)
  - —Protocol integrity test object (models a test that involves transmittal of test protocol data units or messages and verifies appropriate responses)

9.10 FUTURE DIRECTIONS: SELECTED TOPICS

Fault management has come a long way since the line craft drove along a telephone line to see which pole fell down, or which insulator had been shot off by a frustrated hunter. The advances have been primarily in the area of distributed intelligence, deploying more fault detection mechanisms that are specific to potential root causes and reporting more specific detail about the cause of the fault. Future evolution is expected to build on this trend, making more intelligent use of detection check information closer to the root cause and combining the total information from subnetworks of increasingly larger scale in stages.

9.10.1 Role of Subnetwork Operations Controllers in Fault Management

One strong influence on the techniques of fault management is network topology. The trend has been and continues to be that the number of choices is increasing: linear, star, ring, mesh, hierarchical, etc. Topologies are used in combination, some in the local area network, some in the access network, and some in wide area networks. The generic problem of handling all aspects of network management in a fully centralized operations system is growing ever more complex.

A very promising approach to the management of a network composed of several different topologies is to decompose the network into subnetworks of single topologies. Each such subnetwork is then managed by one or more separate subnetwork fault management application functions. Such subnetwork functions can be implemented either in standalone equipment specific to the subnetwork (sometimes called a subnetwork operations controller or an element manager) or in software systems with a distinct subnetwork identity in a central processing environment.

Such a subnetwork fault management application should include, or have access to, a database structured specifically for the subnetwork and algorithms optimized for the specific topology and technology of the subnetwork. The application should be able to combine information from all relevant information sources, including alarm surveillance, fault localization of both equipment and software, and circuit testing, to determine the root cause and specific location of the fault.

Information from outside the subnetwork should also influence the process of finding the root cause if it is within the network, and information should be provided for external fault management functions if the root cause is external or cannot be determined within the subnetwork.

9.10.2 Impact of Larger-Scale Integration

The scale of integration is an important influence on the process of identifying the root cause and specific location of a fault. The larger the scale, the less resolution is needed because the objective is to determine the location within one replaceable module. Ideally, fault detection mechanisms within each chip [35] and each module would be able to unambiguously detect faults within the module. Clearly, such detection mechanisms would have to survive any likely fault in the module. Such survivability might imply duplication of power feeds and communication interfaces. One hundred percent internal detectability and detection survivability is a challenging goal for future design but, if achieved, would greatly simplify the process of finding and locating a root cause of a fault.

Larger-scale integration will also present new challenges to fault detection. For example, as the area available to store a data bit decreases, the stored charge decreases, making the value of the bit more vulnerable to a hit by an alpha particle [36]. This challenge is being met by designing error-correcting codes into chips [37],[38].

9.10.3 Cooperation among Administrators

Trends in providing more competitive alternatives for communications are expected to continue into the next century. Successful world-wide communications will require the interconnection of public networks, private networks, and networks formed by cooperation among specific users. The value of a network depends not only on the number of its internal users and the services available to them but also on the number of external users and services that are available through that network and others. This central fact will promote the interconnection of networks of all types. It follows that a fault in any network may decrease the value of all networks.

Therefore, the open exchange of information concerning fault management is an important component in upholding the value of all networks. Examples of such information are:

- Trouble reports on trails at the circuit, path, and transmission media network layers and on services
- Alarm records on trails and services
- Test results and localized faults
- Service restoration records, where protection switches cross administrative boundaries
- Repair schedules
- Requests for cooperation in fault localization and circuit testing

ITU-T recommendations are a very helpful source for establishing the principles and tactics of cooperation among administrations. One important point is the need for a uniform structure of the management information base across all administrations. Each trail that crosses administrations must receive a universally unique name and a well-structured set of attributes that describe not only the service, characteristics but also information needed for fault management. One very important attribute is the one that identifies the serving test centers for each end of the trail, the points of contact for cooperation.

The provision of test access points as close to the boundary between administrations as possible is also important in promoting cooperation, as are agreements on cooperative work between both craft and automated processes among administrations.

Security considerations (specifically authentication and access control) are important for ensuring the free and open exchange of data relevant to the fault management function.

REFERENCES

[1]. CCITT Recommendation M.3400, TMN Management Functions, 1992.

[2]. A. Bouloutas and P. M. Gopal, "Clustering schemes for network management," *Infocom '91*, vol. 1, 1991.

[3]. S. Rabie, A. Rau-Chaplin, and T. Shibahara, "DAD: A real-time expert system for monitoring of data packet networks," *IEEE Network,* vol. 2, no. 5, 1988.

[4]. J. Malcolm and T. Wooding, "IKBS in Network Management", *Computer Communications,* vol.13, no. 9, November, 1990.

[5]. H. Koga, T. Abe, K. Takeda, A. Hayashi, and Y. Mitsunaga, "Study of fault-location expert systems for paired cables," *Electronics and Communications in Japan, Part 1,* vol. 74, no. 10, 1991.

[6]. G. T. Vesonder, S. J. Stolfo, J. E. Zielinski, F. D. Miller, and D. H. Copp, "ACE: An expert system for telephone cable maintenance", *Proc. 8th Int'l Joint Conf. on Artificial Intelligence,* 8, 1983.

[7]. P. E. Spencer, S. C. Zaharakis, and R. T. Denton, "Fault management of a fiber optic LAN," *SPIE* vol. 1364, *FDDI, Campus-Wide, and Metropolitan Area Networks,* 1990.

[8]. J. Giarratano and G. Riley, *"Expert Systems: Principles and Programing,"* Boston: Boyd and Frazer, 1989.

[9]. T. J. Laffey, P. A. Cox, J. L. Schmidt, S. M. Kao, and J. Y. Read, "Real-time Knowledge-based Systems," *AI Magazine,* vol. 9, no. 1, 1988.

[10]. A. Mahmood and E. J. McCluskey, "Concurrent error detection using watchdog processors—A survey," *IEEE Computer,* vol. 37, no. 2, 1988.

[11]. S. N. Demidenko, E. M. Levin, and K. V. Lever, "Concurrent self-checking for microprogrammed control units," *IEEE Proceedings, Part E: Computers and Digital Techniques,* vol. 138, no. 6, 1991.

[12]. F. F. Tsui, *LSI/VLSI Testability Design,* New York: McGraw-Hill, 1987.

[13]. N. H. Roberts, W. E. Veseley, D. F. Haasl, and F. F. Goldberg, "Fault Tree Handbook," Systems and Reliability Research, Office of

Nuclear Regulatory Research, Nuclear Regulatory Commission, Washington DC, USA, NUREG-0492, January, 1981.

[14]. R. B. Worrell and D. W. Stack, "A SETS user's manual for the fault tree analyst," SAND 77-2051, Sandia Laboratories, Albuquerque, NM, November, 1978

[15]. R. T. Hessian Jr., B. B. Salter, and E. F. Goodwin, "Fault-Tree Analysis for System Design, Development, Modification, and Verification," *IEEE Trans Reliability*, vol. 39, no. 1, April, 1990.

[16]. D. Marvel, G. Hurst, and J. Sturbois, "T-1 maintenance takes a new path, *Telephony*, June 15, 1992.

[17]. P. K. Lala, "Fault tolerant and fault testable design," Englewood Cliffs, NJ: Prentice-Hall, 1985.

[18]. J. Nash, "USL to bolster Unix by filling security, fault-recovery gaps," *Computerworld*, May 18, 1992.

[19]. CCITT Recommendation Z.315, Input (Command) Language Syntax Specification, 1988.

[20]. CCITT Recommendation Z.316, Output Language Syntax Specification, 1988.

[21]. Operations Technology Generic Requirements (OTGR): Operations Application Messages—Network Maintenance: Network Element and Transport Surveillance Messages, Bellcore, TR-NWT-000833, issue 5, December, 1992.

[22]. Operations Technology Generic Requirements (OTGR): Operations Applications Messages—Network Maintenance: Access and Testing Messages, Bellcore, TR-NWT-000834, issue 4, December, 1992.

[23]. Operations Technology Generic Requirements (OTGR): Operations Application Messages—Memory Administration Messages, Bellcore, TR-NWT-000199, issue 1, May, 1991.

[24]. M. T. Rose, *The Simple Book—An Introduction to Management of TCP/IP-Based Internets,* Englewood Cliffs, NJ: Prentice-Hall, 1991.

[25]. J. Galvin, K. McCloghrie, and J. Davin, "SNMP administrative model," Internet draft, Internet Engineering Task Force, Network Information Center, SRI International, Menlo Park, CA, April, 1991.

[26]. J. K. Case, M. Fedor, M. L. Schoffstall, and C. Davin, "Simple Network Management Protocol," RFC1157, Network Information Center, SRI International, Menlo Park, CA, May, 1990.

[27]. ISO/IEC 10040, Information Technology—OSI—Systems Management Overview.

[28]. ISO/IEC 10164-4, CCITT Recommendation X.733, Information Technology—OSI—Systems Management—Part 4: Alarm Reporting Function.

[29]. ISO/IEC 10164-12, CCITT Recommendation X.745, Information Technology—OSI—Systems Management—Part 12: Test Management Function.

[30]. ISO/IEC 10164-14, CCITT Recommendation X.737, Information Technology—OSI—Systems Management—Part 14: Confidence and Diagnostic Test Categories [for CCITT Applications].

[31]. CCITT Recommendation M.3200, TMN Management Services: Overview, 1992.

[32]. CCITT Recommendation Q.821, Stage 2 and 3 Functional Description for the Q3 Interface, 1992.

[33]. CCITT Recommendation M.3100, Generic Network Information Model, 1992.

[34]. ISO/IEC 10164-2, CCITT Recommendation X.731, Information Technology—OSI—Systems Management—Part 2: State management function [for CCITT Applications].

[35]. A. P. Jayasumana, Y. K. Malaiya, and R. Rajsuman, "Design of CMOS circuits for stuck-open fault testability," *IEEE J. Solid-State Circuits,* vol. 26, no. 1, January, 1991.

[36]. M. Franklin and K. K. Saluja, "Pattern sensitive fault testing of RAMS with built-in ECC," *Digest of Papers, Fault-Tolerant Computing: 21st Int'l Symp.,* Los Alamitos, CA: IEEE Computer Society Press.

[37]. H. L. Kalter et. al., "A 50-ns 16-MB DRAM with a 10-ns data rate and on-chip ECC," *IEEE J. Solid-State Circuits,* vol. 25, October, 1990.

[38]. E. Fujiwara and D. K. Pradhan, "Error-control coding in computers," IEEE Computer, July, 1990.

CHAPTER 10

Network Performance Management

Subhabrata Bapi Sen
Bellcore
Piscataway, NJ 08855

10.1 INTRODUCTION

Network performance management is a critical part of a telecommunications management network (TMN) that ensures that the network is operating efficiently and delivers the prescribed grade of service. Performance management operational measurements are key inputs to the planning process to track growth of demand and to the resource provisioning process to flag actual or impending resource shortages. Performance management, like fault management that ensures that the network resources operate in a trouble-free and error-free manner, depends on the configuration management[1] resource inventory in referencing its monitoring measurements and directing its network control messages. The perspective adopted here addresses operations as they might apply to an operating company, a common carrier, private network operator, or an enhanced service provider. Here they will be collectively referred to as the *operating telephone company* and their network/services will be referred to as *telecommunications network / services.*

Figure 10-1 shows the mapping between traditional telecommunications management functions, i.e., operations, administration, main-

---

[1]Configuration management includes the processes of network planning, resource provisioning, and service provisioning.

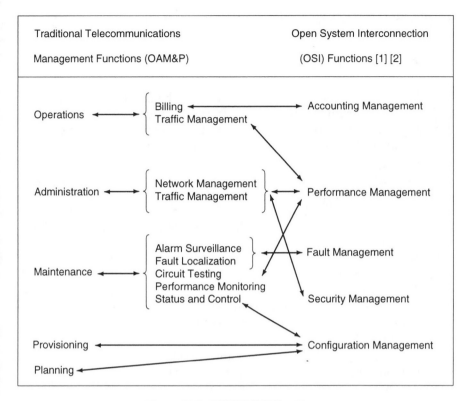

Figure 10-1 OAM&P/OSI Functions

tenance, and provisioning (OAM&P[2]) along with planning, and open system interconnection (OSI) functions.

Performance management encompasses the traffic management functions of operations, both network and traffic management functions of administration, and the performance monitoring functions of maintenance. There have been many standardization efforts for network management [3] and performance monitoring among standards bodies [4], which provide a conceptual functional framework for OSI management including performance management. Among the five functional areas seen in Fig. 10-1 under OSI, performance management enables the behavior of resources in the OSI environment and the effectiveness of communication activities to be evaluated. Performance management includes functions to gather statistical information, maintain and examine logs of system state histories, determine system performance under natural and artificial con-

---

[2]In Chapter 1 this traditional function was called OAM (Operation, Administration, and Maintenance), provisioning was considered under operation. In this chapter, OAM will be called *OAM&P*.

ditions, and alter system modes of operation for the purpose of conducting performance management activities.

10.1.1 Chapter Organization

This chapter starts with a brief review of the mapping of traditional OAM&P functions into OSI functions. These OSI functions are used mainly to manage three interrelated parts of a telecommunications network, namely, switching, interoffice facilities, and distribution (or local loop), as shown in Fig. 10-2.

The performance management activities of these parts can be in three different levels: (1) level of physical design as in the case of inadequate transmission through interoffice facilities and local loop; (2) level of traffic (not physical path) as in blocking in switches, and (3) level of fault analysis or failure analysis mainly due to lack of maintenance. Throughout this chapter, focus is on different aspects of performance management; goals, process, implementation, operations system (OS) support, and evolution/future direction of performance management are discussed in chronological order. However, examples of these different aspects are derived from three different parts in three different levels in a somewhat ad hoc manner.

In Section 10.2, where the goals and aspects of performance management function are discussed, the examples are brought out in the first two levels for two different parts of the network, namely, physical design of transmission and traffic characteristics of digital switching. In this section, network characterization, setting performance objectives including models of performance and customer opinion, determining grade of ser-

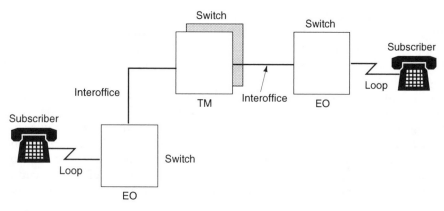

EO: End-office switch
TM: Tandem switch

Figure 10-2 Three Parts of a Telecommunications Network

vice, and measurements and control of service and performance, are discussed.

In Section 10.3, the performance management process is covered by describing performance monitoring and performance management control. Section 10.4 is devoted to an example of a more recent technology: performance management for packet networks using an OS system. Section 10.5 describes the issues of using OS systems in performance management and Section 10.6 elaborates on the major issues of performance management.

10.2 GOALS AND ASPECTS OF PERFORMANCE MANAGEMENT

The goal of performance management is to maintain the quality of service (QOS) in a cost-effective manner. The QOS is maintained by using a service evaluation process that starts with network characterization, followed by setting performance objectives and measurement. Control of network parameters is then exercised to improve service and performance [5]. An important aspect of QOS is that the time for service provisioning is becoming a QOS measure, thus intertwining the realm of configuration management and performance management. The next subsections describe network characterization, setting performance objectives, and performance measurements and control of QOS along with a set of examples.

10.2.1 Network Characterization

This step involves several substeps that describe, analyze, and measure customer-perceivable service characteristics and related network performance parameters. Here customer expectation of service is analyzed in terms of network characteristics to determine what can be provided in terms of current and evolving network performance. It is customary to perform network characterization studies on existing networks to set performance objectives, to determine whether performance is adequate for existing and proposed new services, and to indicate where changes in the network design and objectives could lead to improved performance or lower costs. Moreover, new services and technologies, e.g., integrated services digital network (ISDN) and advanced intelligent network (AIN), need such characterization to provide critical data in the introductory phase. New network structures, e.g., separation of inter-LATA and intra-LATA networks or a multilayered network like signaling system 7 (SS7), necessitate such characterization to establish separate performance criteria for different parts of the network.

The current trend is to mechanize performance characterization and reuse the characterization components as much as possible. Figure 10-3 shows an overview of the characterization process that produces performance-related documents and includes a monitoring system.

Performance planning of a new service, technology, or network structure begins with the identification of key parameters in such a way that it reflects service performance (i.e., customer perception) on one hand and measurable characteristics (i.e., the network side) on the other. The characterization studies are designed using up-to-date measurement technologies and analysis, e.g., statistical sampling, etc. The characterization studies are conducted to create performance-related documents as well as produce interim[3] performance (including fault management[4]) monitoring systems.

10.2.2 Setting Performance Objectives

Performance objectives must ensure a satisfactory level of service to customers while being cost-effective for the service providers. Moreover, the customers' needs and perceptions with respect to service change with time. Also, the provider cost structure changes with technology. These factors dictate that setting objectives for the performance of elements of the telecommunications networks can be an iterative process. The major steps in this process are:

- dividing the network into components of manageable size and describing and quantifying the operation or performance of each component using some mathematical models based on measurements made during the characterization process
- determining user opinion and acceptance levels by subjective testing (or laboratory testing if the "users" are equipment such as computers)
- determining distribution of quality-of-service to the customer
- formulating performance objectives

Iterations in this objective-setting process are generally stimulated by the availability of new technology, which leads to proposals for new

---

[3]The word *interim* is used here to convey that the performance monitoring system is first (interim) field-trialed and then the final system is established.

[4]The phrase *fault management* relates to monitoring of alarms and deals with event data rather than statistical data. The term *statistical data* refers to masses of data or rates of event as opposed to events created individually.

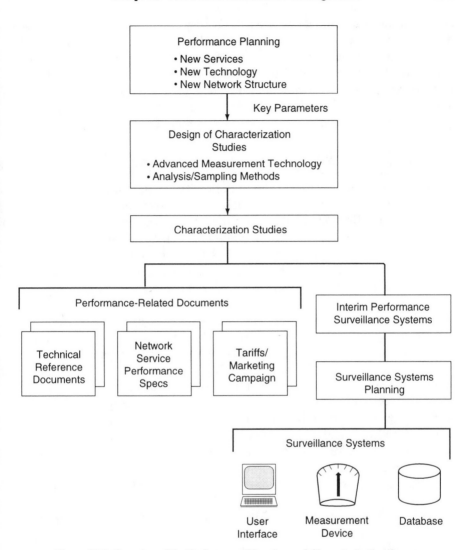

Figure 10-3 Overview of the Performance Planning and Characterization Process

services or cost-saving or revenue-stimulating changes to the network. A proposal is first evaluated by sensitivity studies of models of the existing network. These studies develop estimates of operation or performance effects of the proposal, the impact on service, and the cost of achieving those effects. Given acceptable results, a trial consisting of a small-scale application of the proposal may be conducted. Here, quality-of-service ratings are determined and performance objectives are formulated. If the results of the trial application are satisfactory, the proposal may be implemented throughout the system. Once the changes have been incorporated in the telecommunications network, performance or operational measures are again characterized, mathematical models are refined, new grade-of-service ratings are determined, and objectives are reformulated. In this stage, integration of service standard, component service objectives, engineering objectives, and administrative control takes place.

Performance Model. Performance models facilitate analysis of proposals for new services or new network technology because they are more tractable and less expensive than field trials for evaluating alternatives. Generally, mathematical models allow better understanding and more precise definitions of processes and the characteristics of facilities than do word descriptions. The models must be sufficiently detailed to account for all factors that significantly affect or contribute to the performance of the telecommunications network. When general models do not exist (for example, for new services under consideration as public offerings), they may be created as described below.

The first step is to specify the nature and features of the proposed service or technology, the facilities that will be involved, the performance parameters that must be controlled in order to render satisfactory service, and the characteristics of the geographical area where it would be implemented. The next step is to collect all information pertinent to the cause-and-effect relationships among these factors. This step may include searching sources, referring to operating telephone company measurements of facilities, and, if necessary, planning and conducting a new characterization study. Mathematical expressions are then generated for each of the parameters under consideration. These expressions are defined in statistical terms based on standard techniques of data analysis, for instance, correlation analyses relating performance parameters and facility descriptors. The set of mathematical expressions derived constitutes the mathematical model.

An example of this process can be found in the transmission (through local loop and interoffice facility) of the public switched telecommunications network (PSTN). Here the performance model is for transmission of voice telephone service (loss, noise, and echo). A call using the PSTN involves the telephone sets on the customers' premises, the communication paths (the loops) to local class end offices, and a number of interconnecting links (interoffice trunks) and switching systems. The actual route chosen for a particular call through the network could be one of several possible combinations of links. Transmission parameters are determined primarily by path length; therefore, a performance model for transmission parameters requires a way of expressing the expected route length for a call and combining that with the relationship between trunk length and each parameter.

Computer simulation is used to generate routing data weighted by the probability of occurrence of a given route. These routing data provide information as to the number of trunks and the length in miles of each trunk in a connection between two end offices. Transmission characteristics are then derived by combining this information with trunk statistics based on company-wide measurements of transmission parameters.

Customer Opinion Model. Customer opinions and perceptions of the quality of a telecommunications service influence the demand for the service. Given the increasing role of competition in telecommunications, customer perceptions may also influence the customer's choice among alternative services. Therefore, it is important to appraise customer satisfaction with offered services.

The models of customer opinions quantify subjective opinion of performance conditions by discrete ratings (e.g., excellent, good, fair, poor, unsatisfactory). These ratings, which are obtained primarily from subjective tests, are first tabulated and then converted to mathematical expressions of the form: $P(R \mid X)$.

In quantifying customer satisfaction with the transmission quality of a telephone call, for example, $P(R \mid X)$ is the conditional probability that a customer will rate a call in opinion category R, given a value X of the stimulus (noise, loss, echo, etc.). This function can be estimated from measurement of the responses of groups of subjects to various controlled levels of the given stimulus.

As Cavanaugh, Hatch, and Sullivan (1976) [6] found, opinions depend on various factors such as the subject group, the time, and the test environment (that is, whether the test was conducted during conversations in a laboratory or during normal telephone calls). Furthermore, the effects of various stimuli may be interdependent. In testing the effects of noise and loss, for example, both the noise and loss stimuli must be varied so that different combinations of noise and loss can be observed. By taking into account the type and size of the sample and also the influences of other pertinent factors and stimuli, customer opinions that reflect the distribution of telephone users can be tabulated.

10.2.3 Determining Grade of Service

Mathematical models of performance and models of customer opinion are combined to obtain grade-of-service ratings for specific aspects of telecommunications service.

Grade of service (GOS) is defined (for a single parameter, X) by the weighted average[5] of opinion curve $P(R|X)$ (defined in the last section) where weighting function $f(X)$ is the probability density function of obtaining stimulus X.

Both subjective tests results via opinion curves and performance characterization via distribution of performance enter in the process to determine grade of service. Controlled subjective tests provide the needed opinion curves, $P(R|X)$, and characterization studies, together with a mathematical model, provide the required performance distribution, $f(X)$.

Grade-of-service ratings are essential criteria used in formulating objectives for various performance parameters. Specific instances of the transmission quality discussed in Section 10.2.2 are shown in Fig. 10-4, which shows that as loss in a telephone connection increases, the quality of service for loss and noise (in terms of the number of people who rate transmission as good or better) decreases (*dotted curve*). On the other hand, this same increase in loss increases the echo path loss, thus improving the grade of service for echo (*dashed curve*). The result is an optimum value of loss for which the highest values of the combined loss-noise-echo grade of service (*solid curve*) are achieved. The value of optimum loss is a function of connection length because the delay is longer on long connection and, thus, more loss is required to control echo.

The data[6] for Fig. 10-4 were gathered by measuring calls in a network with a realistic distribution of echo loss and noise. For the connection length illustrated, an end-to-end accoustic loss of approximately 18 dB (the *asterisk* in the figure) results in the optimum loss-noise-echo grade of service.

Deployment of digital transmission facilities, digital switching, and fiber optics have, in the past decade, greatly improved the grade of service for voice communications. However, early attempts to mix voice, video, and data in packet-switched networks have introduced variable delays in the voice signal that have renewed interest in the topics of evaluating grade of service over the voice network.

10.2.4 Examples of Various Performance Objectives

The previous subsection discussed an example of analog transmission; in this subsection two major examples are discussed—one from traffic characteristic of switching and the other from digital transmission.

---

[5]Weighted average is calculated by an integral: $GOS = \int_{-\infty}^{+\infty} P(R|X) f(X) dx$. Multiple impairments are handled similarly with multiple integrals and joint density functions.

[6]Data is for analog transmission. With digital transmission, amount of noise decreases considerably; however, relationships described in Fig. 10-4 still hold. In digital transmission model, delay is introduced which makes echo more prominent.

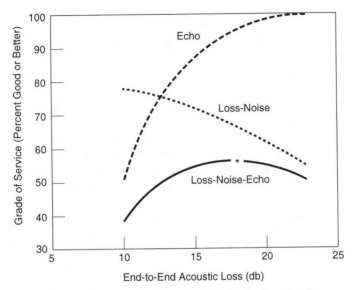

Figure 10-4   Example of Loss-Noise-Echo Grade of Service

Performance Objectives of Switches. The switching example is quite different from the transmission example because it deals with another part of network (switching instead of interoffice or loops) and at another level (traffic[7] rather than physical characteristic). The LATA switching systems generic requirements (LSSGR) [7] specifies requirements for various service criteria (performance) for data collected in average busy season busy hour (ABSBH),[8] ten high day busy hour (TDBH),[9] high day busy hour (HDBH), average bouncing busy hour load (ABBH),[10] and expected high day (EHD).[11] These criteria are shown in Table 10-1. Tandem network service is measured both as probability of first failure to match (FFM) (i.e., call fails to get the network path in the first attempt) and as the probability of final failure to match [matching loss (ML)], (i.e., inability of an originating call to find any link path through the system).

---

[7]Traffic load is different from how traffic is physically carried. So, performance objectives that deal with traffic are quite different from performance characteristic that deal with physical path in transmission.

[8]ABSBH is the average time consistent (same busy hour each day) busy hour traffic over the busy season.

[9]TDBH is ten highest busy hour, during busy season.

[10]ABBH is average busy hour traffic when traffic is measured at different times each day at the daily peak rather than time at a consistent hour.

[11]It is an estimated load using ABBH and variation of ABBH from day to day.

Dial tone delay is a measure of the time it takes a telephone system to return dial tone to an origination subscriber after the subscriber goes off hook (i.e., $t>3''$ indicates more than 3 seconds delay and $t>0$ indicates any delay at all).

These detailed performance objectives cannot be derived directly from customer perceptions, but are based on system engineering judgement. The intent is to distribute allowable degradations in such a way as to achieve an acceptable degradation at minimum cost.

As seen in Table 10.1, there are usually a few numbers that specify the service standard. However, there are usually a large number of rules and procedures that must be be specified to adequately size the network so that it meets the limited number of service standards.

TABLE 10.1 Busy Hour Criteria for Switching in PSTN

	Criterion Type	ABSBH	THDBH	HDBH	ABBH	EHD
Network Service						
Line-to-Trunk	Blocking (ML)	1%	—	2%	1%	2%
Trunk-to-Line	Blocking (FFM)	2%	—	#	2%	#
Line-to-Line	Blocking (ML)	2%	—	#	2%	#
Trunk-to-Trunk	Blocking (FFM)	2%	—	—	2%	—
Trunk-to-Trunk	Blocking (ML)	0.5%	—	2%	0.5%	2%
Dial Tone Delay	Delay($t>3''$)	1.5%	8%	20%	1.5%	20%
Dial Tone Delay	Average	0.6	—	—	0.6	—
IR Attachment Delay	Delay ($t>3''$)	1.5%	8%	20%	1.5%	20%
Service Circuits						
Customer Digit Receivers	Delay ($t>0$)	—	—	5%	—	5%
Interoffice Receivers	Delay ($t>0$)	1%	—	—	1%	—
Interoffice Receivers	Delay ($t>3''$)	—	—	0.1%	—	0.1%
Interoffice Transmitters	Blocking	—	1%	—	—	—
Ringing Circuits	Blocking	—	—	0.1%	—	0.1%
Coin Circuits						
Coin Control	Blocking	—	—	0.1%	—	0.1%
Overtime Announcement	Delay ($t>0$)	1%	—	—	1%	—
Announcement Circuits	Blocking	1%	—	—	1%	—
Tone Circuits	Blocking	1%	—	—	1%	—
Reorder Tone Circuits	Blocking	—	0.1%	—	0.1%	—
Conference Circuits	Blocking	0.1%	—	—	0.1%	—
Billing	Error	—	—	2%	—	2%

#There are no explicit HDBH or EHD performance standards for the (#) items, but suppliers should use the 20 percent HDBH or EHD dial tone delay ($t>3''$) criteria to define concentrator congestion objectives for their switch.

Performance Objectives of Digital Transmission. Besides the switching area example, performance objectives can be found in digital transmission [8] areas as well. In the digital transmission area, detected impairment events serve two main purposes. First, they are used as key elements in defining more severe impairment events or failure criteria, e.g., loss of frame. Second, they form the basis for deriving a fundamental set of performance parameters that are accumulated and stored in a monitoring network element (NE).

For the most part, performance parameters are raw counts of impairment events or seconds containing impairment events. Performance parameters are normally accumulated over quarter-hourly (i.e., every 15 minutes) and daily intervals, and are maintained in designated storage registers. Some of the error measurements are shown below:

- Code violations (CVs)—a single detected transmission error. Any parameter beginning with CV_ is a count of bit error (for line and paths) that occurred for the monitored interval within the scope of measurements, in the direction of measurement.

- Errored seconds (ESs)—count of seconds containing one or more CVs or other error detection events (for lines and paths).

- Severely errored seconds (SESs)—count of seconds containing more than a particular quantity of CVs or other bit error detection (for lines and paths).

- An unavailable second (UAS)—any second in which the monitored facility was unavailable due to loss of signal or frame.

Interpreting performance monitoring (PM) data can be done in the following manner.

- Code violations: If number of code violations (any parameter beginning with CV_) per hour or per day is extremely high, then check the corresponding ES and UAS counts (the one with the same scope of measurement, direction, and time interval). If ES count is zero or very low, and the UAS count is high (>200/hr or 1000/day), then equipment doing the monitoring is counting CVs during unavailable time. In this case only, ignore CVs as measure of error performance and use ES and SES as a guide.

If there is no reason to conclude that a significant number of the CVs are being counted during unavailable time, then note the general level of CVs per hour for use in comparison with ES level.

- Error seconds are usually caused by an impairment of some sort. If the ESs occur for just one or a few hours and then stop permanently, then the cause is transient impairment and likely will not reappear. If the ESs are persistent or intermittent without a clear pattern, then use the general level of ESs and type of service supported to decide whether proactive maintenance is necessary.

Comparison with the corresponding CV counts and computing the ratio of CV/ES gives a measure of small-scale burstiness of the errors. A high ratio indicates bursty errors.

- Severely errored seconds occur when the CV/ES ratio mentioned above exceeds a vendor set threshold. SES indicate an extreme burstiness of errors. Such burst would manifest themselves as chirps or other interface in a voice transmission or static in a video transmission. For some data transfer applications, especially encrypted data, a SES is no worse than a regular ES as long as it does not cause the application to drop the line.

10.2.5 Performance Measurements and Control of QOS

Performance measurements and control of QOS is the process of monitoring and controlling conformance to the performance objectives and determining plans to rectify the situation if objectives are not met. The measurements start with customer perception as reflected in trouble reports and questionnaires, e.g., the telephone service attitude measurement program (TELSAM). Performance measurements for the installation of service include the probability of proper provisioning and the probability of meeting the promised schedule. Availability (dial tone delay, network blockage, etc.) and billing integrity are also important aspects of QOS that are measured.

The measured results are then compiled and categorized (e.g., high, objective, low, unsatisfactory bands) and distributed through a management information system for local action. Local action quite often uses diagnostic aids to identify the source of the problem and generating means of correcting the problem.

10.3 PROCESS OF PERFORMANCE MANAGEMENT

Performance management [9] provides functions to evaluate and report on the behavior of telecommunications equipment and the effectiveness of the network or NE. There are two primary aspects of performance management: *Monitoring and Control.*

Performance monitoring involves the continuous collection of data concerning the performance of the NE. Very-low-rate or intermittent error or congestion conditions in multiple equipment units may interact, resulting in poor service quality. Performance monitoring is designed to measure the overall quality of the monitored parameters to detect such deterioration. It may also be designed to detect characteristic patterns before service quality has dropped below an acceptable level. The basic function of performance monitoring is to track system, network, or service activities to gather the appropriate data for determining performance.

Performance control involves the administration of information that guides or supports the network management function and the application or modification of traffic controls to help provide network management.

The most operations environment currently is based on the traditional OS/NE systems view of operations control in the network. The OSs tend to have centralized functionality, a span of control that covers a large portion of the network, and tend to provide most of the operations functions. The network elements are distributed, have a span of control limited to themselves, and have a relatively limited set of operations functionality. In this model, the OS/NE interface is the prime focus of interface specification work. To further elaborate the two aspects of the performance management process, however, it is important to discuss the proposed functional operational view of three layers, namely, the network management layer (NML), the element management layer (EML), and the element layer (EL). Descriptions for these layers follow:

Network management layer (NML). This layer contains those functions used to manage an end-to-end view of the telecommunications network. NML access to the network is provided by the element management layer, which presents network resources either individually or in aggregation as a subnetwork. The NML controls and coordinates the provisioning or modification of network resource capabilities in support of services to the customer through interaction with other layer functions such as those in a service management layer. It also provides other layers with information such as the performance, availability, and usage data provided by the network resources.

Element management layer (EML). This layer contains those functions used to manage resources within the network individually or in aggregation as a subnetwork. These functions provide and use standard or open information models to define interactions with NML, element layer, and other layer functions. Individual functions in this layer may be very similar to those in the network management layer but have a more limited span of control.

Element layer (EL). This layer contains functions that are linked to the technology or architecture of the network resources that provide the basic telecommunications services. These functions are accessed by the EML functions using standard or open information models that may hide vendor-specific functions within network resources.

The next two subsections discuss the following two management applications[12] using EML performance management.

- Performance monitoring
- Performance management control

10.3.1 PERFORMANCE MONITORING

Performance monitoring[13] is concerned with nonintrusive (not service-affecting) statistical measurements of errors that may degrade service if they exceed an acceptable rate. Transport performance monitoring measures transmission error rates detected by line, path, or section terminations. Examples of transmission error rates that are monitored include counts over periodic time intervals (typically 15 minutes or one hour) of cyclic redundancy check errors, frame slips, and severely errored seconds. These events are detected by element layer (EL) functions.

In draft International Telecommunications Union—Telecommunications (ITU-T—formerly CCITT) Recommendation M.3400, the functions of performance monitoring are subdivided into *generic functions, control functions,* and *analysis functions.* These distinctions are not made in this section, but for purposes of cross reference, the functions that deal with PM events, PM exceptions, and current data are related to line items in

---

[12]The actual process of performance management can also be grouped into performance monitoring, traffic and network management, and quality-of-service observation. Performance monitoring is described here. Network traffic management constitutes the following activities: requesting traffic data from an NE, receiving report from an NE due to exceeded threshold, modifying network thresholds, setting traffic data collection attributes, setting error analysis parameters, receiving error analysis from NE, and modifying real-time network controls.

Quality-of-service (QOS) observations constitute determining a provisioning sequence to maximize long-term network flexibility and utilization and assessing statistical data trends to avoid overload and maintain QOS.

[13]The term *performance monitoring* in many documents, including the 1988 issue of Annex B of ITU-T Recommendation M.30 [10] refers to the performance monitoring of transport resources. In the current draft of M.3400, which replaces Annex B of M.30, the term has been generalized to include performance monitoring of switches, including traffic and quality of service measures. However, this document considers only the narrow topic of measurement of transport resources.

One example of another area of performance monitoring is the statistical measurement of protocol errors in packet, frame, or cell data services. The principles of this section may apply to this topic, but the subject has not been examined in detail.

M.3400 that are listed under generic functions. Schedule and threshold data entities are related to the control functions. Functions that have the terms *correlate* and *analyze* in their names relate to the analysis functions.

Performance monitoring is closely related to alarm surveillance. Both are concerned with the detection of problems in equipment and transmission media. While alarms are designed to detect failure events, PM exception events indicate the rate of errors measured over a time interval that has exceeded a threshold. Since PM exception event data are rate of event, rather than events created individually, this data is considered statistical data rather than event data.

Figure 10-5 shows a data flow diagram of EML [11] transport performance monitoring. Performance monitoring is also related to fault localization and to testing. Performance monitoring is used to monitor statistics concerning the health of equipment that is in service or on standby as a spare. Applications of performance monitoring are exemplified in the following scenarios:

Scenario 1: Proactive Servicing. Transport PM detects and quantifies degradation in circuits providing services. It is a particularly powerful tool for the following:

- Quantifying subtle problems in circuits whose components pass all diagnostics and measurements.
- Locating intermittent causes of degradation.
- Anticipating failures that are preceded by gradual increases in error rates.
- Verifying quality of transmission over a long period.

When the statistical performance of a circuit drops from an acceptable level to an unacceptable, or significantly toward an unacceptable level, repair can be started, even if a customer has not reported the condition. If a customer has reported trouble, PM can be helpful in verifying the trouble and in guiding repair.

Scenario 2: Acceptance Testing. In the course of acceptance testing of new transport equipment, performance monitoring can be used to verify the quality of newly installed facilities. In this case, the success of the test may be determined by reading the current data or by reading exceptions in the surveillance condition entities. In the latter case, the threshold may be set to a lower value than usually used for service.

Scenario 3: Contract Fulfillment. In some cases, a customer guarantees a level of performance for a transport service. In such cases, performance management exceptions may be used for billing rebates, for

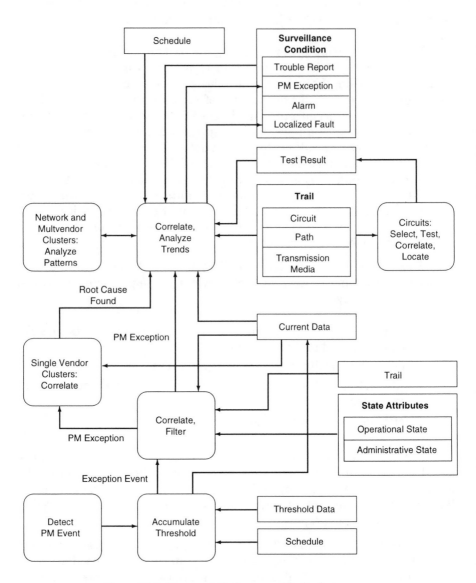

Figure 10-5 Transport Performance Monitoring

customer information, or for verification of the contractual requirement. In this scenario, the appropriate threshold values may depend on the customer's application of the service.

Note that the trail data entities (circuit interconnection topology) are managed by the provisioning functions of configuration management. The state attributes[14] are managed by the status and control functions of configuration management and by the behavior of the managed objects. The threshold data and schedule data entities are managed by the data access function of common operations management.

In a wire center subnetwork (the distribution area of a switch), many of the NEs are located at remote locations. It can be difficult to gather PM exceptions detected at these remote locations and bring them to the central points where corrective action can be started. Therefore, one aspect of EML functionality is to gather exceptions from throughout the domain and make them available to the operations system functions of the network management layer (NML) [11] or higher layers.

Additional potential EML functionality is to remove unnecessary exception reports in a manner very similar to the removal of unnecessary alarms. One example is to refrain from reporting PM exceptions concerning terminations that are already in a disabled state, or that are supported by equipment in a disabled state.

10.3.2 PERFORMANCE MANAGEMENT CONTROL

Performance management control is the monitoring of traffic flow within the telecommunications network, the optimizing of the traffic utilization of network resources, the preservation of the integrity of the network during high usage periods [12], and the surveying of an NE's traffic processing behavior for network engineering and administrative purposes (i.e., network data collection). The traffic data collected and processed is also used to support other network operations functions such as fault management, network capacity provisioning, and configuration management. This data is collected and maintained over specified time intervals—such as 5 minutes for near real-time management—and 30 minutes, 1 hour, and 24 hours for general analysis and patterning.

A traffic manager relies on having these functions provided by some operations application (e.g., an OS or element manager) and the complementary functions being accomplished in NEs in order to perform the network traffic management activity. Currently, traffic performance management is accomplished mainly in OSs (e.g., a network traffic

---

[14]Note that state attributes are properties of objects in a database, not objects per se.

management OS and a network data collection OS). However, the data collection aspect of traffic management can be performed at the EML as a means to accommodate the shorter analysis periods required by some services (e.g., switched multimegabit data service [SMDS]) and to perform data concentration and correlation to aid the OSs in accomplishing their tasks. The activities that comprise performance management control can be generalized to the following management functions:

- Accessing and administering local management data in an NE
- Requesting management data from an NE
- Accumulating and "preprocessing" of traffic data
- Forwarding of management data

Each of these functions, as depicted in Fig. 10-6, has some data and/or functionality that it provides to the other functions or external procedures.

Application of Performance Management Control. Application of management control during performance monitoring can be exemplified in the following scenario of switching equipment:

Scenario: Switching Equipment. Data may be generated within the EL over a five minute time interval. This data, obtained through periodic scans of switching equipment and maintained in equipment registers, can be forwarded at any designated time during the five-minute interval period to the EML for thresholding, storage, and filtering needed to provide traffic data reports at the NML. The EML organizes the data from several switching elements based on established report criteria and makes it available to the NML according to the data schedule. Hence, an EML function is the gathering of five-minute traffic counts for the purpose of interface consolidation, thresholding, initial calculations, and data routing to the NML.

Access and Administer Local Data. The Access and Administer Local Data function is a combination of a data gathering function and a data administration function. This function is vital to the traffic management and data collection applications in that it constitutes the primary operations data link to the network elements being monitored. The administrative functions update the local data that guide the various traffic management and data collection functions. These updates (e.g., resetting local thresholds) can be initiated by input from external applications, users, or known equipment states

Request Data. This function is responsible for originating polls to the network for the data needed in monitoring, controlling, and reporting on subnetwork traffic and facility utilization conditions. The polls are

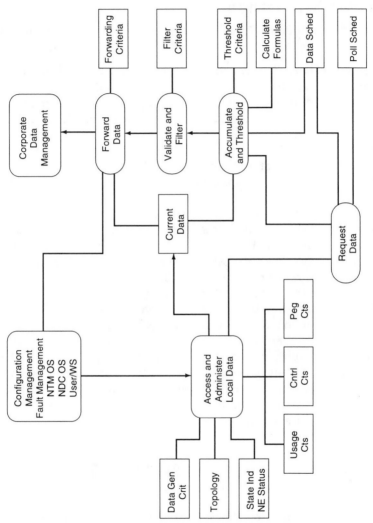

Figure 10-6 Performance Management Control

321

generated based on schedules or ad hoc data requests from other functions or applications.

Accumulate and Threshold Data. This function reads and finalizes the collection of traffic data from NEs and compares this data with various thresholds to help identify critical traffic conditions.

Validate and Filter Data. This function accomplishes activities that verify the data collected and separates it for further handling.

Forward Data. This function is responsible for routing or forwarding data reports of subnetwork traffic and utilization data to corporate data repositories and upstream or downstream processes or users, in accordance with requests of system users, network customers, or other operations applications.

10.3.3 Example of the Process of Switch Performance Management

The switch performance management provides an example of the performance management process. There are three parallel processes: fault management[15] with seconds to 5 minutes data, performance management with 5 minute data, and capacity or configuration management with 30 minutes data or mostly busy, hour data. In fault management the focus is on immediately correcting anomalies of switch behavior, e.g., maintenance, in real time. In performance management the focus is to correct unusual traffic behavior using real time controls like rerouting, etc. In capacity management the focus is a delayed reaction through load balancing, using proper load assignment, and proper addition of future capacity using present measurements and forecast of access line growth. In all these processes, a specific grade of service as in Table 10-1 needs to be specified. The action is quite diverse but the aim is to provide a specific grade of consumer service.

10.4 PERFORMANCE MANAGEMENT OF PACKET SWITCHING: A CASE STUDY

This section elaborates on the concepts developed in the earlier sections by using an example of a plan for performance management of packet switching [13], in particular how an OS architecture can be deployed to support the process. Packet switched data communication networks pro-

---

[15]In time domain the performance starts where fault management leaves off. Fault management deals with event data whereas performance management deals with statistical data.

vide the means of convergence of areas of information gathering, processing, and distribution and, thus, helping provide a more coherent view of collecting, transporting, storing, and processing information.

Asynchronous transfer mode (ATM), which is a form of packet switching, is stated in ITU-T Recommendation I.121 as the target transport mode for future broadband services and architectures such as broadband integrated services digital networks (BISDNs). Three types of packet switched networks are currently available or being deployed: public packet switched networks (PPSNs), common channel signaling networks (CCSNs), and metropolitan area networks (MANs).

PPSN (e.g., X.25 and basic and primary rate ISDN) is intended to provide economical common user switched data/information transport for bursty data. PPSN can be used to transport services such as electronic mail, reservation services, point of sale, order entry, credit authorization, credit card verification, electronic funds transfer, information retrieval, time-sharing, automatic teller machine, facsimile, telemetry, education, home banking, home shopping, library, brokerage services, stock quotation, advertising, bulletin board, and so on. Terminals, host computers, and gateways constitute the sources and sinks of the data messages to be transported by the network.

CCSN, on the other hand, is intended to provide signaling message transfer for participating stored program control (SPC) switches, databases, and operator systems in Public Switched Telephone Networks (PSTNs). It allows call control messages from the PSTNs to be transferred on separate communications paths from the voice connections, thus permitting a great reduction in call set-up time, not only by its inherent speed but also by signaling the next office in the route before an office has finished switching. These advantages stem from the fact that the signaling function is fully dissociated from the voice path and is handled in a highly concentrated form through the CCSN. The signaling system uses the SS7 protocol over its signaling links for transfer of signaling messages between exchanges or other nodes in the telecommunication network served by the system. The messages in the CCSN not only provide for call control signaling between network nodes but also for transaction-based services, e.g., 800 service and alternate billing service (ABS). In the CCSN, the sources and sinks of messages are the signaling points (SPs). A signaling point is any node with the necessary hardware and software to interface to the CCSN, for example central office switches, tandem switches, service control points (SCPs), and service switching points (SSPs).

SMDS [14] is a proposed high speed packet switched data service that many operating telephone companies plan to offer in the near future. SMDS provides performance and features similar to those found in local area networks and it is regarded as the first phase of BISDN. It will furnish subscribers with high speed access (1.5 Mbit/s and 45 Mbit/s initially and 150 Mbit/s eventually) as well as multipoint connection by today's private line facilities. SMDS will initially be supported by network elements using metropolitan area network (MAN) technology. Subscribers would access the service via a dedicated interface, the subscriber network interface (SNI), the function of which is based on the IEEE 802.6 standard for access to MANs. The interface function is defined as the SMDS interface protocol (SIP), which is based on the DQDB (distributed queue dual bus) protocol.

As more and more packet networks, such as the aforementioned PPSN, CCSN, and MAN, are being deployed, appropriate network management systems are required to support all functions, including performance management. The performance monitoring function provides statistical data on utilization problems that indicate that a network change is needed. To achieve a total integrated management architecture and system such as the International Standard Organization's (ISO) OSI network management architecture, a generic interface [network management protocol, e.g., OSI's Common Management Information Protocol (CMIP)] will have to be defined between the network elements and the supporting operations systems. With this goal in mind, the performance function will be discussed as a part of the overall scope of network management.

10.4.1 Performance Management and Network Planning Functions

Since performance management and network planning (part of configuration management) are complementary activities, they will be discussed together here. The performance management and network planning functions can be depicted in a three-dimensional model shown in Fig. 10-7. For the public switched telephone networks, different systems exist to support different disciplines such as strategic planning, tactical planning, traffic engineering, and traffic data analysis. These are shown as the dimension Y in Fig. 10-7. Systems are also separated across different network components; for example, different systems support switch node engineering, inter office trunk engineering, and outside plant (loop) engineering as shown as the dimension Z in Fig. 10-7. Also, as shown along dimension X, discrete systems often serve to isolate network technologies. Instead of building a separate operations system for each individual network technology, planning and engineering discipline, network component, the Integrated Network Planning System (INPLANS™—a trademark of Bellcore) offers an integrated system to support the traffic-sensitive planning and engineering capabilities shown in Fig. 10-7.

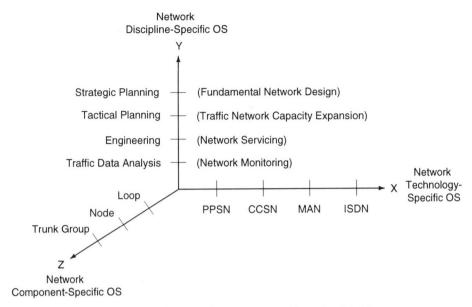

Figure 10-7   Performance Management and Planning Model

For traffic-sensitive packet networks, a performance management and network planning model as shown in Fig. 10-8 can be used to describe the functions supported by INPLANS. Each function in the model has a one-to-one correspondence with an INPLANS module:

• The integrated network monitoring (INM) module supports both planning and performance monitoring functions by monitoring the ability of in-place networks to meet both grade-of-service and utilization objectives. The purpose of this type of study is to identify situations where corrective planning action is necessary because objectives are not being satisfied and to provide, where possible, early warning of problem situations to allow corrective plans to be formulated before service problems result. Combining the functions of utilization measurement and grade-of-service measurements is conducive to efficient management of high-growth services because growth in access lines and growth in rate of usage can be monitored coherently.

• The integrated network servicing (INS) module provides an integrated environment for the network traffic engineering and servicing decision support process. With the integrated network monitoring capability to monitor the current network continuously, the network planner/servicer will be able to make necessary adjustments to absorb changes in demand before service is affected.

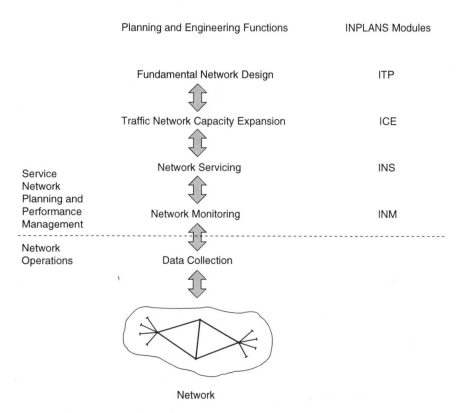

Figure 10-8 INPLANS Performance Management and Planning Model

- The integrated capacity expansion (ICE) module provides a 3- to 5-year forecast of network demands and requirements (e.g., trunk and equipment requirements). It computes the network element growth rates, estimates future network demands, times (epoch at which resources exhaust) and sizes network elements, and generates a 3- to 5-year capacity expansion plan.

- The integrated technology planning (ITP) module provides cost-effective backbone and network access design plans. These plans identify the placement and interconnection of network nodes as well as an optimal homing arrangement plan for new customers.

The ICE and ITP modules provide mainly planning functions. The INS module serves the functions of both network performance management and planning, including two categories of algorithms. Synthesis algorithms provide solutions for network element congestion problems identified through a network monitoring process such as the INPLANS INM module. Analysis algorithms evaluate the performance of a synthesized network or any user-specified hypothetical network. Specifically, to support PPSN traffic engineering and servicing, the synthesis algorithms include network element sizing algorithms, a processor load balancing algorithm, a network layer window size adjustment algorithm, and a traffic rerouting algorithm. The analysis algorithms include network element utilization and delay estimation as well as end-to-end delay estimation. To support other packet networks, such as CCSNs and MANs, only synthesis and analysis algorithms require network-specific modifications; the underlying software remains the same.

The prospective users of INS are network traffic engineers who are responsible for appropriately sizing and configuring packet networks based on service objectives such as end-to-end delay. The existence of such service objectives presumes that measurements will be taken to determine whether the network is meeting them. The interplay between traffic measurements and service criteria is important: there is no point in establishing service objectives that cannot be adequately verified by the measurement process. In fact, the service objectives must be defined such that measurements can be developed to pinpoint the corrective action that must be taken when objectives are not met. This process of converting the raw traffic measurements into performance measurements is referred to as *traffic data analysis*. Typically, raw traffic measurements are collected by a network operations data collection system and stored in a database in batch mode. The monitoring process then performs traffic data validation, converts the validated raw measurements into meaningful network loads (e.g., trunk group packets per second transmitted), and calculates a statistic to characterize the load for engineering purposes (e.g., average busy season busy hour (ABSBH) load). The process then computes network element performance (e.g., throughput or delay) based on the characteristic engineering loads. Finally, the process compares the calculated performance results with the user-defined service objectives to identify any service or performance exception.

The monitoring process (INM) identifies network elements that do not meet their performance objectives. The servicing process (INS) is then used to investigate ways of alleviating those performance problems. The servicing process includes two main subprocesses: network synthesis, which provides solutions to network element congestion, and network analysis, which evaluates the performance of proposed networks. To carry out the servicing process, the system iterates between these two subprocesses until a satisfactory result is obtained. The output of the servicing process is a servicing order that, depending on the action it suggests, will drive other operations systems such as the network operations, link (line and trunk) provisioning, and equipment and facility engineering systems for actual implementation of the synthesized results.

In summary, INM processes the raw traffic measurements, estimates network performances, and identifies performance exceptions; INS synthesizes servicing options, analyzes the synthesized network performance, and issues appropriate servicing orders. Although the underlying software structure remains the same no matter what type of network is involved, the servicing alternative synthesis and analysis algorithms are network specific.

10.4.2 Critical Performance Measure

A critical performance measure for PPSNs, especially those that support voice and video data through ATM, is the length of data transfer delay (end-to-end delay) experienced by end customers[16]. Network performance objectives are usually stated in terms of quantitative averages such as the average end-to-end delay through the network and probability of scoring above a threshold. Given the available traffic data and the end-to-end delay objective, the question would be: What is to be done if the objective is not met? Long delay could result from a problem in one or more of many trunk groups from which an end-to-end path might be formed and/or from a problem in one of several switching systems. The specific location(s) and cause(s) of the problem would have to be determined.

10.4.3 Traffic Measurements and Point-to-Point Traffic Estimation

To properly engineer a PPSN, a performance evaluation of the network needs to be conducted periodically with the purpose of monitoring whether the network is meeting its stated performance objectives (IN-PLANS INM capability). It is not currently possible, however, to directly measure the levels of end-to-end delay sustained on all paths through the PPSN. Therefore, limitations exist in the available traffic measurements. For example, utilization and delay are not directly measurable nor are the point-to-point traffic loads. Table 10.2 shows an example of the available measurements for a link set. To obtain utilization and delay performance for engineering purposes, models or algorithms have to be set up to convert the raw measurements into the meaningful performance measures. Note that the performance analysis algorithms used in INS can also be used in INM and ICE.

---

[16]Most PPSNs use an X.25-like protocol within the network, although the implementation details (e.g., acknowledgment scheme) could be quite different.

Table 10.2 Example of Available Traffic Measurements for a Link Set

ITRAN	—Number of information frames transmitted
IRECV	—Number of information frames received
IRTRAN	—Number of information frames retransmitted
DBYTET	—Number of data bytes transmitted
DBTTER	—Number of data bytes received
ABORT	—Number of abort frames transmitted
ERROR	—Number of frame check sequence (FCS) error frames transmitted
FRMR	—Number of frame reject (FRMR) frames transmitted
RR	—Number of receive ready (RR) frames transmitted
RNR	—Number of receive not ready (RNR) frames transmitted

To plan and engineer a communications network, the point-to-point traffic loads are vital input to several processes such as demand forecast, network servicing, and network design. For the synthesis algorithms, the point-to-point loads are required for the window size adjustment and traffic rerouting algorithms. There are three possible approaches to obtain the point-to-point loads:

- measure point-to-point traffic directly,

- extract from other systems, such as the billing system, that have the data or the data that point-to-point loads can be derived from, and

- estimate from the available measurements

The first approach is difficult, because the equipment manufacturers would have to make changes. This would increase the cost and degrade the equipment performance (since some processing time is now required to collect the point-to-point data). The second approach is expensive because an interface would have to be built to extract the point-to-point data from the source system and, furthermore, the data is usually incomplete. Hence, the third approach is used and an algorithm was developed to estimate the point-to-point traffic loads from the available trunk group loads. The next section evaluates the major issues of using the OS for performance management.

10.5 OPERATIONS SYSTEMS SUPPORT
FOR PERFORMANCE MANAGEMENT

This section discusses major issues that need to be resolved before an OS architecture can be effectively deployed to address performance management. These are mainly issues of data flow and OS system integration, as discussed in the next two subsections.

10.5.1 Data Flow

The past history of collection of data for surveillance and monitoring of telecommunications networks has resulted in a fragmented data flow. Typically, alarm surveillance, transport performance monitoring, traffic performance monitoring for the purpose of planning and engineering, and traffic performance monitoring for the purpose of traffic control have been specified by different user groups. As a result, separate parameter definitions, data protocols, and OSs have come into being for each purpose.

In the future, there is a need to collect the data stream in a more coherent fashion, thereby simplifying the interface requirements of NEs, reducing the number of separate OSs, and making the total data more available to high-level analysis purposes. For example, a process that determines traffic control action should be aware of known faults in the network as well as congestion. Coherence in data definitions and collection strategy will encourage flow-through by eliminating the need for human translation of data from one functional domain to another.

An example of this adverse flow of data can be seen in a TMN architecture that was evolving in telephone companies as late as 1993. In this situation there are at least six different interfaces for data flow for six OSI functions (or part of a function) as seen in Fig. 10.9 and explained in Table 10.3

TABLE 10.3 Explanation of Six Interfaces

No.	Description of Interface	TA & TR[a]
1	Configuration management data are primarily sent (secondarily collected) to NEs (for switches) or a mediation device [for intelligent transport elements like digital cross-connect system (DCS), integrated digital loop carrier (IDLC), add-drop multiplexer (ADM), etc.]. Provisioning systems usually work through interface 1.	TR-303
2	The fault management function is performed by interchanging messages between another OS and NE. Part of the performance management function that is associated with monitoring transport errors may share this interface.	TA-1030
3	Currently an OS like engineering and administration data acquisition system (EADAS) collects traffic data for both performance and fault management.	TA-495
4	EADAS along with signaling engineering administrative system (SEAS) collects traffic data for configuration management.	TA-376
5	Customer interaction with the network as required by a platform like SONET switched bandwidth networks (SBN), will have to be through an OS like FLEXCOM™/linc,[b] which in turn needs a direct interface with the network.	

TABLE 10.3 (continued)

No.	Description of Interface	TA & TR[a]
6	Accounting management (billing) currently has a separate interface that involves delayed processing of billing tapes (called AMA tapes). This interface needs to be integrated with other operations to reflect effect of customer control and increase the speed of obtaining the billing information.	TR-1039

[a]The *Technical Advisory* (TA) and *Technical Reference* (TR) are published by Bellcore to inform the industry of Bellcore's view of proposed generic requirements for various systems. TA and TR mentioned here are for interfacing systems.

[b]FLEXCOM is a trademark of Bellcore. FLEXCOM/linc provides services such as customer direct connectivity and control of telecommunication network.

As seen in Fig. 10-9, data communications between network and OSs takes place in six distinct places, as described in Table 10.3.

An ideal scenario will be to have an integrated data pooling and processing for configuration management, fault management, performance management, etc. One important advance in the right direction will be to achieve logical integration [17] of the data collection and processing functions while they may continue to be physically separate. Moreover, these segmented data flow and separate OSs make it very difficult to implement a platform like dynamic routing (DR) or an architecture like AIN or to set up SONET switched Bandwidth Networks (SBN), as seen in Fig. 10-9.

10.5.2 Integration of Operations Systems in Performance Management

OSs have long been used for performance management. As depicted in Fig. 10-10 (where "●" indicates separate OS's for the OSI function and types of network(s), however, there was no attempt to integrate the OSs across functions or types of network. Lack of OS integration is a liability for performance management functions to be adequately implemented in the new integrated services supported by integrated networks.

In the future, integration of OSs (at least logically if not physically) must be achieved both horizontally (across functions) and vertically (across types of network) to meet the performance management (as well as other functions) requirements of new services and architecture.

---

[17]This logical integration could be partially achieved (because of existing systems) through standardization of interfaces as seen in TA & TR specified in Table 10.3.

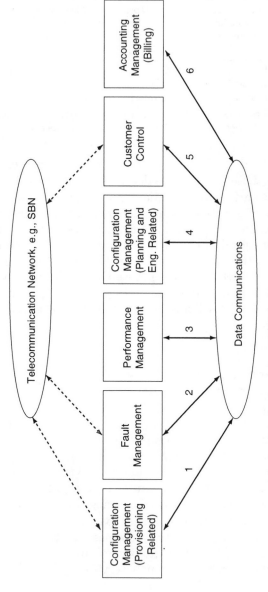

Figure 10-9 An Example of TMN Architecture: Multiple information flow path to OSs

TYPES OF NETWORK	TYPES OF FUNCTIONS			
	Network Traffic Data (Analysis & Distribution)	Configuration Management (Planning & Engineering)	Fault Management (Survivability & Restoral)	Performance Management (Monitoring & Control)
PSTN (Circuit Switched)	•		•	
Channel Switched	•		•	
CCS Network		•		•
(AIN+)	•		•	
PPSN, SMDS	•		•	
INA[a] etc.	•	•	•	•

[a]Indicates future networks like information networking architecture (INA).

Figure 10-10 Example of OSs Supporting Various Types of Network/Functions

This integration of OSs also will accommodate the following future trends:

- OSs were previously in the support role and not in the critical path of the delivery of the service to the customer. Increasingly they are becoming part of the critical path.
- Architectures like AIN demand more OS flexibility and better OS performance.
- Platforms like SBN move provisioning directly toward the customer, thus requiring more integrated overall operation between functions like performance management, configuration management, account management, etc.
- Most other responsibilities, except bulk provisioning and exception action, are shifting from OSs to NEs.
- Current functions are now merging and are being redefined very rapidly and OSs are required to play an ever-increasing, critical role.
- Finally, fault management in the network has traditionally been concentrated on failed elements. However, it has become increasingly clear that degraded performance must be considered as a significant parameter in the network performance. This degraded service as measured by performance parameters is then used by fault management. This type of application is already available in many intelligent transport elements and other hardware designs.

All of the issues of integration above affect all OSI functions to some extent but are critical for performance management.

10.6 MAJOR ISSUES OF PERFORMANCE MANAGEMENT

A number of performance management issues are being settled or are becoming better defined. In this final section, some of these issues are described.

10.6.1 Definition of Metric of Measurements

The OSI management framework assigns the gathering of statistical data as one of the duties of performance management. The discussion of useful metrics, e.g., mean, variance, minimum, and maximum [15], has always been a major controversy. The issues of the definition of metrics for the managed objects, specification of a monitoring function, specification of QOS measurement functions, and the efficiency of calculation of the metrics were addressed by a proposal. This proposal contains collection policy, calculation policy, and delivery/reporting policy [16].

10.6.2 Integration

As discussed previously, performance management must accommodate rapidly evolving network management trends such as the way separate functions (e.g., performance monitoring, fault management, and configuration management) are converging into one integrated function. Moreover, within the performance management function, increased interdependence of performance of various services (predefined/static path/ capacity for different services are becoming state-dependent/dynamic) requires joint performance measures. Lastly, performance management will require an integrated approach to collecting and analyzing data and responding—using control schemes in all phases.

10.6.3 OS/NE Interaction

Transition from OAM&P to OSI network management functions will aid in focusing the OS evolution and better serve the customer needs using emerging technology like enhanced NE capabilities (e.g., intelligent transport elements). The performance management process definition will be based upon the analysis of the changing relationship between NEs and OSs as in increasing function of NE (e.g., data screening by NE and existence of more intelligent NEs in the network) or proper resolution of conflicting levels of control between OS and NE. Lastly, changes in the relationships between NEs and OSs (NEs more capable of taking over traditional OS functionality) are taking place; therefore, the evolution plan must be based on an analytic framework using three interrelated components: OS/NE function, data, and control.

10.6.4 New Services and Technology

Rapid deployment of new services will be essential to meet customer demands for those services in a very short interval. At the same time, new approaches to network management will be offered increasingly by switch and transport vendors. New traffic characteristics and traffic capacity management procedures are evolving with emerging technologies, so both performance measures and processes must be revised. As a result, performance management must be dynamic to accommodate customer demand for more control and more flexible service offerings.

In summary, though performance management is well defined and well established (evolving from the OAM&P framework), it must resolve a number of issues like that of integration (flow-through of data), OS/NE relationship (management of conflicting level of control), demands placed by new services and technology (new performance concepts due to state-dependent dynamic path for services), etc.

REFERENCES

[1]. American National Standard for Telecommunications—Operations, Administration, Maintenance and Provisioning (OAM&P) "Principles of functions, architectures and protocols for interface between operations systems and network elements," ANSI T1.210-1989.

[2]. TE&M Special Report, "Network management standards—standardizing standardization, May 1, 1990.

[3]. P. J. Brusil, and D. P. Stokesberry, "Toward a unified theory of managing large networks," *IEEE Spectrum,* pp. 39-42, April 1989.

[4]. Information Processing Systems—Open Systems Interconnections—Basic Reference Model—Part 4: Management Framework, USA, ISO 7498-4:1989 and ISO/IEC JTC 1/DIS 74984-4, April 1998.

[5]. R. F. Rey, Technical Editor, *Engineering and Operations in the Bell System,* ed. 2, 1984.

[6]. J. R. Cavanaugh, R. W. Hatch, and J. L. Sullivan, "Models for the subjective effects of loss, noise, and talker echo on telephone connections." *Bell System Technical Journal,* vol. 55, pp. 1319–1371, 1976.

[7]. Service Standards: LSSGR, Section 11, Bellcore TR-TSY-000511, July 1987

[8]. OTGR, Section 5.1, Bellcore TR-TSY-000820, issue 1, June 1990.

[9]. Framework generic requirements for element management layer (EML) functionality and architecture, *Framework Technical Advisory,* FA-TSV-001294, issue 1, August 1992.

[10]. CCITT Recommendation M.30, Principles for a telecommunications management network, 1988.

[11]. CCITT Recommendation M.3010, Principles for a telecommunications management network, 1992.

[12]. TA-TSY-000753, Network Traffic Management (NTM) Operations Systems (OS) Requirements, Bellcore, issue 2, December 1988.

[13]. J. L. Wang, and E. A. White, "Integrated methodology for supporting packet network performance management and planning. *Computer Communications,* vol. 13, no. 9, November 1990.

[14]. J. Shantz, "SMDS: An alternative to private network?" *Telecommun. Mag.* pp. 52–54, May 1989.

[15]. ISO/IEC JTC 1/SC 21 N3313, Information Processing Processing Systems—Open Systems Interconnection—Performance Management Working Document (Third Draft), USA, January 1989.

[16]. Y. B. Kim and A. G. Vacroux, *Real-time packet network analysis for ISO/OSI performance management,* GLOBECOM '90: IEEE GLOBECOM Telecommunications Conference and Exhibition, vol 1. pp. 397–401, 1990.

CHAPTER   11

Distributed Restoration of the Transport Network

Dr. Wayne Grover, Senior Member, IEEE
Associate Professor (EE), University of Alberta
Director of Networks and Systems Research, *TRLabs*
Edmonton, Alberta, Canada T6G 2M7

11.1 INTRODUCTION

11.1.1 Background and Motivation

At 1.2 Gb/s (SONET OC-24) a single optical fiber carries the equivalent of over 16,000 voice circuits—and there may be over 48 fibers in a cable. This large capacity, plus advantages in cost, payload flexibility, and transmission quality have made digital fiber-optic technology the medium of choice for the future transport network. Nonetheless, fiber optical transmission is a cable-based technology that is subject to frequent damage. Failure reports since 1987 include damage from backhoes, power augers, lightning, rodents, craftsperson error (wireclippers), fires, train derailment, bullets, vandalism, car crashes, ship anchors, trawler nets, and even shark attacks. Revenue losses can be as high as $75,000 per minute of outage [1]. The traffic-dynamic impact of major outages on the trunk (inter-exchange circuit) switching layer of the network and in customer data networks is also important. In sum, fiber-optic cable cuts are more frequent and result in more serious consequences, than did nonfading outages in past networks.

The technological change to fiber-based networks also coincides with a time when telecommunications is a major part of the national infrastructure. Business, medicine, finance, education, air traffic control and reservations, and police and government agencies are among those critically dependent on telecommunications integrity. Automated restoration has therefore become an essential adjunct to the deployment of large-scale fiber networks.

This has motivated research on an elegant and exciting technical development in network management: networks that autonomously reconfigure themselves to protect services from physical failures. Such networks go beyond the dedicated-spare switching of today's automatic protection switches (APS) or survivable-ring equivalents. These are networks that react organically, as a whole, self-organizing their resources in a general fashion as required to accommodate failure or unforeseen traffic shifts.

The architecture is based on digital cross-connect systems[1] (DCS), which provide grooming, provisioning, testing, bandwidth management, new facility cut-ins, broadcasting, and support of customer virtual networks [2]. It is potentially economic to add distributed restoration to the functions performed by the DCS. This may be either as a backstop to other methods (APS and self-healing rings) or, with suitably fast DCS implementations, as the sole survivability strategy. The latter is especially attractive when it is considered that a mesh-survivable network may be 3 to 6 times more efficient in terms of capacity than its ring counterparts for the same service base [3, 4, 5]. Commercial equipment supporting distributed restoration does not exist yet, but the necessary theory and algorithms are evolving rapidly. Leading manufacturers and the research arms of Canadian, US, and European network operators are studying implementation.

The concepts behind distributed restoration also have the potential for automated service provisioning, traffic adaptation, and network surveillance [6, 7, 8, 9]. Ideas such as the network constituting its own database (i.e., the data structure within which the algorithms execute is the current state of the host equipment hardware) and self-organizing distributed interaction among network elements engender a view of the network with almost biological intelligence. This type of intelligence is complementary to that which the network derives from databases and centralized computation, as in the formal advanced intelligent network (AIN) development (e.g., Bellcore SR-NWT-002247). Both forms of intel-

---

[1]For restoration purposes the DCS can be perceived as a switching machine that operates directly on the multiplexed carrier signals of the network to manage the logical configuration of the transport network, without regard to the individual calls or other services borne by those signals. The DCS-3/3 (or SONET equivalent) is considered the primary vehicle for distributed restoration of the backbone transport network but the methods can be adapted to any level at which a restoration switching function is implemented, e.g., DS1, or STS-n, or OC-n on an optical cross-connect. DCS machines are discussed further in [57, 58, 59].

ligence can coexist in the future network. As in humans, autonomous brain-stem functions and higher cerebral functions complement each other.

11.1.2 Chapter Objective

The chapter introduces the basic concepts, research progress, and current issues in distributed restoration. The coverage is primarily based on published sources from 1987–1992 and the author's own research. It includes problem definition, terminology, basic theory, performance measures, and a review of some distributed restoration algorithms (DRA). It also offers approaches or alternatives for a number of implementation issues. Not all topics on distributed restoration can be covered. Optimized capacity planning for distributed restoration [3, 10, 11], asynchronous transfer mode (ATM)-based distributed restoration [12], economic assessment of distributed restoration [13], and coordinated operation ring and mesh survivability techniques [5, 14] are beyond the scope of this chapter.

11.2 RESTORATION SPEED AND NETWORK CAPACITY

What speed and level of restoration coverage is required? Some customers are critically dependent on uninterruptible service and will pay a premium for 100% restoration in not more than 50 msec. This is the characteristic time for a dedicated 1:1 automatic protection switching (APS) action. All services continue uninterrupted except for a momentary "hit." Bank teller machines and large financial trading houses are examples of users that need this level of service. A large class of other applications (e.g., data networks, 1-800 services, and some voice business users) need 100% restoration but can tolerate interruptions of seconds to minutes. Another large sector, residential customers, can tolerate outages of up to 30 minutes, especially if it lowers the cost of service. Nonetheless, if large numbers of active subscribers are simultaneously disconnected, their collective attempt to reestablish a connection can cause undesirable network transients and congestion.

A detailed assessment of the impact of various outage times is found in [15]. With an outage of between 50 msec and 200 msec there is a less than 5% probability of connection loss in switched services and a minimal impact on the CCS-7 signalling network and on cell-relay type services. From 200 msec to 2 seconds of outage there is an increasing chance of switched-circuit disconnect where older channel banks are still employed. At 2 seconds of outage a major event occurs—all circuit-switched connections, private line, nxDS0, and DS1 dial-up services are disconnected. If the outage continues to 10 seconds, most voice-band data modems time out, connection-oriented data sessions may be lost depending on the session time-out, and X.25 sessions may time out. Other disruptive effects on electronic funds transfer and high priority and high speed data communications are outlined in [15]. With outages greater than 10 seconds all sessions are considered to be disconnected. Data networks and voice callers begin attempts to reestablish connections and applications, with significant transient traffic implications for the network. If the outage lasts 5 minutes or more, the congestion experienced by digital switching machines can be severe. An outage greater than 30 minutes may require reporting to the Federal Communication Commission (FCC) and noticeable social and business impacts are considered to result at 15 minutes of outage.

It is easy to see that a significant goal for restoration systems is at the 2 second level, which is called the connection-dropping threshold (CDT).[2] There is a quantum improvement in service quality for all users and in the stability of the trunking and signalling networks if restoration can be achieved in under 2 seconds. Voice users experience a gap in conversation rather than being returned to dial tone. Almost all data session protocols can survive an outage of less than 2 seconds, so dial-up data connections and data network applications do not collapse and terminal users are not involuntarily logging out.

11.2.1 Centralized Control

Until recently facility restoration in many networks was a manual process of patching up rearrangements at passive DSX-3 cross-connect panels. The reroutes used the "extra-traffic" inputs to the spare-span input of transmission systems where possible, but also used agreements to access satellite transponders and certain amounts of precommissioned spare capacity [16]. Average restoration time was 6 to 12 hours, much of which was in travel time to unmanned transmission hubs [17].

Electronic DCS 3/3 cross-connects (and SONET equivalents) have replaced the DSX-3 rearrangement process under centralized control. AT&T's FASTAR [18] and DACScan [19] systems are the first fully automated, large-scale mesh restoration systems operational in a real network and represent the state of the art in centralized control. FASTAR can restore 100 DS3s in 5 minutes (mid- 1992 target) [20]. Other central-

---

[2]Readers may ask why the network is so eager to disconnect everything. After 2.5 ± 0.5 seconds of trunk group outage, it is standard for the switch to busy-out all failed trunks. This avoids new call setup attempts into failed trunk groups and prevents the switching machines from processing noise-like signalling states on the affected trunks. With older equipment this occurred with as little as 320 msec of outage.

ized control systems are described by Vickers [21], Fluery [22], and Sutton [23].

11.2.2 Distributed Control

It may, however, be unfeasible to advance the technology of centrally controlled restoration to the 2 second target. In this regard, distributed control is an alternative that can enter the domain of split-second restoration speeds while retaining the networking efficiency of centrally controlled mesh-restorable networks. Several studies have confirmed the initial findings that efficient multipath rerouting plans can be developed through distributed control in under 2 seconds [24, 25]. Distributed control also offers architectural simplifications. It does not require centralized computation or a real-time database image of the network, nor is there a need for a database at the nodal sites or for physically diverse telemetry arrangement. Networks with these properties are called *self-healing networks* (SHN) [24, 25, 26]. The nodal elements of an SHN are most often considered to be DCS nodes, but add-drop multiplexers (ADMs) can, in principle, interwork with DCSs in an SHN [5,37][3].

11.2.3 Automatic Protection Switching and
Self-healing Rings

Automatic protection switching (APS) systems restore service by switching to a dedicated standby system. To protect against cable cuts, 1:1 APS is required and the standby must be diversely routed. Self-healing rings (SHRs) vary in detail, but can be visualized as an extension of either 1:1 or 1:N APS. One class of SHRs, unidirectional SHRs (U-SHRs), use counter-rotating duplicate signal feeds to create a 1:1 receive selection situation at each location. The second class, shared protection (SP) rings, are analogous to 1:N APS systems. SP rings share a common protection ring onto which traffic is substituted in the reverse direction at nodes adjacent to a failure.

1:1 APS and U-SHRs can realize 50-msec restoration times, which is the time typically required to detect signal failure and close a tail-end transfer relay. 1:1 and U-SHRs effectively have a permanent head-end bridge. 1:N and shared protection rings (SP-rings) more typically realize

---

[3]An ADM is the network element of which rings are comprised. It can be thought of as a functionally specialized form of DCS machine. Its specialization is that it interfaces to exactly two logical spans. It provides cross-connection between these two spans and provides local add-drop access, as the DCS does for a generalized number of spans. This implies that ADMs need not necessarily work only in rings. Rather, as DCS machines for two-span sites, they can interwork directly with DCS nodes in a single self-healing network [5, 14, 61].

100- to 150-msec switching times because of the need for signalling and head-end bridge setup. SHRs are treated extensively in [27].

11.2.4 Capacity Efficiency

A basic motivation for mesh restoration is to exploit the physical route diversity of real networks for capacity-efficient survivability. For full restoration with APS systems 100% redundancy is required. Rings require 100% or more redundancy: in the U-SHR, the bandwidth in each counter-rotating direction must exceed the sum of all end-to-end demands in the ring. In the SP-ring, sparing for all spans of the ring is set by the largest aggregate demand routed between any two adjacent nodes (except, strictly, for the sparing on the span that has the largest working capacity).

In contrast, mesh-restorable networks under centralized or distributed control are based on generalized rerouting over all diverse routes of the network. Figure 11-1 illustrates the generality and flexibility of the rerouting patterns in mesh restoration. Spare capacity on each span contributes to the survivability of many other spans. Network redundancy is not dedicated to restoration of one span or ring. Such networks are called *mesh-restorable*, not to imply that the network is a full mesh, but to reflect the ability of the routing mechanism to exploit a mesh-like topology.

Figure 11-1 also shows node-oriented considerations that give a lower bound on the redundancy (total spare to working capacity ratio) needed for full restorability of any span cut. Egress for rerouting of all failed capacity on any span is possible if the sum of the spare capacity on the $(n-1)$ surviving spans equals or exceeds the working capacity on the failed span. If the remaining body of the network is not limiting in path number between this node and the other end of the failed span, and if the same nodal conditions apply on average at the adjacent node, then full restoration is possible in the limit with network redundancy of $1/(n-1)$. Studies of survivable mesh-network design have confirmed that on a range of study networks [3, 4,], the practical redundancies achievable for full restorability are often within 10 to 30% of this lower bound.

By comparison $n = 2$ for every element in a logical ring, regardless of how many spans physically exist at the nodes transited by a ring. Redundancy of 100% is therefore the lower bound for survivable rings. Redundancy is, however, typically much higher than 100%. The 100% minimum requires ideal working capacity balance in each span of a ring. Even in this case, the 100% redundancy is with respect to working capacity whose routing follows the defined ring(s). It is not with respect to a shortest-path mapping of demands through the native route structure of the network, prior to the ring overlay. Mesh networks not only achieve lower redundancy levels than rings but this is with respect to a working base that is already an efficient shortest paths placement for working capacity. The

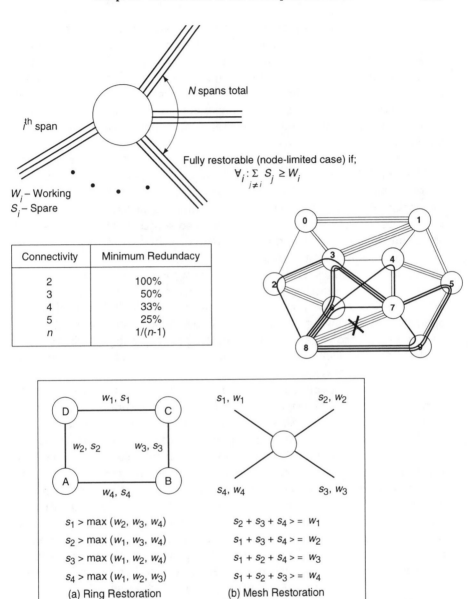

Figure 11-1   Illustration of Mesh-rotation and Node-based Lower Bounds on the Network Redundancy Required

total capacity (working plus spare) for mesh networks is therefore generally several times less than that of a ring-based network for the same total service base [3, 4, 5].

This does not necessarily mean better cost efficiency for mesh networks, particularly with current electrically interfaced DCS machines, because the DS3 interface costs are high. The capacity difference is, however, of such a magnitude (three to six fold in typical networks) that it signals the opportunity for developers to convert the intrinsic efficiency of mesh networking into cost savings for telephone companies. A key to this conversion will be development of integrated optical terminations on the DCS [5].

Many dense metropolitan and national long-haul network examples can be found that have average nodal degrees of three or more in existing route topologies. Viewed in the perspective of mesh restoration, this topological diversity is a real, though intangible, asset.[4] To benefit from this diversity, the telephone companies need solutions that are designed to exploit connectivity as a natural resource of their networks.

In continental or national scale networks, distance places a premium on capacity efficiency and mesh networking can economically prove-in against rings. In metropolitan networks, distance-related line costs are less of a leverage, but the relative simplicity and flexibility of planning, operating, and growing a single mesh network may become more attractive as experience is gained with operating multiple logical-ring subnetworks and as the DCS platforms are optimized for interface cost reduction and speed of restoration.

Mesh networks are also inherently flexible in that traffic need not be accurately forecast. Survivable mesh networks can conform to arbitrary patterns of service growth with a single algorithm to update the survivability design in the least change manner [3, 11]. Mesh networks can also support service growth and survivability from a common pool of network spare capacity. Algorithms are under development to support this dual use management concept by exploiting routing and capacity modularity effects to defer the time at which restorability updates are triggered [3, 11]. In this way, survivability and service provisioning may be jointly managed and efficiently supported from a single investment in network spare capacity. In contrast, SP rings require unused capacity on the working fibers for service growth, and survivability requires a second reservation of spare capacity on another fiber or channel set.

Figure 11-2 summarizes the unique position of distributed restoration in terms of survivable network efficiency and restoration speed. It has the potential to offer the best of two existing worlds; the capacity efficiency of meshes and the CDT-beating speed that only SHRs and APS systems can currently provide.

---

[4]An analogy is that route diversity is a natural resource in the same sense that synchronous orbital slots are also a valuable, although intangible, natural resource.

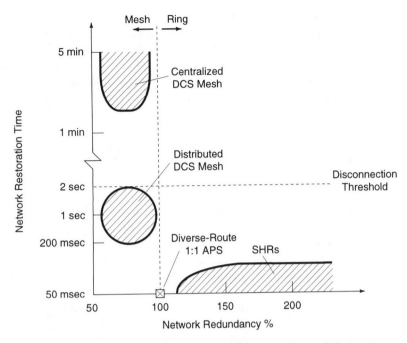

Figure 11-2   Position of Distributed Mesh Restoration in Terms of Restoration
Time and Survivable Network Redundancy

11.3 THEORY AND CONCEPTS

11.3.1 Philosophy and Key Ideas of Distributed Restoration

The central idea of distributed restoration (DR) is to recognize each DCS machine as a computer of considerable processing power embedded in a fabric of multiple communication links. Viewed this way, the DCS-based transport network is a distributed, irregularly interconnected, multiprocessor on which to implement a specialized computational problem, the computation of path-sets within its own topology. This allows computation without any conventional database requirements. Rather, the network in which nodes are embedded forms the database within which the distributed computation is performed. Every node will perform in an apparently isolated manner, with no global knowledge of the network in which it is part. The independently deduced cross-connection decisions of each node will, however, collectively constitute efficient multipath rerouting plans.

At the most general level of abstraction, mesh restoration requires three conceptual steps: (1) accessing a network description, (2) computing a rerouting plan, and (3) deploying cross-connection actions to put the plan into effect. Abstractly, this remains true for DR, except that; (a) the network is the database that is (indirectly) accessed at the instant of failure, (b) the network is the computer on which the reconfiguration algorithm runs and, (c) the result is in place, i.e., the solution in each node is exactly the command set that would have been downloaded to it had the plan been centrally computed. The ideal properties for such a scheme are:

(a) no node has, obtains, or depends upon, a global network description;

(b) every node acts autonomously and has no dependency on external telemetry to a supervisory or coordination center;

(c) each node uses only data that are obtained locally from links impinging on that node; and

(d) cross-connections computed in isolation at every node coordinate with the connections made autonomously by all other nodes, collectively synthesizing a complete and coherent routing plan at the network level.

The aim is not, however, to replace centralized supervisory control over the network. Rather, the DR system becomes a real-time assistant to centralized administrative systems. Network elements will react autonomously to the emergency but then use normal telemetry to a network operation center (NOC) for ratification of restoration actions and for all other normal provisioning and maintenance purposes.

11.3.2 Terminology

A *link* is an individual bidirectional DS3, STS-1, or STS-n carrier signal between two adjacent nodes. For example, one bidirectional OC-12 fiber system implements 12 STS-1 links. In general, a link is whatever unit of capacity that cross-connection is performed to make restoration paths. This could be an OC-n optical carrier in the case of an optical cross-connect switch. A *span* is the set of all links in parallel between two adjacent nodes, whether borne on the same transmission system or not. A span could be comprised of 12 DS3 links carried on an asynchronous 565 Mb/s system in parallel with 24 STS-1 links borne on a SONET OC-24 system. This admits the possibility of partial span failures if only one of the parallel transmission systems fails and the prospect of restoration of mixed signal types. The terms *line* and *channel*, in [9] and elsewhere, are analogous to span and link, respectively.

Spans are identified by the names of their end nodes, e.g., (A, B), or, when spans of a network have been numbered as $S[i]$ where S is the set of spans of the network. Spans are the basic elements that are concatenated to form a *route* through the network. A concatenation of links through the network is called a *path*. Every path has a route, which is the series of spans over which it is realized. A route can support several paths, but every path has only one route. Note that a link is to a path as a span is to a route, the former identifying individual capacity units of the network, the latter referring to the topology of the network only. The logical length of a path and length of a route is the number of spans in series along the route or path.

A *working link* is a link that carries live traffic and is part of some working path through the network between origin and destination nodes. Every working path originates at an origin node where an individual multiplex signal is first created by equipment that sources that signal type (e.g., M13 mux, DS3 video codec); working paths terminate at a destination node where the working path signal is demultiplexed or otherwise terminated. A *spare link* is a link that is in a fully equipped and operational condition but is not in service and is not part of any path through the network. In all transmission respects a spare link is identical to a working link and can enter working status by cross-connection to a live signal within a DCS.

Adjacent nodes are those that are directly connected by a span. The *custodial nodes*[5] for a given span cut are those that were adjacent to a span before failure of that span. A *restoration path* is assembled from spare links in the network to logically substitute for part or all of one working path that was severed by the failure. A *simple graph* has at most one link entity between nodes. A *multigraph* can have any number of links in parallel between nodes.

11.3.3 Span and Path Restoration

In span restoration, replacement paths substitute for the lost direct span between two custodial nodes. In path restoration, replacement paths are effected via origin-to-destination (end-to-end) rerouting for every working path affected by the span cut. The difference is illustrated for a single path in Fig. 11-3. Figure 11-3 also shows the range of rerouting options that exist between pure span and pure path restoration, depending on how far back from the failed span the route of the restored path departs from the prefailure routing. In general, the further one backs away from the failure, the more rerouting options exist and the more likely one is to avoid the back-hauls apparent in Fig. 11-3(a).

On the other hand, span restoration currently offers greater speed, ease of implementation, and fault isolation capabilities because the problem is to find a replacement path set in which all paths have the same end-nodes—i.e., to find k paths between two nodes where k was the number of working links on the failed span. By comparison, path restoration requires identification of the different origin–destination nodes of each

---

[5]This term arises because these nodes play special roles on behalf of the network in initiating and (indirectly) coordinating the actions of remote nodes for restoration.

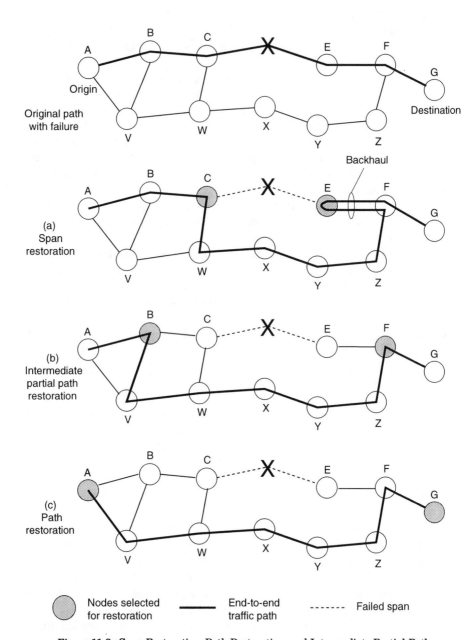

Figure 11-3 Span Restoration, Path Restoration, and Intermediate Partial-Path Restoration Concepts

path that transited the failed span and production of $k_{x,y}$ end-to-end replacement paths for each demand pair (x,y) that was routed over the failed span.

Path restoration therefore converts a single node-pair routing problem into (worst case) an all-pairs routing problem still requiring several paths per node pair, although each is individually less than k. In return for greater complexity, however, path restoration will often require less spare capacity for a given level of survivability. (It is not generally known yet, however, how much less capacity will be required compared with an optimized span restorable network, but studies are underway.) To obtain the best possible efficiency with path restoration, the placement of spare capacity must reflect optimal path restoration. This implies finding the optimum permutation of path restorations for each end-node pair, i.e., considering all orderings of individual node pairings. This makes optimal path restoration considerably more complex than optimal span restoration. In practice, a nonoptimal form of path restoration can be effected by applying a process that achieves span restoration in an arbitrary order to each individual end-node pair in the path recovery formulation.

At this time, research on optimized path restoration and design of path-restorable networks is just beginning. References [2, 4, 47] refer to path restoration but this is in the same nonoptimal sense as multiple reprovisioning events using a span restoration protocol [48]. The optimization issues to be solved for path restoration are outlined in Section 11.9. We concentrate on span restoration for the present chapter as it is the most developed of the two topics and many concepts from span restoration will later transfer to path restoration.

11.3.4 Network Recovery from Node Loss

There is another context in which path restoration is of interest—network recovery from node loss. Because path restoration operates end-to-end for every disrupted demand entity, it can restore paths that transit a failed node or react to arbitrary combinations of node and span failure. DR for node recovery is the primary orientation in [2, 28]. There it is assumed that span failures are addressed by APS or SHR systems and that DCS-based restoration is a secondary response to protect against outages of the DCS nodes that connect rings but are not protected within them.

Despite the famous Hinsdale central office fire, node destruction actually occurs far less frequently than cable damage. A consequent issue is whether to optimize DR for maximum speed and routing efficiency in response to the much more frequent cable cuts, or to compromise performance in that role to some extent to incorporate node recovery. The degree of physical protection is inherently greater for nodal equipment than for cable plant. Moreover, services can be entirely protected by rerouting in response to cable failures, but cannot be for node failures. Only the through traffic can be protected by restoration. Good site security and software reliability are far more important and effective for defense of nodal equipment than is reliance on network rerouting. This philosophy would lead us to optimize performance and cost of DR for span restoration and simply ensure that the restoration rerouting process can also be applied in a path recovery context for a possibly slower, but still automated, network response to a node loss.

11.3.5 Relating Restoration to Call and Data Routing

Many communications engineers are familiar with packet data or call routing processes, so it is useful to compare DR with those problems. There are a number of routing algorithms for packet data networks and alternate-routing call networks that realize distributed routing but use essentially centralized computational methods to derive the routing plan. These schemes either (1) perform routing calculations in a central site using a global view followed by the downloading of results to each node [29, 30] or, (2) include a scheme of mutual information exchange (or topology orientation) through which all nodes of the network first build a global view, then execute an instance of a centralized algorithm and extract the part of the global routing plan that applies to their site [29, 31]. In either approach the final routing policy is, indeed, put into effect in a distributed manner, but there is no truly distributed *computation* of the routing plans. Thus, there is a distinction between distributed implementation and distributed *computation* of a routing plan. The real-time demands of the restoration problem motivate the search for a scheme where computation and routing deployment are both part of a single, truly distributed process involving only isolated action by each node.

The algorithms that determine routings that optimize delay or blocking in packet and call routing can be quite sophisticated. Ultimately, however, the solution is specifiable for any origin, destination pair (AB), in the form of one (least cost) route in a simple network graph:

$$r_{AB} = (A, u_2)\,(u_2, u_3) \ldots (u_{n-1}, B) \tag{1}$$

where u_i is a node of the network. r_{AB} specifies a route in a simple graph that is the sequence of nodes through which packets will be relayed. There is no (direct) constraint on the simultaneous use of any link in this route by different node pairs. All routings can reuse and access any link through delay queuing. In dynamic call routing, the call attempt replaces the data packet; trunk groups, accessed in blocking contention rather than delay queuing, replace the links of the packet network problem. In this case, r_{AB} specifies the sequence of trunk groups over which a new call attempt should be successively offered for call establishment. Otherwise, the situation is the same in terms of the information required to specify a routing solution and the lack of constraints on other simultaneous routings. The call attempt is routed nondisjointly with other call attempts through a simple graph comprised of all the logical trunk groups in the network.

In contrast, for restoration, the link entities for routing are numerous links in parallel in each span, and each bears one multiplex carrier signal. Restoration of a span-cut therefore requires multiple unique replacement paths between the custodial nodes. The target number of paths is the number of working links failed. Each restoration path must be a completely closed transmission path dedicated to logical replacement of one failed link. The routing of the replacement paths as a group must be consistent with discrete capacity limits on each span. Unlike packets or calls, there is no equivalent to blocking-based, contention-based, or delay-based shared access to links or spans. A carrier signal completely fills a restoration path in space, time, and frequency. The restoration path-set must therefore be: (1) mutually link-disjoint throughout and, (2) collectively consistent with the whole number link capacities of each span of the network. In contrast to packet and call routing where a suboptimal routing state may mean some excess delay or blocking (relatively graceful forms of degradation), the penalty for failure to satisfy (1) and (2) in restoration is abrupt and complete outage for the carried services.

These attributes are relatively atypical of traditional routing problems. Using a transportation routing, call routing, or packet routing analogy, the equivalent requirement would be that once a truck (or a packet, or a call) is routed over lanes x, y, z, of streets r, s, t, then none of those street lanes can be used again by any other truck (packet or call). That is absurd for those problems but it is a true analogy for the restoration problem. The difference is that packets, calls, trucks, letters, trains, etc. do not fill the medium of the routing path simultaneously in space, time, and frequency, as carrier signals do. Restoration of the transport network is therefore more like routing multiple fluids (or flows) through a network of pipelines, but without allowing any mixing (and without any reservoirs). But even this is not a complete analogy because flow omits any aspect of the need to also stipulate precise interconnections between discrete links at the flow merging points. Restoration must accordingly be formulated in a discrete multigraph network representation, not a simple graph with flow quantities on each span. A summary of comparative aspects of call, packet, and restoration routing is in Table 11.1.

TABLE 11.1 Comparative Aspects of Packet, Call, and Restoration Routing Problems

	Routing in Packet Networks	Alternative Traffic (Call) Routing	Span Restoration
"Link" Resource	logical "pipe" between nodes	logical trunk group	digital carrier signal e.g., DS-3 or STS-1
# "Links" between nodes (type of network graph)	1 (simple)	1 (simple)	$k \geq 1$ (multigraph)
mode of access to links	queued (delay engineered) (time shared)	full availability (traffic engineered) (space shared)	nonshared (dedicated in time and space)
sharing of links among paths?	yes	yes	no (fully link disjoint paths)
grade of service metric	path delay	call blocking	network availability
application	data communication	voice telephony	physical bearer network for all voice and data services
impact of failure	increased delay	increased blocking of originating calls	total loss of all calls in progress plus service outage

11.3.6 Routing Formulation for Span Restoration

A solution to the restoration problem for span (A, B) must comprise the following information, and the solution is subject to the constraints that follow thereafter:

$$k_{AB} = min \ (W_{AB}, T_{AB}) \tag{2}$$

$$\mathbf{R}_{AB} = [\mathbf{r}_1 \ . \ . \ .\mathbf{r}_k] \tag{3}$$

where W_{AB} = number of working links on span A–B,

T_{AB} = number of link disjoint paths that are topologically feasible between A and B over spare links of the network,

$\mathbf{R}_{AB}$ = a set of k_{AB} restoration paths, of various lengths, comprised as follows:

$$\mathbf{r}_1 = [A, u_{<1,2>}, x_{<1,1>}], \ [u_{<1,2>}, u_{<1,3>}, x_{<1,2>}] \ . \ . \ . \ [u_{<1,n(1)-1>}, B, x_{<1,n(1)>}]$$

$$. \tag{4}$$

$$\mathbf{r}_k = [A{<}u_{<k,2>}, x_{<k,1>}], \ [u_{<k,2>}, u_{<k,3>} x_{<k,2>}] \ . \ . \ . \ [u_{<k,n(k)-1>}, B, x_{<k,n(k)>}]$$

where

$$[u_{<i,q>}, u_{<i,q+1>}]$$ = the qth span in the ith restoration path.

$x_{<i,q>}$ = the individual link in the span $(q, (q+1))$ allocated to the ith restoration path.

$n(i)$ = the logical length of the ith restoration path

$[u_{<i,q>}, u_{<i,q+1>}, x_{<i,q>}]$ specifies an individual link of the network in terms of its span, $[u_{<i,q>}, u_{<i,q+1>}]$, its individual link number within the span, $x_{<i,q>}$, and its use as the qth link in the ith restoration path. $\mathbf{R}_{AB}$ is called the *restoration plan* for nodes AB. By detailing all the information required for a restoration plan, equation (4) emphasizes that restoration requires specification of exactly which links in each span to connect together. This is quite important in the actual implementation of DR. Also, because individual cross-connection information must be specified at each node for each replacement path, the terminology and methods in restoration must resolve individual links within spans as the objects of manipulation rather than having only one link between nodes with an associated capacity or flow value. A purely flow-oriented approach to the routing problem does not encompass the detailed cross-connection information that must be generated by a restoration rerouting mechanism. The following constraints must also be simultaneously satisfied by the set of paths in any restoration plan:

Constraint 1: Link Disjointness. Every link used in one path of the restoration plan must not appear in any other restoration paths of the plan:

$$if\ [u_{<q>}, u_{<p>}, x_{<i>}] \in \mathbf{r}_j\ then\ [u_{<q>}, u_{<p>}, x_{<i>}] \notin \mathbf{r}_z\ \forall\ z \neq j \tag{5}$$

Constraint 2: Capacity-consistency. A capacity-consistent set of paths is one in which the sum of all paths that use the same span does not exceed the number of spare links on the span, for every span in the network. Any feasible restoration plan must satisfy S simultaneous constraints on the number of paths routed over any span where S is the number of spans in the network:

$$\forall\ m = 1 .. S\ ;\ \sum_{i=1}^{k_{AB}} |\ [\mathbf{r}_{AB,i}] \wedge [S(m)]\ | \leq C[m] \tag{6}$$

where $[-]$ denotes a set formed out of the named span(s), $\wedge$ is set intersection, and $|-|$ is the number of elements in a set. $C[m]$ is the number of spare links on span m.

Constraint 3: Mapping Preservation. In addition to finding the required number of paths between custodial nodes, those nodes must share a common scheme of ordering the replacement paths to avoid transposition in the traffic streams substituted over the restoration paths. If $M(\mathbf{R}_{AB}) \rightarrow [I]$ represents any function that provides a one-to-one mapping of each restoration path of a restoration plan onto one element of the set of integers $[I]$, then for any set of restoration paths for the relation AB, two mapping functions $M_A(\mathbf{R}_{AB})$ and $M_B(\mathbf{R}_{AB})$ must be obtainable at the nodes A and B, respectively. The constraint that must be satisfied is:

$$M_A(\mathbf{R}_{AB}) = M_B(\mathbf{R}_{AB}) \qquad (7)$$

Constraint 4: Collective Shortest Paths. If the operator L ($\mathbf{r}_{AB,j}$) obtains the number of spans in a restoration path, then the restoration plan $\mathbf{R}_{AB}$ comprised of k_{AB} restoration paths must also satisfy:

$$\sum_{j=1}^{k_{AB}} L\left(\mathbf{r}_{AB_j}\right) = L_{\min} \qquad (8)$$

where $L_{\min}$ is the smallest sum of the lengths of all AB restoration path sets that contain k_{AB} paths.

Constraints 1 to 3 are aspects not found in packet or call routing but are essential for the basic functioning of restoration. Equation (5) says the paths created must be fully link-disjoint, equation (6) says they must coordinate to respect the finite span capacities present, equation (7) says they must share a common end-to-end scheme for traffic substitution, and equation (8) says that wherever several path-sets containing k_{AB} paths exist, the restoration path-set selected must be that with minimal total length.

In an equivalent formulation for path restoration, the path-set equation (4) is still of size k but it is comprised of h subsets of paths having different end nodes. Constraints 1 and 2 have a similar form but now pertain to disjointness and capacity consistency over the set of paths not having common end-nodes. Constraint 3 only applies to each end-to-end relation on which more than one path is restored. Constraint 4 does not change in form but it becomes the source of a major difference between span and path restoration in terms of the complexity to satisfy this constraint.

11.4 ALTERNATIVE ROUTING CRITERIA

Having developed a specification for the path-set required, we now consider known routing algorithms that could serve as reference standards against which to assess the path-sets deployed by a DRA.

11.4.1 Complexity Considerations

Table 11.2 shows a progression in computational complexity as we move through some increasingly sophisticated criteria that could be considered as routing characteristics for DR. Equations (9) through (13) are the theoretical time complexity of the respective problems for computation on a single processor [32]. The complexities listed would not necessarily translate directly into the complexity of a distributed implementation where the parallelism of processors in the network is exploited. Nonetheless, the single-processor complexity of each basic formulation gives some insight into the relative difficulty in attempting to implement DR to satisfy these various approaches.

TABLE 11.2   The Progression of Theoretical Complexity Depending on the Routing Formulation for Restoration

	Problem Formulation	Computational Complexity	
1)	Find a min cost alternate route in trunking network	$0(n^2)$	(9)
2)	Span restoration via k-shortest L.D. replacement paths	$0(k_s \cdot n^2)$	(10)
3)	Span restoration via min cost max flow	$0(n^3)$	(11)
4)	Path level restoration of a span cut via successive shortest paths	$0\left(k_p \cdot n^2 \cdot \dfrac{n \cdot (n-1)}{2}\right)$ $= 0(k_p \cdot n^4)$	(12)
5)	Path level restoration of a span cut with min total path length	$0\left(k_p \cdot n^4 \cdot 2\,\dfrac{n \cdot (n-1)}{2}\right)$	(13)

Equation (9) is the complexity of finding a single shortest replacement route in a simple graph. It can be solved with Dijkstra's shortest path algorithm [33] in $O(n^2)$ (or, in $O(n\log_2 n)$ time using a binary min-heap [34]. For restoration by this criteria, the single route would have to have a capacity at least equal to the failed span. This is the simplest approach but would require the most spare capacity of any mesh restoration scheme because all of the required capacity for each span would have to be in place on some single route.

Equation (10) is for span restoration via k-successively-shortest link-disjoint paths (KSP), found by repeated application of a single-shortest path algorithm. There is no increase in the exponent of complexity for this step. The worst case increase is a multiplier equal to the average number of working links per span. The next step, Equation (11), represents span restoration with min-cost maximum flow routing. Algorithms for max flow are of $O(n^3)$ to identify the maximum feasible flow between two nodes [35] and leave the path-set that realizes this flow to be found as a separate problem.

Computational complexity rises another order when nonoptimal (ad-hoc) path restoration is considered. Equation (12) reflects that one span cut may effect working paths on (worst-case) all demand relations in the network, with each requiring several replacement paths. This is equivalent to $n(n-1)/2$ simultaneous instances of the individual KSP problem in equation (10). However, this is a naive form of path restoration because the total number of restoration paths depends on the order of satisfying the $n(n-1)/2$ KSP subproblems. A result corresponding to equation (12) would be optimal neither in path number or total path length and would not reliably yield a predictable maximum number of paths.

For path restoration with assured maximization of the total restoration capacity, the problem becomes a case of multicommodity maximum-flow (MCMF) and is exponential in complexity. Equation (13) reflects the process of considering all permutations of individual demand relations as necessary to find the composite path-set that is globally optimal. MCMF may be solved in practice with linear programming (LP), but LP is of equivalent exponential worst-case complexity (although average-case polynomial-time LP algorithms exist). The two most pragmatic routing characteristics that could satisfy the requirements just formulated are, therefore, min-cost max-flow (11) (MF) and k-shortest link-disjoint paths (10).

11.4.2 Maximum Flow

The maximum flow routing criterion is based on the min-cut max-flow theorem [36]. In our context, the theorem states that the greatest number of restoration paths that can be found between two nodes (A,B) is equal to the smallest sum of spare capacities on any cutset of the network that separates A and B, excluding the capacity on the failed (A,B) span. A cutset is any set of spans which, if removed from the network, divide the network into two disconnected subgraphs. Some cutsets relevant to the restoration of span (A,B) are shown in Fig. 11-4. The cutset with the smallest sum of spare capacity acts like the minimum cross-sectional area in a pipe: it dictates the maximum rate of flow through the entire network

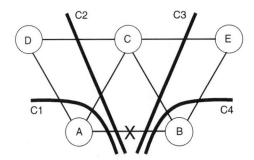

Figure 11-4 Cutsets of a Network
Relevant to Restoration
of Span (A-B)

between (A,B). Thus an MF calculation determines the highest number of restoration paths feasible by any routing mechanism and the MF result is optimal for span restoration. The complexity for computing the MF capacity is $O(n^3)$ [35], however, and does not yield a path-set for use in restoration, only the flow value of the best path-set that can exist. MF therefore requires a secondary computation to find an implementable path-set that realizes the MF capacity.

11.4.3 k-Shortest Link-Disjoint Paths

KSP is computationally simpler than MF and yields a detailed path-set ready for deployment. But the path-set may not always yield the MF capacity. The KSP path-set is comprised of all paths on the single shortest route, all paths on the second shortest route not reusing any links from the first, all paths on the third shortest route not using any links from the first two, and so on. It has been verified that the path-sets deployed by the self-healing network (SHN) protocol are equivalent to centrally computed KSP, [25, 38, 39] although they are formed by a parallel distributed inter-action, which is different from how a centralized KSP algorithm works.

Because KSP yields an implementable path-set and is computation-ally simpler than MF, we are interested in it as a possible criterion for mesh restoration systems. However, we need to know how close KSP rout-ing is to the MF quantity that represents the strictly optimal result for span restoration. In this regard, it can be shown that when KSP fails to match the MF capacity, it is because of an instance of the general situation that is outlined in Fig. 11-5 [40]. At most, two edge-disjoint paths can be created between nodes 1 and 4. One path is 1-5-3-4 and the other is 1-2-6-4. But if the path 1-2-3-4 is chosen, then only one edge-disjoint path from 1 to 4 can be created in total. KSP operating on logical hop distances may make the latter choice because it has no basis for preference between the two equal (logical) length path choices. And if the paths via 1-5 and 2-6 to node 4 are longer than the 1-2-3-4 path, KSP will always make the subop-timal choice in the particular topology because it must include the

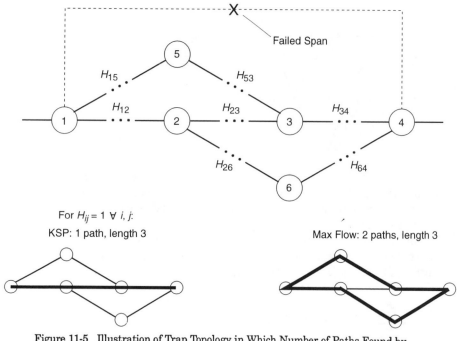

Figure 11-5 Illustration of Trap Topology in Which Number of Paths Found by
Max Flow and KSP May Differ

shortest route. The MF solution requires that we forego the shortest route
to get two individually longer paths. The issue is therefore how often will
this trap arise in realistic networks?

This question was treated in [40]. KSP restoration path-sets were
compared with the max-flow quantities in 300 representative 50-node
transport network models that were randomly generated [41]. The test
networks varied systematically in average nodal degree (D) and in
average span locality (L). L is a measure specific to [40] for reflecting the
propensity of out-of-the-plane spans in a network (i.e., topological cross-
overs), as this has to do with the likelihood of forming a subtopology
equivalent to Fig. 11-5. $L = \infty$ means any node is as likely to have a direct
span to any other node at any distance away; $L = 1$ implies that spans
exist only between nodes that are geographically adjacent in the grid
space in which the trial networks were synthesized. Joint parametric
variations of L and D ensure that the sample space of test networks
covered the range of L and the D characteristics observed in real transport
networks.[6] A representative distribution of nodal degree and some sample

---

[6]Published examples of Canadian, US, UK, German, Indian, and French national
networks, plus several metro networks, are imitated well by random graphs synthesized with
$2.5 < D < 5$ and $1 < L < 1.5$.

networks are shown in Fig. 11-6 to convey that the random networks for these tests are topologically similar to real transport networks.

In trials of over 2,600 span-cuts in the 300 trial networks, the total number of restoration paths found by KSP was over 99.97% of the MF value. The difference is therefore trivial in practice. Given provisioning module effects, no real network will have such a precise provisioning of spare capacity that it would be restorable under MF but not under KSP with a fraction of a percent less routing efficiency.

Because the difference is so small, theoretical substantiation for this result was also sought in [40]. One of the reasons the trap topology in Fig. 11-5 is so infrequent is that the paths for span restoration are always between two adjacent nodes of the network. This delimits the chances of the trap existing. The first-order trap manifestation (i.e., $H_{ij} = 1$ in Fig. 11-5) requires a precise relationship among 7 spans that also must not be short-circuited by surrounding path options, which can defeat the trap effect. Higher-order manifestations are individually even more particular in construction, although the total number of possible formations is greater. The probability of effective trap formation (i.e., not short-circuited) was analytically bounded and is consistent with the experimentally found peak for MF-KSP differences at around $D = 3$. With a lower degree, the density of spans is too sparse to form the trap frequently. On the other hand, at higher D the trap subnetwork is surrounded by additional topology that defeats the trap. KSP is therefore an effective and practical objective for DR systems.

11.5 PERFORMANCE MEASURES FOR DISTRIBUTED RESTORATION

11.5.1 Span Restorability and Network Restorability

In a network operational context, the utmost concern is with the proportion of failed working capacity that is restored. The restorability (or restoration ratio) of an individual span, i, having w_i working links is defined as:

$$R_{s,i} = \frac{\min(w_i, k_i)}{w_i} \tag{14}$$

where k_i is the number of restoration paths feasible by DR. Whenever $k_i \geq = w_i$, $R_{s,i} = 1$. Here, *restorability* is equivalent to the *conditional survivability* used in [13]—i.e., it is the fraction of transport that survives the

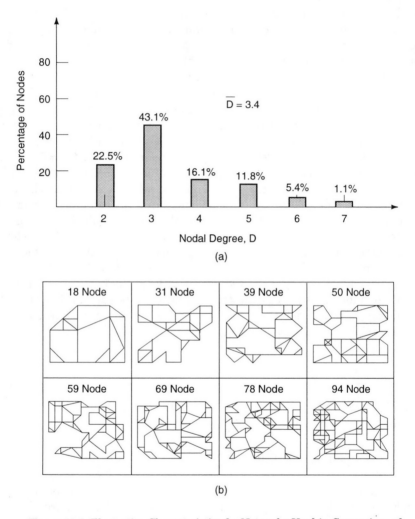

Figure 11-6 Illustrative Characteristics for Networks Used in Comparison of
           KSP to Max Flow

failure, given that a failure has occurred. The restorability of a network as a whole [also called the *network restoration ratio* (NRR)] is:

$$R_n = \frac{\sum\limits_{i=1}^{S}[\min(w_i,k_i)]}{\sum\limits_{i=1}^{S}[w_i]} = \frac{\sum\limits_{i=1}^{S}[R_{s,i}\cdot w_i]}{\sum\limits_{i=1}^{S}[w_i]} \tag{15}$$

where S is the number of spans in the network. Note that R_n is not the average of individual span restorabilities (unless all spans have equal w_i). As defined, R_n weights the restorability of each span by the size of each span so that it expresses the total fraction of working capacity that is protected, not the average fraction protected on each span. This reflects the importance of large spans in overall network performance.

The worst case restorability of a network (to span failures) is defined as the lowest span restorability level of any span in the network:

$$R_{n,wc} = \min_{i\in s}\{R_{s,i}\} \tag{16}$$

In the case of $R_n < 1.0$, $R_{n,wc}$ can be used as part of a determination of whether many spans are slightly below full restorability or, in contrast, one or a few spans are very under-protected.

11.5.2 Path Number Efficiency

$R_{s,i}$, R_n, and $R_{n,wc}$ depend on both the network design and on the restoration rerouting scheme. A given level of restoration may be achieved by a weak routing mechanism in a generously spared network, or vice versa. The intrinsic performance of a DRA as a routing mechanism, therefore, needs measures which are independent of the test network. In this regard, the most meaningful way to compare the routing effectiveness of DRAs is to assume an infinite number of working links in every span cut of a test network and compare the resulting DR path-sets to reference solutions from a KSP algorithm. When all w_is are considered infinite, the DR mechanism will be under maximal path-finding stress in the given network. Results then compared with path-sets from a KSP reference generator are intrinsic measures of routing efficiency, loop-freedomness, and predictability of a DRA. Path number efficiency (PNE) is defined in that manner:

$$\text{PNE} = \frac{\sum\limits_{i=1}^{S} k_{\text{dra},i}}{\sum\limits_{i=1}^{S} k_{\text{ref},i}} \qquad (17)$$

where $k_{\text{dra},i}$ is the maximal-stress number of paths found for span-cut i by the distributed restoration algorithm (DRA) under test and $k_{\text{ref},i}$, is the number of paths in the reference path-set for span-cut i. Failure to realize the number of paths found by KSP appears as a PNE value less than 1. PNE = 1 implies ideal topologically limited performance that is equivalent to centrally computed KSP in conditions of maximum path-finding stress. 100% PNE does not necessarily imply 100% restoration, however, because the restoration ratio depends on the sparing levels in the network, even with PNE = 1 (i.e., an ideal routing capability)). Conversely, a 100% restoration ratio does not imply 100% PNE for the routing mechanism. It may be that redundancy in the test network is higher than necessary, not requiring the longest or most intricate path combinations out of the DRA.

11.5.3 Path Length Efficiency (PLE)

In restoration it is also important to ensure that additional signal path delay is minimized for continental-scale networks and that efficient overall use of network spares is made. In the extreme, poor routing in terms of path length or looping paths will cause a drop in path number, so PLE and PNE are not totally unrelated. PLE is defined as the total distance of all paths in a restoration plans, in a given test network, relative to the total path length of the corresponding KSP reference solution path-set. PLE is defined only on the total length of the path-set, allowing individual path details to differ in two restoration plans that may be equivalent in terms of overall length efficiency. The path length efficiency of a DRA in a given test network is:

$$\text{PLE} = \frac{\sum\limits_{i=1}^{S} \left[\sum\limits_{j=1}^{k_i} L(P_{\text{ref}, i, j}) \right]}{\sum\limits_{i=1}^{S} \left[\sum\limits_{j=1}^{k_i} L(P_{\text{dra}, i, j}) \right]} \qquad (18)$$

where $L\,[P]$ is the length of path P, $P_{\text{ref},j}$ is the jth path in the restoration plan from the reference model and $P_{\text{dra},j}$ the path in the DRAs restoration plan. The number of paths in both DRA and reference solutions is k. PLE should not be used in comparisons in which PNE is not identical because it is only meaningful to compare path-set length between solutions that have the same number of restoration paths. Note also that PLE is defined such that a path length total *longer* than the reference solution produces PLE *less than* 1 (i.e., PLE = 1 is ideal).

11.5.4 Speed-Related Performance Measures

The following measures have been used to characterize the speed of a DRA executing in a given network:

A. *First path time* (t_{p1}): This is the time of completing the first path of a restoration plan. Its importance primarily is to assess the impact on high priority demands and to characterize speed of a DRA when it acts in a mode imitative of an APS system for single-link failures (Section 11.10.6). One may arrange priority use of the first path(s) found in a restoration plan so as to have minimal outage for priority services. The first path time indicates the minimum interruption that a priority service would see and gives an immediate indication if any portion of the affected traffic was able to beat the CDT in any given restoration event.

B. *Complete restoration time* (t_R): This is the time to complete the last path that can be realized for restoration of a given span-cut. "Complete" does not imply 100% restoration. Rather, it is the time at which the transient restoration event is complete, regardless of the outcome.

C. *Individual and mean path outage times* ($t_{p,i}$), (t_{pav}): $t_{p,i}$ is the individual outage experienced by the ith working path restored in a span-cut. When working links fail simultaneously, the path outage times are identical to the sequence of restoration path completion times. But in general, the onset of individual link failures may be time-dispersed within the failure event. Then $t_{p,i}$ is the elapsed time from failure of the ith path until its individual restoration. This is relevant to the discussion of time-dispersed random failures in Section 11.10.5.

The average of $t_{p,i}$ over all working links restored in a restoration event is t_{pav}. This is not necessarily the restoration time t_R divided by k, the number of paths found. In a strictly iterative DRA, $t_{pav} * k = t_R$ may be expected. Because of the high degree of parallelism in some DRAs, however, the restoration time for the whole event (t_R) can be virtually the same as the individual outage time for every path, where link delays exceed nodal delays [25].

D. *95 percentile restoration time* t_{95}: This is defined for a whole network, based on the individual outage times of all working paths over all span-cuts of the network. This is the time by which 95% of all individual links are restored and is the value at which the cumulative distribution function (CDF) of $t_{p,i}$ equals 0.95.

$$\int_0^{t_{95}} t_{p,\,i}\,(x)\,dx = 0.95 \tag{19}$$

where $t_{p,i}(x)$ is the probability density function of individual path outage times (equivalent to path completion times if all link failures in spans are simultaneous). Figure 11-16, (which is associated with Section 11.10.5) serves here, nonetheless, to illustrate the typical form of a CDF arising from $t_{p,i}(x)$ data from all span-cuts in a large study network. The rationale for a 95 percentile level for specifications and measurement of outage times is that such a measure is less influenced (than a pure maximum measure) by single extreme events in a sample.

11.6 THE CAPACITY CONSISTENCY PROBLEM IN DRAs

We are now heading toward an overview and comparative discussion of published DRA proposals. A central issue for any DRA is the manner in which it deals with the capacity consistency issue, i.e., Constraint 2 in Section 11.3.6. We now develop the implications of this constraint to give an understanding of the main technical problem in designing a DRA. This reveals that there are at least three basic approaches to the central problem of capacity consistency. These serve as a basis for categorizing the DRAs proposals that follow.

All DRAs to date use some version of flooding and reverse linking introduced in [24, 25, 26, 38, 39]. The basic idea of flooding to explore all network regions may be familiar to some readers from flooding in data networks to distribute updates on network configuration. In this use, however, the flooding process is not actually the mechanism by which routes between network points are determined. Most often, routing is determined by an algorithm in each node that computes a delay or other cost metric to destinations based on the network data that were distributed by flooding. In flooding for topology update distribution, the sequence of hops by which flooded update packets arrive at a node is not of significance; loops don't matter and redundant arrivals can simply be ignored.

In contrast, a DRA must exploit flooding in a way that derives a complete set of link-disjoint nonlooping paths between two nodes, ideally doing so in only one iteration. The path-set must be collectively the shortest and consistent with the exact quantities of spare links available on each span of the network, i.e., capacity-consistent.

At first glance it is tempting to assume that simple flooding addresses the DRA problem. Indeed, examples in [27] (p. 240) and [42] imply that this is the case. However, simple flooding will only solve the DRA problem if we take paths on one route per iteration. Flooding in a DRA that obtains all paths at once is significantly different from simple flooding. To develop this point, let us consider the network in Fig. 11-7 in which span (6-7) has failed. Overlaid on the network is the set of 10 KSP restoration paths for span-cut (6-7), found in one iteration of the SHN protocol. Now let us assume that a simple flooding process starts from node 6 and develop the information for a flooding tree (as in [27], p. 240), to discover all distinct routes from node 6 to node 7.

Simple flooding functions well at the first hop at which all two-span routes are enumerated. There are three routes and all yield feasible paths:

Route	Paths Contributed to Solution
6-4-7	1 path (limited by span (6-4)), which has only one spare link)
6-3-7	2 paths (limited by span (6-3))
6-8-7	4 paths (limited by span (6-8))

Simple flooding will next discover six routes of length 3. Now we begin to see the breakdown of simple flooding for restoration because no capacity-consistent paths are feasible over four of the topologically distinct routes enumerated:

Route	Paths Contributed to Solution
6-2-3-7	1 path feasible
6-3-4-7	no paths feasible (span 6-3)
6-4-5-7	no paths feasible (span 6-4)
6-8-9-7	no paths feasible (span 6-8)
6-2-8-7	1 path feasible
6-4-3-7	no paths feasible (spans 6-4, 3-7)

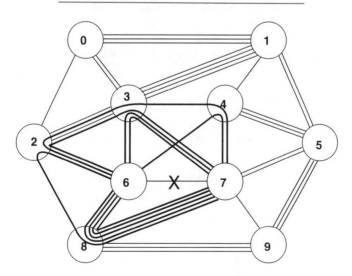

Figure 11-7 Example for Illustration of Capacity-Consistency Requirement in the DRA Problem

The point is that enumeration of all distinct routes, which is what simple flooding accomplishes, is not equivalent to discovery of distinct capacity-consistent paths for restoration. Although the eye easily discerns the feasible paths among the routes above, a DRA must include a means for doing this while at the same time exploring the network by flooding. The output of the DRA must be a set of mutually feasible paths, not a list of all routes. From the fact that all routes are found by simple flooding it does not follow that a restoration path-set is implicit. In fact, having the route-set alone, a large number of path-sets would have to be "tested" to find one that is capacity-consistent.

As we continue the example, divergence between the size of the route-set generated by simple flooding and the number of capacity, consistent paths rises steeply. At the fourth hop, seven more distinct routes are found, but only one offers a restoration path that is capacity-consistent with the rest of the KSP path-set.

Route	Paths Contributed to Solution
6-3-1-4-7	no paths feasible (span 6-3)
6-2-3-4-7	1 path feasible
6-3-4-5-7	no paths feasible (span 6-3)
6-3-1-5-7	" (span 6-3)
6-8-9-5-7	" (span 6-8)
6-4-5-9-7	" (span 6-4)
6-2-8-9-7	" (span 2-6)
6-8-2-3-7	" (span 2-6, others)

And of 21 routes found at hop 5, only one feasible path exists. In addition, 3 *looping* routes appear now:

Route	Paths Contributed to Solution
6-2-3-1-4-7	0
6-2-3-1-5-7	0
6-2-3-4-5-7	0
6-2-8-9-5-7	0
6-2-0-3-4-7	1 path feasible
6-2-0-1-4-7	0 infeasible given (6-2-0-3-4-7) taken
6-2-0-1-5-7	0
6-2-0-1-3-7	0
6-3-0-2-3-7	0 looping, not feasible
6-3-0-1-4-7	0
6-3-0-2-8-7	0
6-3-0-1-3-7	0 looping
6-3-1-5-9-7	0
6-3-1-4-5-7	0
6-4-1-5-9-7	0
6-4-1-0-3-7	0
6-4-1-5-4-7	0 looping
6-4-1-3-4-7	0

Route	Paths Contributed to Solution
6-8-2-3-4-7	0
6-8-2-0-3-7	0
6-8-9-5-4-7	0

We stop the example at 5 hops, but note that capacity-efficient restoration in real networks may require a DRA to realize paths up to 9 hops in length [38]. Simple flooding to this extent will generate a huge number of distinct routes, most of which offer no capacity-consistent paths or are looping. This illustrates the real problem for a DRA—to find the relatively small subset of mutually capacity-consistent paths among the large number of distinct simple routes generated by flooding.

Below 5 spans, looping routes cannot form when flooding messages are not rebroadcast into the span on which the related prior message arrived and the custodial nodes do no rebroadcast. Under these conditions a path of length 5 is the minimum that can exhibit a routing loop: 2 (different) spans are rooted at each custodial node and it takes 3 others to contribute a cycle connected to the ends of the first two spans. A cycle including the custodial nodes cannot arise because neither of these nodes performs rebroadcast. The fact that a feasible path was found at hop 5 in the example shows that restriction of the hop count to less than 5 is not an acceptable means to avoid loops, since it would severely limit restoration in many networks.

The preceding example lets us formally relate the simple flooding process to the DRAs' capacity consistency problem: Let $\mathbf{A}$ be a $[S \times R]$ binary matrix containing all distinct routes enumerated by simple flooding between two (custodial) nodes (x,y), up to a defined limit on hop count. The number of spans in the network is S and R is the total number of distinct routes enumerated by simple flooding. Each column of $\mathbf{A}$ lists one distinct route between x and y. Each row contains a 1 or 0 marking those routes in which a given span is included.

The routing problem in restoration can then be expressed in a form that gives another way of looking at the capacity-consistency constraint, and is more insightful than equation (6): This is:

$$\mathbf{A}\,\mathbf{v} \le \mathbf{c} \qquad (20)$$

where $\le$ is understood to apply on a row-by-row basis, not on a vector magnitude basis. $\mathbf{c}$ is the S-long column vector of spare capacities on each span of the network. $\mathbf{v}$ is an R-long route-compatibility vector that describes a maximal capacity-consistent loop-free path-set for restoration. Each entry in $\mathbf{v}$ is zero or a whole number of paths to be realized over the respective route enumerated in $\mathbf{A}$. Obviously $R \ge k$, where k is the total weight of $\mathbf{v}$ and is the number of paths feasible for restoration. Figure 11-8 presents $\mathbf{A}$ and $\mathbf{v}$ to put the preceding example in this framework. The $\mathbf{v}$

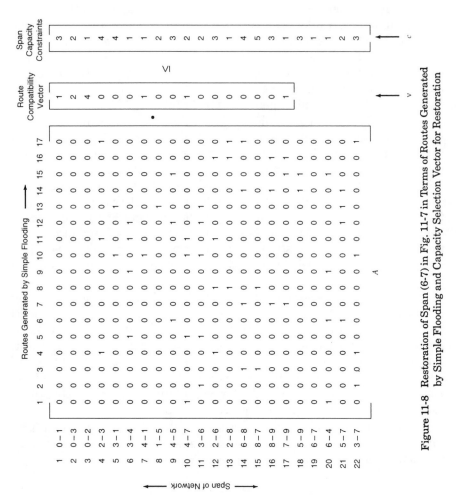

Figure 11-8 Restoration of Span (6-7) in Fig. 11-7 in Terms of Routes Generated by Simple Flooding and Capacity Selection Vector for Restoration

vector corresponds to the KSP path-set in Fig. 11-7 and shows how the path-set required for restoration actually relates to the information produced by simple flooding.

This formulation gives insight about the nature of several existing DRA proposals by relating the simple flooding process to the path-set needed for restoration. Indeed, most of the complexity of DRAs, and their major differences, lie in strategies for inferring **v**. Three basic approaches can be identified from the DRA proposals that follow:

1. *Iterate simple flooding:* Here, simple flooding is iterated, taking only one path (or all the paths common to one route) at a time. Seized links are excluded from the next iteration. In our framework, this can be viewed as generating the full **A** matrix of the network on each iteration but allowing **v** to have only one nonzero entry. This is the trivial case for a capacity-consistent solution of **v**. This is the simplest approach to deal with capacity-consistency but its price is that $O(k)$ iterations of the whole process are required to develop a complete restoration path-set. The FITNESS and Komine DRAs take this basic approach.

2. *Generate* **A** *and solve for* **v** *in backwards search:* In this approach a single stage of simple flooding is executed and a record of all simple flooding messages arriving at each node is built up. Then, in reverse-linking, a search is made through the flooding information recorded at all nodes (which encodes **A**) to find a capacity-consistent path-set by messaging among nodes to coordinate path-set construction. This involves detection of dead ends to the search and restarting the search as local capacity limits are reached in parts of the network. In our framework, this is equivalent to generating the **A** matrix once, storing it in a distributed fashion over the network, and then deducing **v** = $A^{-1}c$, effectively inverting **A** by interacting backward searches through the route tree that is implicit in **A**. This requires only one flooding iteration to create the distributed copy of **A** but a difficult-to-predict amount of time to deduce a **v** vector that satisfies (20) by backward search. The NETRATS DRA works along these principles.

3. *Capacity-consistent modified flooding:* Here, the simple flooding process is modified by rules at every node so that an inherently capacity-consistent flooding pattern is established over links of the network. Implicitly, only the columns in **A** that correspond to nonzero entries in **v** are found. The full **A** matrix of the network is not explicitly generated. This is the way the SHN protocol works.

11.7 OVERVIEW OF DISTRIBUTED RESTORATION ALGORITHMS

To provide an overview of DRAs we begin with the self-healing network (SHN) protocol. Most of what is common among subsequent DRA proposals is embodied in the SHN. The SHN was the first DRA to be reported [24] and to be extensively characterized in terms of PNE and PLE relative to KSP routing characteristics under maximal path-finding stress [25, 38, 39, 43]. With the SHN as a basis, other DRAs can be discussed concisely in terms of details that are different, without repeating common aspects. One proposal for distributed restoration [44] is not included as it assumes that preplans for all failures are downloaded to nodes and only fault identification is done in a distributed manner. Our central interest is in avoiding the need for centralized preplan computation and download.

11.7.1 Self-healing Network Protocol

The SHN protocol is implemented as an event-driven finite state machine (FSM) with three main states: sender, chooser, and tandem. All processing can be described in terms of two basic event types that drive the FSM: a signature arrival (called a *receive signature* (RS) event) and alarm events. *Signature* is the name used for link-borne state indications used in the SHN. Signatures are more like the SONET K1-K2 bytes for APS and ring applications than X.25-style packet message channels. A signature, with a defined application-related set of data fields, is always present on each link; only the contents change. Rationale for signatures rather than messaging is given in Section 11.8.1. Other DRAs have so far assumed messaging through the SONET data communications channel (DCC). Whether state- or message-based, all DRAs employ signalling interactions only between adjacent DCS nodes.

All SHN signatures contain the following basic fields (others may be added for specific extensions of the protocol in following sections):

1. *Custodial node name pair:* This tags all signatures in the network with the failure event to which they pertain. This is a primary basis for handling multiple faults. Also, because each custodial node places itself first in order of inclusion when it emits a signature, intermediate nodes can tell a forward flooding signature (originated by sender) from a reverse-linking signature (originated by chooser).

2. *Repeat count:* Every tandem node increments the hop count on new signatures emitted by it in response to an incoming signature. A hop limit sets the greatest logical distance from the

sender that signatures may propagate and, hence, sets a limit to the maximum restoration path length. Any message arriving at a tandem node with a hop count equal to the repeat limit (RL) is ignored; only messages arriving at a chooser node with a hop count less than or equal to the limit will be considered. The repeat limit can be constant over a whole network or it can effectively be made region- or node-specific within a network by changing the sender's initial repeat value (IRV). Then flooding range = (RL – IRV). If IRV is negative, greater range is effected, so the range can be controlled by the node that acts as sender if it is desired to reflect variations in network topology.

3. *Last immediate node name:* In any signature emitted by a node, the node's name is included so that adjacent nodes have a record of the logical span on which the signature arrived and so that link-to-span associations are always apparent from external data, not from internally stored images in the DCS. Node names are not appended to a growing list but simply written into the dedicated last node ID field.

4. *Index:* A unique index number is assigned to each primary signature emitted by the sender. The indexing is not repeated within each span but runs sequentially over all primary signatures initiated by the sender. Every signature subsequently emitted by a tandem node bears the index value of the incoming signature, which is the precursor for that outgoing signature (*precursor* is defined in the tandem node logic that follows).

5. *Return alarm bit (RA):* Whenever a node observes an alarm on a working link it sets the RA bit in the reverse direction signature. In the case of a unidirectional failure, this ensures that both ends of the span are activated to perform custodial node roles.

The overall operation of the SHN protocol can be described in terms of the following functions: (A) activation, (B) sender-chooser arbitration, (C) multi-index flooding, (D) reverse linking, and (E) traffic substitution.[7] In practice, (B) through (E) are not separate chronological phases. They occur in a concurrent asynchronous manner with updates to the multi-index flooding pattern occurring simultaneously with, and in response to, reverse-linking processes as the path-set self-organizes. However, for simplicity we will describe these processes as if they were separate phases. Later, a single network-level picture of the dynamics of operation is given.

---

[7]The details that follow, plus the use of the DRA in ad-hoc path restoration (called *capacity scavenging*) are covered by US Patent No. 4,956,835 [26].

A. Activation: Before a failure, each node applies a null signature to all working and spare links leaving its site. The null signature contains only one non-null field, the nodes' name (NID). This allows the SHN in each other node to identify all links in the same logical span, without requiring nodal data tables. If it is not feasible to apply signatures to working links, a table of link-to-span associations can be used at each node. However, with signatures on the working links prior to failure, the SHN functions with absolutely no *a priori* stored network knowledge in any node other than its own name. Aside from this use (and the RA bit that follows), the SHN has no other need for signatures on working links. Working links are not involved in any path-finding aspects of the protocol.

An SHN restoration event begins when a failure occurs on a transmission span and causes loss of frame, alarm indication signal, or loss of signal indications on one or more working links. In the most likely case of one failure at a time, the SH task is initially in the normal state. The task first identifies all the working ports that have an alarm state. Alarm persistence testing may be implemented by a delay at this stage, but a preferable strategy is to let the protocol proceed immediately with path creation and subsequently handle individual link alarms as they arise. This style of implementation is called preemptive activation and is described further in Section 11.10.5.

If the node ID (continuously latched until an alarm state occurs) in the failed ports is not common among all failed ports, then failures on multiple logical spans have been detected. In this case a second concurrent instance of the SH task is started to deal with the other span failure(s). Multiple-failure execution of the SHN is described in Section 11.10.4. Our present aim is to describe the basic protocol in a single-failure context.

B. Sender-chooser arbitration: Immediately after the span-cut, two SH tasks are "awake" in the network—those at the custodial nodes. Each task reads the last valid signatures at ports where ALARM = true and SPARE = false to learn the identity of the node to which connectivity has been lost. Each custodial node then performs an ordinal-rank test on the remote node name (NID) relative to its own NID to determine whether to act as sender or chooser. For instance, if span (B-C) is cut, ord (C) > ord(B) so node C will act as sender and B will act as chooser. The outcome of the test is arbitrary but it guarantees that one node will adopt the sender role and the other becomes chooser. The protocol in every node is capable of sender, chooser (or tandem) roles but adopts the one resulting from this arbitration. Any rule that can be conducted independently at each node, with only local information, and ensures a complementary outcome, will suffice.

C. Multi-index flooding: Immediately after role arbitration the sender performs primary flooding, applying a unique index number to each signature transmitted on min (w_f, s_i) outgoing links of each span at its site and setting other fields of the signatures to their initial values. The number of working links on the failed (or failing) span is $k = w_f$. The number of spare links on the ith span at the sender is s_i. The failed span is not included in primary flooding except (optionally) for one signature that, in the case of a single link failure on the span, will result in SHN execution that directly emulates 1:N APS reaction.

The primary signatures from the sender appear as receive signature (RS) events in ports on neighboring DCS machines. This causes the OS of those nodes to invoke their SH task. Nodes awakened by signature arrivals, with no alarm at their site, enter the tandem node state. At this stage the tandem node rules initiate selective rebroadcast of incoming signatures on spare links at the node location. Important details of tandem node rebroadcast rules follow. If, within the range limit, there is any connectivity between sender and chooser in the network of spare links surviving the failure, then one or more of the flooding signatures arrives at the chooser node. This triggers a reverse linking sequence.

D. Reverse-linking: When the first flooding signature on a given index arrives at the chooser, the chooser initiates a reverse-linking process that traces backwards through the family of signatures in the network common to that index. The chooser emits no other signatures. The tracing back, or reverse-linking, collapses the whole family of signatures for that index onto one path through an arbitrary number of tandem nodes. To initiate reverse-linking, the chooser applies a *complementary signature* (to follow) on the transmit side of the port that has the first-arriving signature received on each index. When the reverse-linking signature arrives at a tandem node, the tandem-state SHN protocol makes a local cross-connection, propagates the reverse-linking signature in the direction of the precursor for that index, suspends all other signatures leaving the node on the given index, and, if appropriate, revises the multi-index flooding pattern. (The signatures that remain for that index act as a bidirectional holding thread for that path so that sender or chooser can collapse it at any later time simply by cancelling the remaining signature emitted on that index at their site.)

E. Traffic substitution: After transiting one or more tandem nodes, a reverse-linking signature arrives back at the sender, forming a complement signature pair relative to one of the sender's primary signatures. A bidirectional path is then known to have been established between the sender and the chooser via the port at the sender where the reverse-linking terminated. The sender then cross-connects traffic from one of the working path stubs into the restoration path. Priority schemes can be effected by establishing a priority basis for sender substitution choices at this stage of the protocol.

After initiating reverse-linking in a given port, the chooser expects the arrival of a live traffic signal from the sender as the next event in that port. The traffic signal may bear a network-wide signal ID embedded in it or a locally applied signal ID from the sender. The signal ID is used to map the live traffic substitution by the sender into the surviving stub of the corresponding failed path at the chooser. When the signal ID arrives (with the signal) at the chooser, that port is cross-connected to the surviving portion of the working path on which the same signal ID is found. Use of node-local and network-wide signal ID schemes for sender–chooser traffic mapping is detailed further in Section 11.10.1.

SHN Tandem Node Rules. The key to the SHN protocol's ability to synthesize the KSP path-set in a single pass is in the tandem node rules which implicitly result in capacity-consistency. To explain the SHN tandem node logic, two definitions are needed:

(1) *Precursor:* For each distinct index value present over all non-null signatures incoming to a node, the port at which the incoming link with the lowest repeat count for that index is found is called the precursor for that index. The precursor is always the root of the rebroadcast tree for any index.

(2) *Complement:* A complement condition exists in a port when a flooding signature is matched by a reverse-linking signature, both of which have the same index and valid repeat counts.

With these key definitions, the tandem node rules are as follows. These rules are applied for self-consistency any time an incoming signature changes state or when a reverse-linking event frees outgoing links at the tandem node:

1. For each index at the node, the tandem logic attempts to provide one outgoing signature of the same index on one link in each span, other than the span in which the precursor for that index is presently found. Where rule 1 is not satisfiable for all indexes, the multi-index flooding pattern from the node is determined by competition among indexes based on their repeat values and the spans in which their precursors lie, as follows:

2(a). Find the precursor that has the lowest repeat count (globally) over all indexes present at the node. Assert the target broadcast pattern fully for that precursor. In a span where all links already bear active outgoing signatures, this may require that some index loses its signature in the span. If so, take over the link occupied by the index whose repeat count is the highest among those indexes present on outgoing links of the given span.

2(b). Find the precursor with the next lowest repeat count of all indexes at the node. Apply the target broadcast pattern for that index to the extent possible while only taking over outgoing links whose precursor has a higher repeat count than the present index and is the highest among indexes in the given span.

2(c). Continue adjusting the broadcast pattern for each precursor so that every index present at the node is either fully satisfied (i.e., has one rebroadcast in every nonprecursor span for that index) or is partially satisfied to the greatest extent possible, consistent with the overall rank of its precursor in terms of incoming repeat count at the node and the relative spans in which it and other precursors lie.

A compact implementation that always results in the required multi-index competitive flooding pattern is to maintain an ordered set [precursors], sorted by increasing repeat count. Every time the multi-index pattern is to be revised, the basic flooding pattern per index is applied, working up from lowest repeat precursors to highest, applying no signatures in a span that is already full. An example is shown in Fig. 11-9 for 4 spans and 5 indexes. Precursors are on the incoming links marked with arrows and the respective repeat counts are denoted by "r = ." The transmit side of a port where a precursor is found for one index can be used in the broadcast pattern for another index. The patterns are graphically complex but the underlying principle is that they provide simultaneous maximal, but mutually consistent, flooding patterns for multiple concurrent flooding processes, one per primary signature emitted by the sender. Their multilateral coordination at every node is on the basis of incoming repeat value. Signature arrivals on an index, which are not of the lowest repeat count for that index, are ignored. When one or more precursors have equal repeat counts, the treatment is random as to which gets priority in the process of Fig. 11-9.

Application of these rules results in the following treatment for three frequent events: (i) The precursor location for an index shifts (a lower repeat count appears in another port for an existing index): In this case the broadcast pattern for that index is rerouted (as in a tree) onto the new precursor port and the composite rebroadcast pattern is adjusted for consistency with the overall state rules above. (ii) A new index appears at the node: In this case the composite broadcast pattern is adjusted for consistency with the above rules, fitting the new index into the composite pattern on the basis of its repeat count and anchoring its individual broadcast pattern on the precursor for that index. (iii) An index family collapses from reverse-linking, leaving one or more free outgoing links: In this case, the composite rebroadcast pattern is reviewed, resulting in extension of the pattern for one or more indexes not previously receiving a full broadcast pattern at the node.

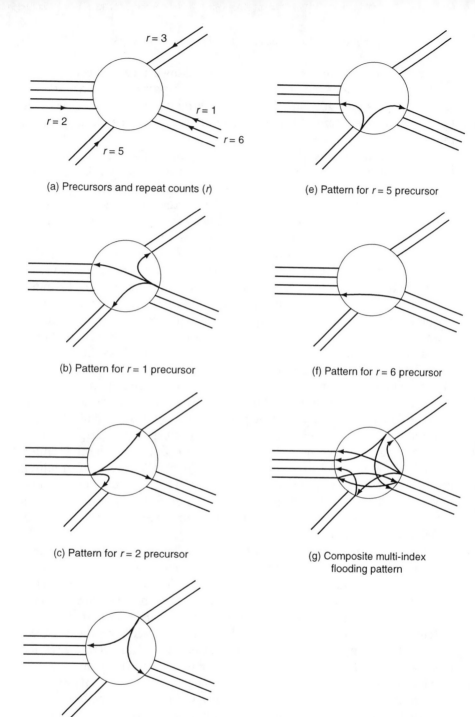

(a) Precursors and repeat counts (r)

(e) Pattern for $r = 5$ precursor

(b) Pattern for $r = 1$ precursor

(f) Pattern for $r = 6$ precursor

(c) Pattern for $r = 2$ precursor

(g) Composite multi-index
flooding pattern

(d) Pattern for $r = 3$ precursor

Figure 11-9 Example of Multi-index Flooding Pattern at the SHN Tandem Nodes

The rest of the tandem node logic pertains to the reverse-linking process, which results in path assembly, and local adjustment of the composite broadcast pattern for remaining indexes. A reverse-linking event is triggered at a tandem node whenever a complement state arises as the result of a new incoming signature in some port. In this case, the following applies:

3. The complement-forming signature is copied to the outgoing side of the port where the precursor of the complemented index is found. This creates a complement condition in that port as well and sets the status of both ports to "working" (or "in use for restoration") status.

4. All other outgoing signatures having that index at this node, but not part of the two complement-pairs that define the new path through this node, are cancelled.

5. The multi-index flooding logic is reapplied to adjust the rebroadcast pattern for all remaining uncompleted indexes at the node. This exploits the collapse of the reverse-linked index to benefit other indexes that may still by vying for complete rebroadcast at this node.

6. A connection is made between the two ports having the complement condition on the given index. Any subsequent appearance of a signature bearing the reverse-linked index is ignored at this node. Diagramatic examples similar to Fig. 11-9 for events (i)–(iii) and reverse-linking are given in [25].

Network Level View of SHN Operation. It is beyond the scope of this chapter to prove the self-organizing effect of the above rules.[8] However, a high-level view of what happens can be given: each index initially expands from the sender in parallel, in contention as in Fig. 11-9, with all other indexes based on the length of the route they followed to arrive at each node. Each index builds a tree of precursor relationships through those nodes at which it is able to obtain rebroadcast. As soon as any index reaches the chooser, its flooding pattern collapses through reverse-linking onto a single path that retraces the precursor sequence for that index. The retracing path is completely predetermined by the precursor relationships established through the network in forward multi-index flooding. (There is no possibility of having to backup, hunt, or iterate in reverse-linking.) Freed links arising from reverse-linking are immediately incorporated into a revised broadcast pattern at each node involved, expanding the surviving index families. As (or if) these indexes reach the chooser, they

---

[8]In the same way that the rules for reacting to neighboring cell states result in pattern formation in the cellular automata game Life, the locally applied rules for sender, chooser, and tandem nodes self-organize the spare capacity of the network into the KSP path-set between sender and chooser.

too collapse onto a single path and make links available for other indices, and so on.

The network-level picture is of multilateral expansion of all index families, in contention with others, punctuated by collapse of successful indexes onto single path reverse-linking chains, exploited immediately by further expansion of unsatisfied indexes. This occurs asynchronously in parallel across the network and is limited only by nodal processing delays and link propagation times. The process exploits the parallelism of the network to the greatest extent possible while still enforcing a mutual self-regulating interaction among expanding meshes so that only the shortest, capacity-consistent path-sets can be formed. Capacity-consistency is implicit because, as in Fig. 11-9, consideration of the logical size of spans is inherently coupled into the rebroadcast pattern logic. The whole process halts any time 100% restoration is achieved because at that point the sender suspends any surplus primary signatures. If 100% restoration is not topologically possible, the whole system freezes when flooding is saturated and no further reverse-linkings occur. In this case some broadcast tree fragments exist for unsatisfied indexes until a follow-up sender time-out suspends uncomplemented primary broadcasts at a fixed time (say, 5 seconds) after initiation. In either outcome, the final state of the network is that only complementary signature pairs persist, each pair running the length of one restoration path.

Discussion: The self-organizing convergence of the above protocol onto a network state in which the KSP path-set is established between the custodial nodes has been extensively verified [24, 25, 38, 39] and, for a range of realistic link and processor delay assumptions, occurs in under 2 seconds. Moreover, when link delays dominate nodal delays, the time for complete organization of the path-set is almost independent of the number of paths found because of the significant parallelism involved. The protocol has the same path-forming effect when executed between any two nodes of the network, as opposed to strictly custodial nodes and, thus, it is also the basis for (nonoptimal) path restoration, by concurrent or sequential execution of an SHN task instance for each demand pair to be recovered.

11.7.2 FITNESS

FITNESS was the second DRA proposal to emerge [45]. It uses iteration to satisfy the capacity-consistency requirement. FITNESS uses the DCC of one SONET link on each span for its messaging requirements. FITNESS works by simple message flooding from the sender to explore all routes of the network, as in the example that introduced the capacity-consistency problem. Because only one path is claimed in each iteration, there is no capacity consistency problem within each step. On each iteration, however, FITNESS finds the largest common module of STS-n capacity over the routes explored by flooding in each iteration.

In FITNESS the first (shortest) path seen at the chooser is not necessarily the one seized in each iteration. The aim is, rather, to find the route with the greatest module of capacity in each iteration, regardless of length. This strategy involves the chooser node waiting for a predefined period before acknowledging any flooding messages that reach it. The point of waiting is to identify the message that bears an indication of the greatest module of capacity that is feasible at this iteration.

Signalling is message-oriented so the initial flooding pattern by the sender needs only one message on every surviving span. The message includes an indication min (w_f, s_i) of the capacity sought on this span in STS-1 units. w_f is the total lost capacity and s_i is the aggregate spare capacity on (nonfailure) span i at the sender site.

Intermediate nodes build a table from messages received in the flooding process. The table records, for each adjacent node from which at least one message has been received, the largest capacity indication seen in any message from that node and the corresponding hop count. Whenever a tandem node receives a message whose capacity number exceeds all previous messages, the message causes an update to the table and is rebroadcast (with hop count increment) to all adjacent nodes (except the node sending the original message). In this way each node, and ultimately the chooser node, learns of the length and largest single common module of capacity that is feasible along distinct routes between itself and the sender.

At the end of the time-out period that defines each iteration, the chooser sends a single acknowledgment to the node adjacent to it from which the largest capacity message was received, regardless of the hop count. If a choice of hop counts exist for the same capacity, the lower will be chosen. The node that directly receives the chooser acknowledgment looks in its table to find the span via which the lowest hop count message that meets or exceeds the acknowledged capacity was received. A message is then sent to this intermediate node and a confirmation is sent back to the prior node. This step repeats until the node reached is the sender, at which point the sender substitutes the largest capacity traffic signal or a group of signals that fits the module found. The payload signal follows the path through to the chooser and the chooser performs traffic mapping based on signal ID, such as in the SONET path trace byte.

FITNESS thereby finds the shortest path at the largest capacity level available in each separate iteration. As iteration proceeds, the path-set will build up to be the shortest path at the highest capacity level, followed by longer paths at the same capacity level, then the shortest path remaining at the next highest common capacity level, and so on.

Discussion: Restoration effectiveness with FITNESS depends on the chooser time-out setting because that determines the length of routing sequence that can be observed by the chooser before time-out of each iteration. Tuning of the chooser time-out for a given network and nodal message delays is required to maximize the restoration ratio achievable before the CDT [45]. If the time-out is too low, many iterations can complete before the CDT, but each is limited to relatively local explorations of the network. On the other hand, if the time-out is too long, high capacity routes of considerable length may be found early on, but then too few iterations may be completed to maximize restoration before the CDT deadline.

11.7.3 NETRATS

NETRATS [4] (renamed TRANS [46]) is a "combined system of self-healing network and spare channel assignment" [4] in which a complete path-set is found in one forward-and-backward process using message-oriented signalling. The forward broadcast phase of NETRATS is a simple flooding exploration of all possible network routes. During the flooding phase, every message is repeated at intermediate nodes into every span except the one it came in on. Intermediate nodes store copies of all messages received and append their name to all messages rebroadcast through their site so that each message carries an accumulating trace of the individual route that it has traversed to its present point. Closed loops are avoided by using this trace information to eliminate rebroadcast to adjacent nodes that already appear in the passing-nodes-list.

Each node looks at the incoming available spare capacity (ASC) indication on each message and reduces it in the output copies of the message on those spans where the outgoing ASC from this node is lower than the incoming span ASC. By replacing the incoming ASC value with its lower capacity amount, the minimum common capacity of all spans in each distinct route is effectively forwarded through to the chooser. This is not by itself NETRATS' means to ensure capacity-consistency, however, because the apparent capacity on a number of nondisjoint simple routes can still represent multiple countings of the spare capacity on spans that are common to several routes.

The chooser acknowledges any forward message with a nonzero ASC. When this response meets the first intermediate node, that node reviews its file of all the messages received in the prior flooding phase to find the lowest hop count message that was received with the lowest compatible ASC. The span on which that message previously arrived is temporarily reserved and a message is passed over this span to the respective adjacent node. At the next node, however, or at any other stage back to the sender, the requested link(s) may have already been allocated to a prior selection response from the chooser. In this case, a negative acknowledgment (NACK) goes back to the earlier intermediate node. That node unreserves the spare link and tries the transaction again to another of its adjacent nodes, presumably based on the second-best hop count choice with the desired ASC capacity. This is repeated until a positive acknowledgment (PACK) is received by the first intermediate node. If no adjacent node can offer to further advance the return path from a given prior intermediate node, then a further level of fall-back to a previous intermediate node is required.

NETRATS uses a third and final confirmation phase between sender and chooser for each restoration path. This confirms to intermediate nodes holding capacity reservations along that path the backward-searching selection process did indeed reach the sender and temporarily reserved links can now be committed to the path. This is necessary because the returning search may have to back up and begin again after any depth up to 1 hop away from regaining the sender. Sender traffic substitution is, presumably, coincident with this final messaging phase in the sender to chooser direction.

Discussion: With reference to Section 11.6, the process of appending every node name visited to messages, means that each message arriving at the chooser conveys exactly one column of the A matrix to the chooser. More generally, the stored copies of all flooding messages at every intermediate node effectively gives each of those nodes a partial A matrix for the network, i.e., it describes a set of distinct routes between the nodes and the sender.

Each chooser selection response initiates a depth-first search (DFS) of the route tree that is effectively stored in the message-trace lists at all nodes from the initial flooding step. The several depth-first searches that the chooser initiates compete with one another on the basis of which one succeeds in seizing certain span capacities (branches in the tree) first. The interacting DFSs are the means by which a mutually capacity-consistent set of paths is built up in the returning direction (i.e., v is found this way).

During the time that capacity is seized by some reverse search that is destined eventually to have to back out of a dead-end, one or several other reverse searches may have already been forced away from that capacity, seizing capacity along suboptimal routes that will be locked into the path-set and may further perturb other paths. It is also explained in [4] that because near-looping routes are encoded in the route messages at every node, serpentine-like paths may arise that consume more spare capacity than necessary. This is addressed by proposing [4] that NETRATS either be tightly coupled with the network design so that the network layout can anticipate and prohibit these routing anomalies, or that tables of steering data be generated for each node from network design information to avoid these effects. In this regard, NETRATS departs somewhat from the ideals of database-independence, and network-independence.

11.7.4 Komine DRA

In 1990 Komine et al. [47] described a DRA that is similar to FITNESS in the sense that iteration of simple flooding finds paths along one route at a time, but the capacity realized at each iteration is that of the first route that reached the chooser. This removes the aspect of a chooser time-out to wait for the largest module of capacity at each iteration. Rather, the Komine DRA iterates as fast as possible, taking the greatest common capacity over the single shortest route found at each iteration. It is paced by sender recognition of reverse-linking responses to identify completion of each forward and backward wave. Messages in each iteration carry an iteration number so that still-propagating messages from a prior wave do not confuse processing for the next.

Komine et al. [47] described the usual sender flooding pattern, i.e., $max(w_f, s_i)$ call for capacity on each span from the sender, followed by simple tandem node rebroadcast on all spans but the incoming span, and a single chooser acknowledgment of the first message to arrive for each iteration number with a single reverse-linking retrace of the flooding pattern back to the sender. The sender then floods again on all remaining spares with a new iteration number to find next shortest route and accept all capacity along that route. Iteration continues until sufficient restoral capacity is found or no more chooser acknowledgments are heard. The amount of restoration before the CDT depends on how many distinct shortest routes have to be explored to build up the restoration path-set (because this sets the number of iterations required), and on link and node propagation delays within each iteration.

Later in 1990, the idea of multiple-destination choosers was added [28] to the algorithm just described. The basic method remains iterative flooding for one route at a time but the possibility is introduced for one of several predetermined nodes to respond as chooser. This is a means for dealing with the uncertainty as to whether an (apparent) link failure is actually due to a span failure or due to failure of the adjacent node. If it is a span failure, then the custodial node at the other end of the failed span will be available and will function as the usual chooser. But in a node failure, the sender flooding messages include the names of other nodes that can act as optional choosers. These nodes are chosen *a priori* to be points logically downstream on paths that normally transit the sender node and the failed node. (Here, paths are implicitly treated as unidirectional entities.) When this is set up in advance for all nodes acting in sender roles with respect to failure of a given other node, node loss is effectively treated as a case of multiple simultaneous virtual span failures, the virtual spans having been between pairs of nodes that were on paths that transited the failed node. The multidestination chooser information is set up in advance, based on knowledge of individual working path routings so that restoration paths are sought that directly replace two or more spans of the affected network paths. This constitutes a form of partial path restoration as in Fig. 11-3. This type of close-in (2-hop) recovery around a node failure is similar to the methods in [48] except that by setting up the multichooser arrangements beforehand, the particular sequence of "scavenges" to knit capacity around the lost node is effectively preprogrammed.

Figure 11-10 illustrates the multidestination chooser arrangement. It shows the setup for node C to restore normally with node B as chooser if span B-C fails or to restore appropriately reduced numbers of links directly to nodes A and D in the event node B fails. Equivalent sender arrangements are established in advance for other nodes D, A, etc. surrounding the prospective failed node. Every node now acts as both sender and multichooser since path finding is inherently unidirectional in this arrangement.

Any nominated chooser that receives a broadcast message responds. A final confirmation message from the sender effects capacity reservation and cross-connections along one path to the chooser whose acknowledgment first propagated back to the sender. Depending on the destina-

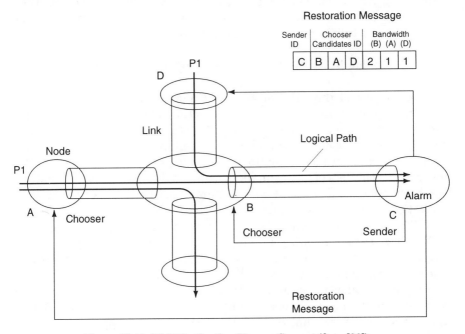

Figure 11-10 Multidestination Chooser Concept (from [28])

tion and capacity found in that iteration, another iteration begins with a revised list of prospective choosers and the associated capacity required if the respective chooser is the first to acknowledge in the next iteration.

Discussion: One aspect not clear from [28] is how two or more chooser acknowledgments are handled when they interfere or if an alternate chooser acknowledgment precedes a primary chooser acknowledgment in getting back to the sender for a span cut. Presumably, tandem nodes will realize if any acknowledgment for the given flooding instance has already impinged at that node and cancel the later-arriving chooser acknowledgment. But returning acknowledgments need not transit any nodes in common. Therefore, it seems to be assumed that the alternate choosers are either always further, in terms of delay, from the sender than is the primary chooser or that alternate choosers may be allowed to assert themselves in a normal span restoration event. Alternatively, a built-in delay at nodes responding in the context of a multichooser could ensure that the primary chooser gets to respond first, if available, thereby capturing the return path unto itself.

11.8 DRA PROPERTIES

Although there is considerable commonality among the DRAs above, their differences are significant in terms of the operational properties they imply for a DR system. We will discuss some basic issues regarding DRA properties and briefly compare DRAs on these terms.

11.8.1 State-Based or Message-Based Signalling

An aspect of the SHN protocol not adopted by other DRAs is the physical-layer, state-based nodal interaction style. In classical interprocessor messaging, all information passes through a serial communication channel between nodes and communication protocol stacks at both ends to handle retransmission for errors and recovery from packet loss, and requires buffers of an adequate size and flow control. These considerations often make it difficult to guarantee real-time performance with packet communications methods.

The state-based interaction introduced in the SHN protocol is potentially very fast because (a) no queuing or communications protocol are involved, (b) the spatial position of links on which signalling arises also encodes information relevant to the problem, and (c) nodal interactions are highly parallel in nature. All of the dynamically changing network state information needed at any time during SHN execution is found in the signatures on the links of the network itself. The protocol action is always defined by the current state of links surrounding the node, not by previous messaging histories. In effect, the network surrounding a node turns into an active memory space within which the protocol executes. Nodes are directly coupled in a shared-memory sense. The effects of protocol execution at any one node automatically updates the memory space within which adjacent nodes are concurrently executing, without any explicit communication between processors. Figure 11-11 illustrates this logical model of nodes connected through external shared memory which, physically, are the links of the network itself with state information stored in their signatures.

With signatures, state changes propagate in the physical layer, are detected in hardware, and are processed on an interrupt-driven basis by an event-driven, finite state machine (FSM) implementation of the DRA. There is no receive message buffering: a new signature overwrites, rather than follows, any previous signature. The host OS has no overhead to support the DRA; it views the protocol simply as an interrupt handler for signature change events.

If each state-change in a signature is counted as a "message," then this signalling paradigm may generate more messages than other means. It is important, however, to see that each spatially distinct state change is not a complete message to be processed in the usual sense. With a DRA that exploits the parallelism of this approach, CPU time may be reduced because communication protocol overheads are eliminated and the DRA protocol is written in terms of simple rules for reacting to changes in signatures. In practice, however, this advantage may be lost if the internal delays for port-to-CPU communications in the DCS are high. In this case, signal grouping or bundling (Section 11.10.7) can be applied to reduce the volume of state-changes to be processed at each DCS.

Signatures are also robust to errors without message retransmission. Simple triplication persistence testing is very effective for the DR application environment because even the alarm activating error rate of 10^{-6} of a typical fiber system is still low enough for single-error correction to be quite effective. If even more protection is desired, a checksum can be embedded in the signature format to validate state changes on the first appearance, with a wait for repetition only when a checksum fails. A thorough treatment of signalling options for distributed restoration (DR) [56] (summarized in [2]), concluded that some form of state-based signalling channel will be required for SONET DCS-based restoration to meet the 2 second objective.

11.8.2 Parallel or Iterative Execution

Another key issue in setting requirements for DR is whether to require that a DRA find all feasible paths in one iteration (parallel path finding) or if it may iterate a single-route algorithm to build up the path-

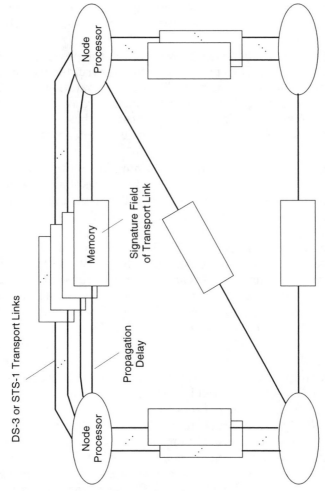

DS-3 or STS-1 Transport Links

Node Processor

Node Processor

Node Processor

Memory

Signature Field of Transport Link

Propagation Delay

Figure 11-11 With Physical-layer State-based Signalling, the Links of the Network Become the Memory Space in Which DRA Executes

set. To compare properties of interactive and parallel DRAs, one must assume the same processor hardware, the same network, and the same ratio of signalling links to spare links (i.e., the bundling ratio). Moreover, one is interested in capacity-efficient mesh-restorable networks, not networks whose spare capacity design is constrained so that all required restoration paths exist solely on one or a few 2-hop routes for each span-cut. In addition, two geographical scale domains should be considered: metro and long haul. In metro networks, propagation delays may be small whereas they can easily match or dominate nodal processing delays in long haul networks.

Comparing the SHN to an iterative scheme such as [28], under these conditions, the first replacement path in a large failure will probably be found by the iterative scheme. This is because its flooding and retrace in each iteration is simpler to execute than the SHN's modified flooding, which is causing a one-time self-organization of the network for *all* required paths. However, the simple flooding process for each stage of an iterative DRA is only slightly less complex than the one-time modified flooding of the SHN. And reverse-linking in the iterative DRAs is essentially the same as in the SHN. The SHN invests slightly more time in its single stage of multi-index flooding but then realizes the entire path-set with concurrent reverse-linking of all paths. There will, therefore, always be some size of failure at which a parallel DRA has returned the entire path-set while an iterative DRA is still cycling. Parallel execution should be especially advantageous in long-haul networks because while nodal delays are processor-dependent there is nothing that can be done to reduce distance related propagation delays. A parallel DRA encounters one set of network propagation delays, as opposed to one per iteration. Another fundamental consideration is that as a mesh-restorable network undergoes service growth while efficiently retaining full survivability, the diversity and lengths of routes in the restoration path-sets will increase. This means that the number of iterations grows, even when all paths common to a route are claimed in each iteration.

11.8.3 Self-Updating Property

A primary objective for DRAs is to remove dependence on centralized databases. The greatest robustness and autonomy arises if a DRA obtains all required information from its own surroundings as they exist at the moment of failure and afterward. In the SHN, the only permanent data that the protocol needs is the name of its host node. All other aspects of the network state are inherently taken into account at the time restoration is needed by the state-driven nature of the protocol. The alarm status, working or spare status, and background NID applied by each SHN node to outgoing links is the database on which the SHN relies. FITNESS uses tables of nearest-neighbor link connectivity information, but needs no other network knowledge, except in the sense that the chooser time-out must be set for the optimum value for each node. NETRATS requires tables of steering information and "tight coupling" [4] between network configuration control and the DRA to ensure desirable routing characteristics. Komine's basic DRA uses only nearest-neighbor link tables. The multidestination chooser arrangement requires network path data for multi chooser setup, but this can be obtained from the network itself through node ID accumulation in the path overhead of each STS-1. Finally, all of these DRAs can mark spare capacity that they use for restoration so that they are immediately self-updating with respect to spares remaining available for subsequent failures.

11.8.4 Topology Independence and PNE

For a DRA to be robust and reliable, the routing characteristic that it deploys should be known *a priori* and this characteristic must be reliable regardless of network topology. For example, the SHN yields a path set that is equivalent to the k-shortest link-disjoint loop-free paths between the restoration nodes through whatever spare links are present in the network. Similar verifications and statistical characterizations against defined reference models are needed for other DRAs. Eventual standardization of a DRA should include adoption of a reference criterion, such as KSP, and a requirement for PNE = 1.0 with respect to the reference in a number of stipulated test networks. Results reported with 100% restoration in a test network do not imply 100% PNE for the underlying routing mechanism, for the reasons given in Section 11.5.

11.9 DISTRIBUTED PREPLANNING WITH A DRA

One of the current practical difficulties for DR implementation is that existing DCS systems tend to have high internal communications delay between port cards and the master CPU. This may be addressed in future DCS designs for which DR will be a primary application. In the meantime, Pekarske and Grover developed an interim scheme for use of a DRA as a distributed preplanning (DPP) tool [49, 50]. The scheme permits fast network restoration with existing DCS equipment that is not as fast as desired for direct execution of the DRA. It uses a suitable DRA to derive preplan information in every node without any centralized database or computational dependency, thereby adhering to a major aim of DR. The motivation was to implement the SHN protocol on the RDX-33 platform [49, 50]. (The important difference of this scheme relative to the system using preplans in [44] is that in [44] the preplan is centrally computed and downloaded to all nodes.) Because of the internal message-passing delays of 200 msec inside the RDX-33 DCS, the SHN was estimated to take about 8 seconds to execute in target networks.[9]

The DPP scheme works as follows:

1. The SHN protocol is executed in a background mode for each possible span failure in the network. This is, in all respects, a full execution of the SHN protocol except that no cross-connections are made.

2. Instead, each node records the cross-connections that *it would have made* for each trial run in which it participated and records the corresponding custodial node name pair that identifies the failure. This results in a table (Fig. 11-12) at each node whose contents are the instructions for that node's portion of the SHN response to each respective span cut of the network.

3. When a real span failure occurs, either (or both) of the custodial nodes emits a single *alerting signature* containing the span identity and an indication equivalent to "this is not a drill!" Alerting can be accomplished either by an activation loop established through all DCS nodes or by disseminating the alert through simple flooding.

4. Any alerted node (including custodial nodes) that has non-null actions in its table for restoration of span (x,y) makes the internal cross-connections between spare ports that are listed in its table and changes the local status of the links used from spare to "in use."

The fastest method to alert all nodes is with an activation loop, created by concatenation of existing spare links in an arrangement that visits all nodes.[10] A loop ensures the span cut cannot isolate any nodes from the altering path. To guard against the case of two outstanding cuts on the activation loop, activation can be backed up by a single-message alerting flood at the same time. With a simple flooding process for alerting, all nodes can still be notified faster than they could execute the complete DRA. All nodes within T logical hops of the custodial nodes will be alerted within $T \times d$ seconds, where d is the time for a node to perform a single-message receive and broadcast operation with no processing. This assumes d dominates over message insertion and span propagation delays,

---

[9]There is little intrinsic reason for such high internal delays in future DCS designs. The particular equipment was designed in an era before it was envisaged that a DCS would perform a real-time embedded computational function in the network.

[10]If a DS3 or STS-1 rate activation path is created it may also be used to support precise time synchronization of all DCS nodes of the network using the method in [60].

Failure Span		Local Part of Network Restoration Plan
1	(x, y)	{cross-point list 1}
.	.	.
.	.	.
.	.	.
.	.	.
S	(k, l)	{cross-point list S}

Figure 11-12 Distributed Preplan Creation with a DRA

as is the case when the original speed problem for DR arises from DCS internal message delays. For example, using 200 msec for a single message forwarding delay through the RDX-33, all nodes for restoration paths of up to 8 hops would be alerted within 1.6 seconds. This conveniently meets the real-time target of 2 seconds with DCSs that may otherwise require about 8 seconds to directly execute the DRA.

A primary advantage of this scheme over centralized preplanning is that it addresses the database integrity issue inherent in centralized planning and avoids the computational and downloading effort in centralized preplanning. The NOC has only to command (or preschedule) a cycle of background trial runs to continually update the distributed preplan (DPP). Alternatively, updates could be commanded selectively for spans within the neighborhood of locations where network changes recently occurred.

One drawback is that there is a finite window of vulnerability between the time when provisioning for network growth or other rearrangements occurs and when the relevant updates to the DPP are effected. The updating frequency of the plan can potentially be quite rapid, however. Moreover, the DRA can also always be run immediately as a backstop to the fast-deployed preplan. The DPP could be continually updated as a background process throughout the day. If the NOC is routinely polling all network elements to confirm communications, that cycle could also serve as the pattern for triggering preplan updates. Alternatively, the update cycle could be prescheduled and left to each node to trigger updates for adjacent spans on its own. New paths and network changes are then incorporated into the restoration plan at most one regeneration cycle later. Assuming that uncompensated time dissemination via SONET overheads can provide DCS nodes with at least ± 100 msec of absolute time coordination, we can obviously pre-establish a rotation schedule to provide regular 10-second slots for each span. Each slot allows for execution of an SHN trial run plus margin before the next scheduled background execution by another custodial node pair. In this case, a 100-span network could have its DPP regenerated entirely every 17 minutes. The window of vulnerability is of less practical concern if, at most, 17 minutes after a network change, the change is incorporated into the network survivability plan.

Automated rules for triggering updates can also be developed. Whenever a new path is provisioned, each pair of custodial nodes along spans of the new path may initiate an update. Conversely, whenever a spare link is taken for growth at one node, the adjacent nodes need to look into their delayed activation tables to identify those (remote) span failures of the network that relied, in part, on the now-used spare link. Those nodes are then advised (via NOC relay) of the need to trigger updates for the span to which they have a custodial role.

To summarize, distributed preplanning offers some trade-offs that could form an acceptable, economic, and technically pragmatic approach for "slow" DCS platforms to support fast DR. The window of vulnerability can be reduced to a matter of minutes and can also be backstopped by follow-up (direct) execution of the DRA. Moreover, one may observe that regardless of how rapidly DCS machines may execute a DRA in the future, DPP with a dedicated activation loop appears to be the fundamentally fastest configuration for deployment of mesh restoration path-sets. An STS-1 or DS3 activation path can alert all nodes in only tens of milliseconds to deploy their components of the DPP. This opens up the possibility that a fast cross-connection DCS could actually match SHRs in terms of restoration speed (i.e., 50 to 150 msec), while still deploying capacity-efficient restoration path-sets that are the main benefit of mesh-restorable networks.

11.10 IMPLEMENTATION CONSIDERATIONS FOR DISTRIBUTED RESTORATION

This section is devoted to selected issues surrounding implementation of DR. These topics do not primarily depend on the DRA used, although properties of some DRAs may help in certain implementation issues. They are, rather, general considerations that pertain to any practical implementation of a DR system.

11.10.1 Traffic Substitution and Mapping

Restoration requires identical mapping of traffic onto the paths available at the two custodial nodes. Regardless of the DRA employed, this can be achieved with two methods outlined in [25]. Let us define [affected_ports] as a list that stores the port locations of the surviving path segment outboard of the custodial nodes. Similarly [alarmed_ports] stores the port numbers where the failures appear. The lists relate 1:1 in that alarmed_port [i] was connected to affected_port [i] through the DCS matrix before failure. The sender selects a member of [affected_ports] each time a new restoration path is available and connects it to the new path. If classes of priority are defined, the sender can exhaust the members of [affected_ports] in priority order. Two mapping schemes are as follows:

1. *With global path_IDs:* If a network-wide Path_ID plan is in place, such as the SONET path-trace function makes possible, the chooser may exploit the fact that, before the failure, there was a Path_ID on every working signal passing through it. By recording the Path_ID from the outgoing direction of each port at its site in normal times, it has information for use in traffic substitution. In the SHN model architecture [25], this information is transparently latched in the DCS port cards. The chooser then matches traffic signals as substituted by the sender with the port in its [affected_ports] list which has the same Path_ID, latched from before the failure. Alternately, latching is not needed if the incoming signal direction in [affected_ports] can be searched for the matching Path_ID. This works as long as the signal incoming from the unfailed stub of the path continues to bear its original Path_ID, i.e., if path terminating equipment does not itself reroute the path or otherwise drop the path_ID when originating AIS, etc. At the sender, forwarding the Path_ID is implicit in the act of signal substitution because the Path_ID is embedded in the traffic signal at its origin.

2. *With local signal_IDs:* If a global Path_ID scheme is not deployed or not available (e.g., to DS3s)., the mapping problem can also be solved by applying arbitrary local numbering before the failure. This allows inclusion of all types of DS3 traffic in restoration as well as SONET. (Except for the C-bit parity DS3 format [51], DS3 signals do not have a Path_ID function equivalent to SONET. However, local numbering can be applied transparently to any framed DS3 format by F-bit modulation [52].) In the local numbering scheme, each node applies its own (arbitrary) number sequence to working signals leaving its site, as outlined in Fig. 11-13(a). At the point of egress, before asserting its own numbers, each node also latches the existing number on each signal that was applied by the previous node in the path of this signal. When the sender connects affected_port [i] to a restoration path, it tags that signal with the local number, x_2, that it had previously applied at alarmed_port [i]. The chooser then connects the arriving signal to the member of [affected_ports] where x_2 was latched before the failure. Because the x numbers are totally arbitrary, the DCS machines' internal port addressing numbers could be used directly.

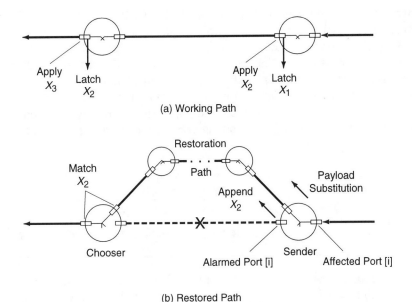

(a) Working Path

(b) Restored Path

Figure 11-13 Payload Mapping in the Absence of a Network-wide Path_ID
Scheme Using Arbitrary Signal Numbering Applied Locally at
Every Node

11.10.2 Restoration of Mixed Signal Types

Let us now consider how a DR system may simultaneously restore a
mixture of DS3, STS-1, STS-3 traffic signals. We proceed in two steps. The
first is to consider how a DRA would deal with mixed incompatible signal
types. The second step is to consider (and exploit) the set of compatibility
relations that actually exist among SONET and DS3 signal types. For
convenience, the discussion is in terms of how the SHN protocol is adapted
for these purposes. The extension to STS-n (n > 3) and other payload
combinations (e.g., CEPT-II, VT1.5, FDDI), and to other DRAs follows
directly.

A. Incompatible mixed signals: If strictly incompatible signal types
are to be restored (i.e., none can become the payload of another), sender
flooding is modified so that min $(w_{<f,j>} \, s_{<i,j>})$ signatures are emitted in each
span where j = {DS3, STS-1, or STS3c}. $w_{<f,j>}$ denotes the number of failed
working links of type j. $s_{<i,j>}$ is the number of spare links of type j in span
i. All signatures bear an additional field, a native format code, which is set
by the sender based on the native format (j) of the spare link into which it
emitted the primary signature for each index. No change is required in the
tandem node logic except that the logic is applied within the naturally
defined subsets of ports that have a common native format at each node.
The result is simultaneous restoration of DS3 failures via KSP routing

through the network pool of spare DS3 links, KSP restoration of STS-1s
through the pool of spare STS-1s, and so on.

   B. *Exploiting SONET mapping control:* But in practice, the flexible
payload mappings of SONET could let us be more efficient than restoring
each signal type with only network spares of strictly the same native
format. For example, a restoration path formed from links whose native
format is STS-1 or STS3c could be configured to restore any of the payload
combinations shown in the following table by forwarding the port mapping configuration through the restoration path.[11]

Native Format of Restoration Path	Compatible Payload Mappings for Mixed Signal Restoration
DS3	DS3 (asynch, Syntran, or Cbit parity)
STS-1	DS3 or STS-1
STS-3c	STS3c, 3@STS-1, (2@STS-1 + DS3), (2@DS3 + 1 STS1), or 3@DS3

Tandem nodes still respect and match native format when creating
restoration paths, but the sender is free to elect the final mapping configuration of each SONET restoration path to suit the priorities and
payload composition of its [affected_ports] list. Obviously the greatest
flexibility in restoring mixed signal types is obtained when raw restoration paths are formed of STS-3c links. A strategy for network management
would therefore be to maximize the amount of spare capacity that is maintained in the form of STS-3c signal units and to stipulate DRA-based
control of the port mappings in specifications for DR future systems.

11.10.3 Reversion after Physical Repair

   Reversion is the process of restoring paths to the prefailure routing,
or to a new working configuration, after physical repair of the failure.
Some particular DCS features and procedures are needed to facilitate
reversion with minimal secondary impact on traffic. One school of thought
is to not do reversion at all, to leave the rerouted paths in place. The
problem with this is that the original routings were (axiomatically) the
most efficient routings. Consequently every rerouting that is left in place
consumes excess network capacity, leaving fewer spares for future growth
and restoration, and eroding the networks' capacity efficiency. Nonetheless, it makes sense to at least inspect the restored routings, especially if
path restoration is performed, because some new routings may be as good
as the original routings and those could indeed be left in place.

---

[11]In this approach, intelligent port cards could automatically recognize and conform to
any payload configuration presented to them by the sender substitution.

Assuming reversion is performed, however, an objective is to make the traffic hit from mesh reversion no greater than that due to reversion in a shared protection SHR. Two considerations that minimize the coordination effort and the traffic impact of reversion from a mesh-restored state are (1) the DRA should leave state information behind in the network that facilitates automated collapse of the restoration path-set, and (2) the DCS nodes should support a make-before-break (in-service rollover) connection mode [53] and fast 1:1 selection feature, which is analogous to the tail-end transfer function in APS systems. With these measures there need not be a reversion outage longer than for reversion in a 1:N APS or an SP-ring.

Use of these features is illustrated by outlining a reversion protocol for the custodial nodes. There are two cases, depending on whether a node is the initiator of the reversion or if it is complying with reversion initiated by the other custodial node. In either case the decision to revert may be delegated by the NOC to the custodial nodes to proceed automatically when they recognize the return of the failed links to a nonfailed status or the NOC may reserve the right to command reversion by a message to either node. Once commanded by the NOC, or self-initiated, all steps are automated:

A. Reversion Protocol for Node That Receives Command From NOC or Initiates Reversion

1. Bridge (i.e., split and copy) each restored traffic signal (i.e., each affected_port[i]), into the corresponding alarmed_port[i].

2. Assert a revert bit in the signature on the (idle) signal on the now-repaired direct link (alarmed_port[i]) to the other custodial node.

3. Wait for the revert bit to also appear on the incoming side of alarmed_port[i].

4. If alarms are indeed absent on alarmed_port[i] at this time, effect tail-end transfer (~ 50 msec), substituting the received signal on the repaired link (on which revert was just asserted) for the received signal from the corresponding restoration path (i.e., connect alarmed_port[i] back to affected_port[i], returning the DCS matrix to its pre-failure state for this path. This is the only traffic interruption during reversion.

5. Remove the head-end bridge connections and, in the SHN, cancel the primary signatures that acted as a static holding thread along the restoration path just transferred. Return the local restoration links to spare status and advise the host-node OS that reversion has been effected.

Step 5 automatically collapses the individual restoration path throughout the network and returns all links to the spare state at nodes along the path. Removing the head-end bridge at Step 5 is safe at this point because the peer protocol at the other custodial node does not return the revert bit until tail-end transfer is complete at its end.

B. Reversion Protocol for Cooperating Node

The peer protocol is activated by the revert bit being set in one of the [alarmed_ports]. This indicates that the other end has set up a head-end bridge.

1. If the alarms on the port where revert is set are indeed clear, effect a 1:1 tail-end transfer operation as above (i.e., affected_ port[i] is switched from its restoration port to alarmed_port[i]).

2. Bridge the signal transmitted into the restoration path from affected_port[i] into alarmed_port[i] and assert the revert indication on the signature leaving in the now repaired alarmed_port[i] link.

3. When the incoming signatures on the restoration path disappear (because of cancellation by the reversion protocol at other end), drop the head-end bridge and cancel the path-holding signatures emitted from this node in the respective restoration port. Return the status of the released link to spare.

The reversion procedure is described for one path but can also be commanded by the NOC for all restoration paths at once, automatically returning the network to the prefailure state. It is also possible for the DCS nodes to use their DRA-related signalling abilities to independently validate the repair work on behalf of the NOC and ensure that no logical channel transpositions have been introduced in the repair before committing to reversion. This is outlined in [8].

11.10.4 Handling Multiple Failures

Structural route diversity guidelines are often used so that transmission facilities going to different destinations avoid sharing the same physical duct or conduit. This is a good discipline no matter what restoration technique is used because it contains the logical complexity of the restoration problem when a conduit is severed, i.e., a single logical span failure occurs, rather than many. With SHRs, this also avoids a simultaneous cut of both sides of one ring and is desirable to avoid simultaneous cuts in adjacent rings.

In networks where the DCS is in the path of all transport capacity that arrives at its node, by definition, single logical span failures always arise. But if capacity from node A, which eventually terminates at node C, is "glassed through" at a node B (i.e., an express route not terminated on the DCS), then one physical cut to the A-B facility will create two logical span cuts—(A,B) and (A,C). There is also a finite probability that two independent failures will overlap in time.[12] The aim for DRA is to behave gracefully and effectively in conditions of two overlapping or simultaneous failures, although overall recovery time may be greater than for the same faults occurring individually.

To illustrate the basic manner in which a DRA may be adapted to handle multiple logical failures, we again use experience with the SHN. The following approach handles multiple failures by spawning concurrent SHN restoration processes. (It should be understood, however, that an alternative is also to fall back to path recovery to deal with multiple span failures, as in node failure. In some respects the latter is simpler than what follows, although it may not be as fast, depending on the number of demand pairs to be recovered when viewing the problem as path restoration, as opposed to dealing with a fewer number of concurrent span restoration processes.) For clarity we consider a dual span failure, but generalization to any number of failures is possible.

The first point is that only when the multiple failures involve a common custodial node does anything have to be done differently from a normal SHN execution. If the two failures involve four distinct custodial nodes, then ordinary independent SHN processes execute automatically and concurrently for both failures and, although they compete for spares in the network regions where they interact, all useful spares are recruited into one or the other recovery patterns.

The more complicated case is when one custodial node is involved in two span restorations. Concurrent task instances are then spawned within a custodial node, when it is apparent that the node is involved in more than one logical span restoration event (see Fig. 11-14). Each SHN task instance is identified by the span failure for which it was invoked and by the role being played by the task for its respective failure, i.e., sender or chooser. The protocol for the single-fault case is used without change by each task instance. Multiple instances of the task are easy to execute because as a pure-code FSM, the basic protocol code is re-enterable in the context of each task instance.

---

[12]If the dynamic phases for restoration of multiple failures don't actually overlap in time, then to any DRA with the self-updating property, such failures are time-successive single failures, each occurring within a network which has a modified spare capacity graph due to the previous restoration(s).

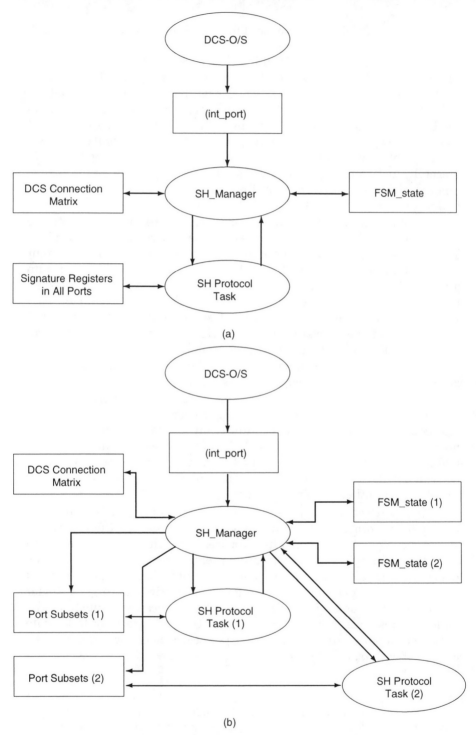

Figure 11-14 Principles of Extending a DRA (in This Case the SHN) to Execute
Concurrently for Multiple Span Failures

When task instances are both senders, however, they are each assigned a subset of ports to which they have access for primary flooding. The subsets are prorated to the size of the failed spans. This provision at the sender is sufficient to ensure that overall network resources will be rationed more or less proportionately between the two failures. When a sender and chooser exist concurrently, all port privileges are granted to the chooser instance first and, when completed, access to all remaining ports is immediately given to the sender task.

At tandem nodes, and when pairs of choosers or senders coexist, the SHN task instances require no mutual exclusions whatsoever. Each task processes only those signatures pertaining to the fault instance for which it was created and treats links in use by signatures for other failures as if those links were not present. At the network level, legitimate interactions for link resources do occur between concurrent SHN processes. It has been verified, however, in the following cases that the overall result is constructive recruitment of all available spare capacity into one or the other span recovery patterns. Specifically, concurrent execution of SHN task instances has been verified for cases of: (a) dual simultaneous span cuts in regions of the network where the restoration solutions do not interact, (b) dual simultaneous span cuts whose restoration solutions interact but share no end-node in common, (c) dual simultaneous failures on spans sharing one node in common as both senders, sender and chooser, and both choosers, and (d) all of the above spatial interaction cases, but where the second span cut follows the first by 1 second, rather than occurring simultaneously [38].

11.10.5 Handling (and Exploiting) Time-Dispersed Failures

In most work on DRAs, the span cut has been modelled as an instantaneous failure event, as if cut by a guillotine. In practice, one expects that when a backhoe pulls a cable apart, a random series of individual fiber system alarms would occur. This is highly relevant to a DR process that acts in the sub-2 second time scale because the recovery process may occupy the same time scale as the failure-causing process. Failures taking less than 50 msec tend to approximate the ideal guillotine failure model. Failure sequences in the 2-second and above regime tend to be equivalent to independent time-concatenated failures to any DRA that is self-updating. Random link failures spread out over a few hundred milliseconds are of special interest, however, because this is the same time scale on which the DR process itself typically acts. Failures of this type plausibly offer great potential for adverse dynamic interactions that may affect DRA performance.

One approach to dispersed failures is to delay protocol activation until after a hold-off time for alarm collection. Assuming a 2-second restoration target, one could budget up to 1 second for alarm collection in a well-developed DR system, and then execute the DRA for the number of failures that had accumulated in the hold-off time as if they were simultaneous alarms occurring at the end of the hold-off period. This is a problematic approach, however, because it can never be ascertained that the collection interval is long enough. And yet the hold-off cannot be increased arbitrarily because all paths suffer the additional delay in restoration time and reduced margin against call-dropping.

Therefore a DRA must somehow start working on the problem almost right away, developing the restoration solution at the same time as the span failure progresses. Two strategies for doing this were studied in [43], using the SHN protocol. In the first, the SHN protocol was directly exposed to a series of randomly timed individual link failures. Nothing is done to simplify, collect, or otherwise organize the failure for presentation to the restoration process. Although preemptive activation (which follows) turns out to be a preferred strategy, the direct exposure tests allowed observation and validation of the SHN protocol and its network-level dynamics, under complex conditions that changed continually while the protocol was executing. As such it constituted a major validation of the SHNs robustness and inevitable convergence (self-organization) onto the final KSP path-set.

In direct exposure to the random failure sequence the SHN protocol is simply reinvoked in response to each new failure, as it occurs, while already active in response to the previous failures. Each reinvocation causes the sender to realize that its primary flooding pattern no longer corresponds to the number of outstanding failures. The sender therefore augments the primary flooding stimulus to the network and these changes propagate through the rest of the network, overtaking and altering the self-organizing dynamics of the network that are already occurring. The tandem node protocol is unchanged and all the chooser does upon new alarm events is increment its target number of restoration paths.

An alternative to direct exposure is to make the working assumption that, as soon as the first link failure occurs, the entire span is doomed. In this case a guillotine-like pseudo-failure is presented to the SHN immediately, with magnitude equal to the total number of working links on the span in jeopardy. This is called *preemptive activation* because at the first sign of trouble, the protocol begins preparing for the worst-case contingency that may follow. But regardless of the number of replacement paths that are found, traffic substitution for any working link will not occur unless that link does subsequently fail. This avoids unnecessary hits on links that may ultimately be unaffected by the fault. The preemptive implementation is as follows. All changes are put into effect at the sender and chooser nodes only: (a) as soon as one working link alarm arises, the SHN protocol is invoked without delay. (b) Each custodial node counts the total number of working links, whether failed or not, on the affected span. This is *max-paths-required*. (c) Each custodial node marks the spare links on the affected span as not available for restoration, except (optionally) for a reserved number of spares on the affected span to provide for APS-like responses on the direct span for single link failures. The reserved number of spare links can be set to match the modularity of the fiber transport to emulate 1:N APS for single (unidirectional) failures (see also Section 11.10.6). (d) The sender applies the primary flooding pattern, as appropriate for *max-paths-required* in the local environment of spans and spare links. (e) Sender, chooser (and tandem) act as usual except that the sender only performs traffic substitution for working links that do eventually fail.

The sender performs a continual matching between accumulated failures and replacement paths available. When path completions outrun failures, the surplus is stored in a set [paths_ready] and allocated to new alarm events as soon as the latter arise. When failures outrun the accumulation of restoration paths, outstanding failures accumulate in [affected_ports] and new restoration paths are used immediately. Network-wide path organization proceeds at its own pace, once triggered by the worst-case sender primary flooding, accumulating all of the replacement paths that could potentially be required (or are feasible in the given network). At the same time, the process that is physically causing the damage continues. The two independent processes occur in parallel, as illustrated in Fig. 11-15. One is accumulating replacement paths, the other accumulating failures on working links. Note the implication that in areas where the number of paths accumulated exceeds the failures accumulated, new failures will see essentially zero path-computational delay for restoration. Their outage will be only to the time for cross-connection combined with signal propagation and reframing time. But even when new dispersed failures do not find paths immediately ready for them, they are still better off than those in a simultaneous mass failure scenario because the nth failure has only to wait the incremental time to the completion of the nth path, as opposed to the total time elapsed from $t = 0$ until organization of the nth path of the restoration pattern.

A prediction inherent in Fig. 11-15, is that the average individual path outage time should improve with preemptive activation in the presence of staggered alarms. Indeed, this was confirmed by an all-spans-cut series of experiments with randomly timed failure sequences in [43]. The resulting CDF of individual path restoration time is shown in Fig. 11-16 (c), in comparison with the CDFs arising from (a) simultaneous failure and (b) staggered alarms with direct exposure. With preemptive activation there were 242 cases, out of 1296 paths that were restored, where a replacement path was available before the next link failure occurred, giving essentially zero outage time for 19% of the failed paths. The

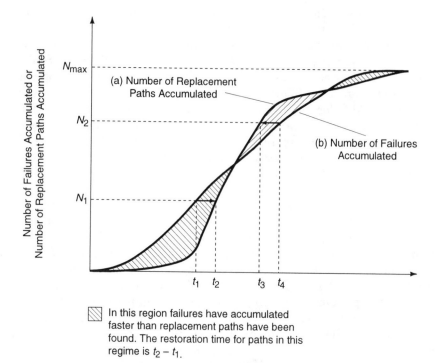

In this region failures have accumulated faster than replacement paths have been found. The restoration time for paths in this regime is $t_2 - t_1$.

In this region replacement paths have accumulated faster than failures have been detected. The restoration time for paths in this regime is nearly zero.

Figure 11-15 Path Synthesis and Failure Processes with Preemptive Activation for Time-dispersed Failures

staggered alarm model had a 75 msec standard deviation centered about a maximum failure density at 188 msec. This is still a relatively fast failure process, compared with path creation times in the test network. As failure durations increase, a greater fraction of paths would find that rerouting is ready for them before they fail. All paths also benefit to a degree from preemptive activation because the average waiting time for all replacement paths is reduced.

11.10.6 Relationships Between Distributed Restoration and APS Systems

There are three options to interwork an existing APS system and a DR system.

1. The first approach leaves the APS system in place to handle single-fiber or electronics failures or, if the APS is a diverse routed 1:1, to handle

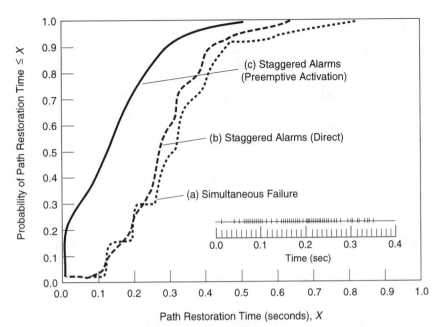

Figure 11-16   Cumulative Distribution of Individual Path Outage Time with
and without Preemptive Activation Using the SHN

cable cuts as well. The DR system is then installed with a, say, 150 msec
hold-off on activation to allow an APS reaction when appropriate. The
DRA executes only if failure indications persist beyond the time for APS
reaction. Thus, the DRA acts as a backup for the APS, while possibly also
being available on a no-delay basis for non-APS spans terminating at the
same node. This is the simplest, least change architecture in terms of
introducing DR to the existing network. But it forgoes the opportunity for
network protection against cable cuts simply by allowing DRA access to
existing spares.

2. The second approach recognizes that the APS system is not strictly
necessary when a suitably fast DR system is deployed. With a simple
arrangement to trigger the DRA for unidirectional fiber failures (e.g., the
return alarm (RA) bit in the SHN signature), a DRA will inherently emulate an APS system for single fiber failures, assuming at least one fiber-
module of spare links on the same span are available to the DCS. In such
circumstances the SHN's normal sender flooding and chooser response
results in a single exchange of signatures over intact spare link(s) on the
direct span, followed by traffic substitution in about 40 to 50 msec (computation time only). In this approach the DRA actually provides a generalized s_i: w_i APS equivalent where s_i is the total number of spare links on a
span with w_i working links.

To allow the DRA to take over the APS function, the spare span lockout can be asserted on most APS systems, and access to the spare span provided to the DCS through the "extra traffic" input(s) of such systems. When this is done, the whole network survivability increases without adding any new redundancy because previously dedicated APS spare spans become available for network-wide protection. This offers an interesting survivability upgrade strategy in some networks: In one highly connected metropolitan network (average nodal degree ~ 4) studied by the author [24, 25], existing APS spares were made accessible for DR. No other spares were added. Assuming 1:7 APS protection ratios to start, the network became 67% restorable against any span cut based on KSP rerouting characteristics. With 1:1 APS as the starting point, the network becomes 100% restorable (and can grow about two-fold while remaining fully restorable) just by giving DRA access to the existing APS spares. There was no cable cut survivability at all in the baseline network, only APS protection against single fiber and electronics failures. This strategy also allows cost reductions on the purchase of new optical line systems by eliminating the need for APS subsystems when DR is deployed.

3. A third possibility is a more complicated cooperative interworking between APS and DRA. This is to operate the DRA with a hold-off for APS reaction, but also allow the DRA to access the spare span via the extra traffic input for use in network-level protection of other span failures when the APS system does not need its spare span. This offers little theoretical advantage over strategy 2 and it requires coordination between the fiber system and DRA to share spare span availability information. It would, however, be of possible interest if the DCS nodes were too slow to meet APS speed objectives, but the network survivability advantage of giving DRA access to existing spares was still to be obtained.

11.10.7 Grouped Signalling Bundles

In mesh restoration, the most efficient use of spare capacity is made when every failed link is replaced by an individually found replacement path of the same unit capacity. This is the case of maximum resolution in the sense that a DRA has greatest effectiveness in threading paths through all avenues of the network. If larger units are rerouted as a block, the granularity is increased with a possible loss in total restoration capacity. But in networks where the cross-sections of spans are sufficiently large, a DRA's speed of execution may be increased (or maintained in the face of growth) with negligible loss of routing efficiency by executing the DRA for a smaller number of logical rerouting bundles. For instance, in networks of 10 to 20 DS3s per span, direct restoration in unit DS3s may be appropriate. But with several hundred DS3s per span, it would make sense to limit signalling or signature volumes by executing the DRA with

$B = 10$ bundled signalling (and rerouting). Each bundle is comprised of B failed links that will be replaced by B replacement paths that are found as a group with common routing through the network. An analogy is that unbundled restoration threads k thin wires through the network maze using the thinnest openings if needed, and B-bundled execution guides ceil (k/B) thicker ropes (of up to B threads each) through the network. Only if B is too large relative to sparing levels will the quantization effect noticeably reduce the number of feasible replacement paths.

An adaptation for the SHN to execute in bundled mode is as follows. Equivalent adaptations follow for other DRAs:

1. Each node arbitrarily numbers all working and spare ports in each span at its site, starting at 1. These numbers may be repeated in each span.

2. Every outgoing spare port with number x, such that $x \bmod(B) = 0$, is a designated signalling link for the bundle of spare links numbered max $(1, x-B+1)$ to x, inclusive. Each signalling link is numbered separately among the subset of signalling links for a given span. This defines a thinner virtual network of "meta-spans" for which the SHN will logically execute. Each meta-span is roughly $1/B^{th}$ as deep as the real span it corresponds to. All links in the meta-span are signalling links to be processed by the regular SHN protocol.

3. To effect bundled signalling, the SHN signature gets extra fields: (i) to designate signalling links to adjacent nodes, (ii) to number signalling links within meta-spans, and (iii) to tag nonsignalling links with the signalling link to which they belong (i.e., a bundle number). The bundle numbers are used in making ganged cross-connections but nonsignalling links are otherwise ignored from a protocol processing viewpoint.

4. When a tandem node reaches a connection decision between links in the meta-network, all real ports in the bundles associated with the two given meta-links are cross-connected. No particular number matching within the bundles is necessary since later sender-chooser mapping implicitly "buzzes out" the actual connection pattern. With random span sizes, one bundle per span may be smaller than B. When bundles of unequal sizes are to be connected, only the feasible connections to the smaller bundle are made. The leftover on the larger bundle is part of the quantization penalty due to bundling.

5. Sender-chooser traffic substitutions are done in groups of B working signals from the real network (except, obviously, for the fractional bundle that may exist on each span). The chooser

relies on the individual signal_IDs of each signal forwarded by the sender link to do traffic mapping in the usual manner at its end (i.e., the chooser makes no use of bundling information for the final step of traffic mapping).

Consequently, a network with several hundred links per span could set $B = 10$ and execute the SHN protocol with about 1/10th the signature event processing. The speed-up is not exactly 10 because the int$(k/B)^{th}$ bundle on each span may have fewer than B members. Note that the bundling scheme is self-coordinating. All numbering to implement the scheme is local to each node. There is no centralized setup or number assignment requirements from the NOC except to say what bundling ratio to use. The numbering process that establishes the meta-link groupings may be executed any time before failure or be included dynamically as an initialization step when the protocol is invoked for a failure. Finally, any time restoration is incomplete with a given bundling granularity, nothing stops the protocol from immediately re-executing in direct mode ($B = 1$), seeking fine granularity routings for outstanding failures.

11.10.8 Importance of a Prompt Connection Specification

The present generation of DCS machines were not designed, nor were industry specifications for them set, with fast restoration in mind. The role for 3/3 (or STS-n/n) cross-connects was initially seen to be in provisioning and transport network rearrangement only. This is reflected in the current specification of 1 second cross-connection time [53]. This allows for remote communications with internal test and external verification before cross-connection. Philosophically, there is little reason to hurry, and every reason to be cautious, in routine provisioning and rearrangements on DS3 / STS-1 streams.[13] But in DR, connections are deduced locally within each node and the context is that of an emergency. Specifications for normal remote provisioning applications need not transfer into specifications for a restoration context. The idea of a prompt-connect specification is necessary, distinct from a regular cross-connection operation.

A prompt connection may be considerably faster than 1 second with only software changes in a DCS, primarily to waive testing. The rationale for waiving routine cross-point tests are: (a) it is an emergency situation, (b) we are operating exclusively on the spare links of the network, (c) follow-up testing to verify the success of each cross-connection can begin

---

[13]As a point of interest, DCS systems were often called *slow switches* by the community of designers that built them (about 1984–1988). The new vision of DCS machines suitable for DR would be to call them *infrequent-but-fast switches*.

immediately after the recovery action is complete, (d) spare links may be continually under test prior to failure by virtue of the idle signal on them and, (e) unused paths in the DCS matrix could be kept under continual or periodically applied local testing prior to failure. If DCS resources are continually or periodically tested during nonfailure times, then testing during an emergency only protects against failures in the DCS matrix over a very small window of time. And yet, one known failure already exists (the span-cut) and outage is inevitable if restoration is delayed.

The importance of a technically based objective (e.g., based on hardware capability) for prompt connect times to DR is clear from a recent study assessing DR speed with existing DCS systems [54]. With data from [54] and [9], one can estimate that a prompt cross-connection is technically feasible in 20 to 40 msec if spare port testing and connection verification is excluded. In this case the conclusion in [54] would be that DR *is* achievable in under 2 seconds. But with testing and verification time assumed at 110 msec per cross-connection, the outcome of the study in [54] is that 2 second restoration *is not* feasible.

11.10.9 Path Restoration and Network Recovery from Node Loss

Here we aim to provide some clarifications and insights into the related topics of path restoration and node recovery. Path restoration of a span cut and node recovery are similar problems because in each case the problem presents an arbitrary set of demand pairs from the demand matrix to be reprovisioned between origin and destination for each pair. First, regarding path restoration, any DRA that performs span restoration can also be used in path restoration as meant in [4, 9, 28, 47] and prior to those works, as explained in [25, 26, 38] based on *capacity scavenging* which is the use of a DRA in a path provisioning context. The name is apt because one is asking the network (using a DRA) to scour itself for all feasible paths between two named end-points considering only existing spares of the network. A prime use for this is in on-demand service provisioning and fast reprovisioning for path-level restoration for node recovery or if span-level restoration were to prove incomplete. Multiple concurrent instances of the SHN executing in the scavenging context will result in "ad-hoc" node recovery or path restoration of a span cut in the same sense in which [4, 28, 47] allude to a DRA being the basis for path restoration and node recovery applications.

It is, however, misleading to conclude that DR technology is available for true (optimal) path restoration. The main point is that the recovery patterns obtained from uncoordinated concurrent (or arbitrary sequential) execution of one DRA instance for every demand pair affected by a node or span failure will not yield an optimal allocation of recovery level among demand pairs. Optimized path restoration is a problem of multicommodity maximum flow with lower bounds, which is an NP-complete problem [55]. Each demand pair is a commodity and the total flow of all commodities is to be maximized subject to lower bounds on each flow in order to effect some proration of the individual commodity flows to the relative prefailure demand levels. In practice, maximum flow can be approximated by KSP for each demand pair but the problem requires a search over all permutations of the order of allocation of network links to each demand pair for an optimal result.

In practice, when path restoration is performed in response to a span failure, the suboptimality of ad-hoc path restoration is probably not crucial. Assuming sparing levels similar to that required for span restoration, suboptimal composite path-sets may reasonably still provide full recovery for single span failures. But optimality is a significant issue for node recovery because of the generally severe shortage of spare capacity for recovery from node loss in a network that has been efficiently provisioned for span recovery.[14] A DCS node failure will typically be equivalent to 3 to 6 simultaneous span failures, causing a general starvation of capacity for recovery. In these circumstances it is most important that path restoration be near-optimal. The aim is not simply to throw out an ad-hoc set of paths in response, nor to maximize the simple percentage of restored (bulk) capacity. It is most important to obtain some nonzero apportionment to every demand, which is ideally prorated to the prefailure levels.

With ad-hoc path restoration for node recovery it is common to have results in which some relations get high restoration and others remain disconnected. Indeed, such a solution can maximize the total percentage restoration while leaving some relations disconnected. Figure 11-17 illustrates this with some experimental results using the SHN for node recovery. An independent instance of the SHN is triggered at the origin–destination nodes of all affected demands. The star plot[15] representations show demand levels routed through node 13 of the network before failure and after various ad-hoc node recoveries via path restoration. Although executing concurrently, some SHN instance will always be the first to use any given link in the network. Given this, all of the different recovery patterns (a) through (e) are obtainable by introducing slight perturbations in the initial conditions and exact launch timing of each concurrent SHN instance for node recovery. Each pattern corresponds to a composite path-set that uses all spares in the region of the network affected so that in one

---

[14]It is economically unlikely to expect that networks should be 100% survivable to node losses. Such networks would be extremely overbuilt in anticipaiton of a still extremely rare event.

[15]The starplot is a way of representing multidimensional data. Here, it shows the connectivity for each demand pair by the length of its line along a radius, all demands being displayed at equal angles around the circle.

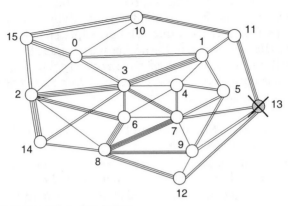

Node 13 Connectivity before Failure

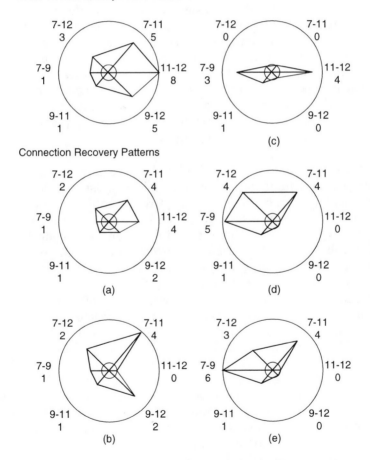

Connection Recovery Patterns

Figure 11-17   Different Patterns of Network Recovery from Node Loss Obtained by Ad-hoc Path Restoration with the SHN

sense they are all equally efficient recoveries. But marked differences in recovery patterns arise from detailed considerations of which demand pair each link was first recruited into. The principle shown in Fig. 11-17 will apply to any DRA executed in this ad-hoc path restoration manner. Ad-hoc recovery will effect some form of node recovery, but highly asymmetric recovery ratios may be common, including disconnected demand, depending on the exact order demands are restored.

11.10.10 Issues in Simulation and Comparison of DRAs

For practical reasons, DR must be developed, and possibly even standardized, based on computer emulation of the DRA on which it is based. Therefore, the utmost attention must be paid to the emulation methods used and to the environment and modelling assumptions whenever DRAs are compared or considered for standardization. The basic methodology for testing DRAs is concurrent execution with interactions defined by the network topology using discrete event simulation (DES). Chapter 6 covers DES in a data network context. A number of special considerations for applying DES to DR modelling are given in [25, 39]. One issue is important enough to single out for emphasis here. This is the modelling of nodal processing delays. In addition, a recent study [56] raises several points of basic discipline in attempting quantitative comparison of DRA performance and properties.

Modelling Nodal Delays. If all nodal processing delays are small with respect to the link delays, then the following issue is not a concern. In practice, however, 64 kb/s or higher DRA signalling channels may be used. Link delays are then on the order of 1 to 10 msec, depending on message length (insertion time) and link distances (propagation time). This is similar to practical estimates of nodal processing time. In practice, therefore, nodal processing delays will not be insignificant with respect to link delays, and the following issue is important.

If the network emulator does not dynamically reflect processing time of the protocol code on each invocation, in each node of the network, the emulator may not predict what the DRA will do in reality. The point is as follows: DRAs represent asynchronous logic systems and employ either simple or modified flooding processes to explore the network. If the DRA is to result in either geographical or logical shortest-path properties, link delays must dominate the dynamic variation in nodal processing delays. It is the variation in processing delays at nodes, not any common absolute nodal delay time, that is at issue. Any fixed component of delay per node is equivalent to a fixed extra length on all links.

A commonly used modelling method is to assume nodal processing delays that are random on each message, with a predetermined average.

The effect is similar to assuming a constant nodal execution time, however, because there is no reflection of the instantaneous context-dependent variation in processing times during the restoration event. Reliable prediction of DRA behavior requires that elapsed processor time is measured (or calculated based on a detailed execution trace) for each DRA invocation during the dynamics of the restoration event. The dynamically measured processor delays are then reflected in scheduling of events arising from the DRA execution within each node in response to each incoming event as the restoration proceeds.

Evidence to support this point, and insight into the difference it makes, was obtained in development experience with the SHN. From 1986 to 1987, a v.1 network emulator was built in which processing delays were constant at each node. This lead to the first proof-of-concept results for DR (in [24]), while initially avoiding the complexity of dynamic modelling of nodal processing delays. However, recognizing the importance of this issue to predicting the behavior of a totally asynchronous protocol with confidence, a v.2 emulator was developed in which event scheduling includes context-specific instantaneous measurement of the execution times of the protocol itself. The predicted dependency was confirmed with the v.2 emulator for the SHN (v.2 was used for [25, 38, 39]). Specifically, when the nodal processing speed is made low enough that average delay becomes high relative to link delays, larger nodes then become relatively slower than small nodes and have higher variation in execution time, entry-to-entry. Path leakage can then occur, resulting in path-sets that are not the shortest path set. Path leakage arises when the flooding process (of any DRA) is able to explore, via longer routes, around a slow node faster than it can propagate through the node. This can be observed in the SHN emulator if nodal processing speeds are made deliberately slow with respect to link delays ($P = 1$ or less for networks in [26, 38]), and nodal sizes vary ~ 10:1. Dynamic measurement of nodal delays is therefore important to verify the behavior of a DRA with confidence. Side effects other than path leakage may arise with any given DRA when emulated in this more accurate way. Until tested under these conditions, with parameters of node and link delays also varied to find the limits of performance, the deep understanding of a given DRA on which to predict real-world performance may not be obtained.

The existence of the path leakage phenomenon makes sense. By intent, DRAs exploit the opportunity for self-paced asynchronous network-wide execution. It is reasonable, therefore, that we have to engineer so that variances in nodal delays do not dominate over the distance-related delays in the network. These conditions can be met whenever nodal delays are either (a) relatively invariant, or (b) negligibly small relative to link delays. Whenever (a) or (b) is true, the restoration path set can be predicted in advance from the network topology and is uninfluenced by nodal speed variations due to DCS size or manufacturer differences.

If neither (a) nor (b) is true, however, three approaches may be considered: (1) minimize signalling volume in DRA design by bundling or by making the processing action for each individual event be as primitive as possible. (The basic SHN takes the latter approach, with bundling as a further option. Other DRAs have stressed message reduction.) (2) Design to explicitly ensure that link transfer times dominate processing time variations. This means that, notwithstanding overall speed objectives, the slowest feasible message transfer times are actually desired. This ensures link-delay dominance for shortest path properties. For example, if 64 kb/s is used to transfer 64 byte messages, then link delay is 8 msec (for insertion) plus a distance-dependent delay. If the minimum to maximum message processing times vary from 0.5 to 2 msec, then shortest path properties will be assured. If necessary, link delays apparently can be increased by enforcing repetition before accepting a new message or signature. (3) Alternatively, a constant component in the nodal execution times can be added so that the variation in execution times is again made small relative to the constant component of delay per logical hop. For example, if the intrinsic variation in execution for the DRA was 2 to 6 msec and link delays were 4 msec, then all message processing delays could be deliberately built out to 5 msec. The link delays (now effectively $4 + 5 = 9$ msec) dominate over the variation in nodal delays ($6 - 2 = 4$ msec). In this case the shortest path characteristic will reflect logical hop length, not geographical length.

Consequently, to standardize DR for implementation on a variety of manufacturers' equipment, it may be advantageous to specify a *minimum*, as well as maximum, per message (or signature) processing time, based on the principle of ensuring link delay dominance over nodal delay variations.

Principles for Comparative Studies. A comparison of four different DRAs running in four different emulators, has been reported [56]. The emulators used different methods for timing nodal execution and different constants representing conservative or optimistic views of DCS processing speed. The DRAs were also compared under different bundling ratios. Some of the DRAs included small nodal databases and/or control of the network spare design. Some were run in a path restoral context, others in a span restoration context. There are several obviously important principles to be observed from this.

A meaningful quantitative methodology for comparing DRAs requires at least the following: (1) the emulator should dynamically measure execution time on each DRA invocation and schedule future events based on measured execution time on a common host processor or processor model. (2) DRAs must be compared in the same engineering context, i.e., path or span restoration, same bundling ratio, same signalling channel speed. (3) Reference path-sets should be computed separately and DRA performance summarized in PNE, PLE over all span-cuts in several test networks. This should include less than fully restorable networks to create maximum path-finding stress and should require creation of paths up to 8 or 9 hops long.

11.11 LOOKING AHEAD

Distributed restoration (DR) is a network operations and management technology poised for maturity and deployment in the early twenty-first century. It is technically and economically attractive by virtue of its operational autonomy, flexibility, network efficiency, and ancillary applications that can be derived from it. Span restoration is presently well developed in terms of theory and algorithms required. Strategies for many of the implementation issues in DR have been covered in this chapter. Investigations on path restoration are beginning and may ultimately leapfrog span restoration in terms of the technical solutions deployed in real DR systems. A challenge to developers is to evolve the DCS platform to economically support mesh-restoration principles, primarily through development of integrated optical terminations and fast (internally commanded) crosspoint operations. Centrally controlled mesh-restoration systems will likely provide the first operational framework for development of field confidence with DR, eventually delegating real-time reconfiguration control down onto the network elements themselves via DR. Hybrid restoration mechanisms such as distributed preplanning may also be deployed as an intermediate step toward fully autonomous real-time distributed restoration. This chapter is expected to be only one of many ongoing works arising from the new technical domain of distributed DCS-based network computation and control of network configuration.

REFERENCES

[1]. J. G. Nellist, "Fiber optic system availability," *Proc. FiberSat Conf.*, Vancouver, BC, pp 367–372, 1986.

[2]. Digital Cross-Connect Systems in Transport Network Survivability, SR-NWT-002514, issue 1, Bellcore, January 1993.

[3]. W. D. Grover, T. D. Bilodeau, and B. D. Venables, "Near optimal spare capacity planning in a mesh-restorable network," *Proc. of IEEE GLOBECOM'91*, Phoenix, pp. 2007–2012, 1991.

[4]. H. Sakauchi, Y. Nishimura, and S. Hasegawa, "A self-healing network with an economical spare-channel assignment," *Proc. GLOBECOM '90,* pp. 438–443, 1990.

[5]. W. D. Grover, "Case studies of restorable ring, mesh and mesh-arc hybrid networks," *Proc. IEEE GLOBECOM '92,* Orlando, December 1992.

[6]. M. H. MacGregor, "The self-traffic engineering network," PhD thesis, University of Alberta (TRLabs), Fall 1991.

[7]. M. H. MacGregor, W. D. Grover, and U. M. Maydell, "The self-traffic engineering network," *Canadian J. Electrical and Computer Engineering,* vol. 18, no. 2, pp. 47–57, 1993.

[8]. M. H MacGregor, W. D. Grover, and J. H Sandham, "Capacity scavenging in the Telecom Canada network: A preliminary assessment," *Telecom Canada* (Stentor) CR-89-16-03, November 1989, reissued by Telecom Canada as Committee T1 Contribution T1Q1.2/91-031, October 9, 1991.

[9]. Restoration of DCS mesh networks with distributed control: Equipment framework generic criteria, FA-NWT-001353, issue 1, Bellcore, December 1992.

[10]. B. D. Venables, W. D. Grover, and M. H. MacGregor, "Two strategies for spare capacity placement in mesh restorable networks," *Proc. IEEE ICC '93,* Geneva, May 1993.

[11]. B. D. Venables, "Algorithms for the design of mesh-restorable networks," MSc. thesis, University of Alberta (TRLabs), Fall 1992.

[12]. K. Sato, H. Ueda, and N. Yoshikai, "The role of virtual path cross-connection," *IEEE LTS Magazine,* vol. 2, no. 3, pp. 44–54, 1991.

[13]. Y. Kane-Esrig, G. Babler, R. Clapp, R. Doverspike, et. al. "Survivability risk analysis and cost comparison of SONET Architectures," *Proc. IEEE GLOBECOM '92,* pp. 841–846, 1992.

[14]. Y. Okanoue, H. Sakauchi, and S. Hasegawa, "Design and control issues of integrated self-healing networks in SONET," *Proc. IEEE GLOBECOM '91,* Phoenix, 1991.

[15]. J. Sosnosky, "Service applications for SONET DCS distributed restoration," *IEEE J-SAC Special Issue on Network Integrity* (in press), 1993.

[16]. D G. Doran, "Restoration of broadband fiber facilities," *Proc. Fiber-Sat Conf.,* Vancouver, BC, pp. 298–301, 1986.

[17]. D. F. Saul, "Constructing Telecom Canada's coast to coast fibre optic network," *Proc. IEEE GLOBECOM '86,* pp. 1684–1688, 1986.

[18]. C. W. Chao, P. M Dollard, J. E. Weythman, L. T. Nguyen, H. Eslambolchi, "FASTAR–A robust system for fast DS3 restoration," *Proc. IEEE GLOBECOM '91,* Phoenix, pp. 1396–1400, December 1991.

[19]. D. K. Doherty, W. D. Hutcheson, and K. K. Raychaudhuri, "High capacity digital network management and control," *Proc. IEEE GLOBECOM '90,* San Diego, 1990.

[20]. "AT&T submits report on network reliability," *Communications Week,* March 23, 1992.

[21]. R. Vickers and T. Vilmansen, "The evolution of telecommunications technology," *IEEE Communications Magazine,* vol. 25, no. 7, pp. 6–18, July 1987.

[22]. B. Fleury and S. Fleming, "Digital broadband restoration (DBBR) implementation with digital broadband cross-connect systems," *Proc. IEEE Montech Conf. Commun.,* pp. 169–172, 1987.

[23]. P. J. Sutton, "Service protection network," *IEEE GLOBECOM '86,* Houston, pp. 862–865, 1986.

[24]. W. D. Grover, "The selfhealing network: A fast distributed restoration technique for networks using digital cross-connect machines," *Proc. IEEE GLOBECOM '87,* Tokyo, vol. 2, pp. 1090–1095, 1987.

[25]. W. D. Grover, "Selfhealing networks - A distributed algorithm for k-shortest link-disjoint paths in a multi-graph with applications in real-time network restoration, Ph.D. thesis, University of Alberta (TRLabs), Fall 1989.

[26]. W. D. Grover, "Method and apparatus for self-healing and self-provisioning networks," U.S. Patent No. 4,956,835, September 11, 1990.

[27]. T. H. Wu, *Fiber Network Service Survivability,* Norwood, MA.: Artech House, 1992.

[28]. H. Komine, T. Chujo, T. Ogura, K. Miyazaki, and T. Soejima, "A distributed restoration algorithm for multiple-link and node failures of transport networks," *Proc IEEE GLOBECOM '90,* pp. 459–463, 1990.

[29]. M. Schwartz and T. E. Stern, "Routing techniques used in computer communication networks," *IEEE Trans. Comm,* vol. 28, no. 4, pp. 539–552, April 1980.

[30]. D. M. Topkis, "A k-shortest path algorithm for adaptive routing in communications networks," *IEEE Trans. Comm,* vol. 36, no. 7, pp. 855–859, July 1988.

[31]. E. C. Rosen, "The updating protocol of ARPANET's new routing algorithm," *Comput. Networks,* vol. 1, pp. 11–30, February 1980.

[32]. A. Aho, J. E. Hopcroft, and V. A. Ullman, *The Design and Analysis of Computer Algorithms,* Reading, MA: Addison-Wesley, 1974.

[33]. E. W. Dijkstra, "A note on two problems in connection with graphs," *Number Math.* vol. 1, pp. 269–271, 1959.

[34]. M. H. MacGregor and W. D. Grover, "Optimized k-shortest paths algorithm for facility restoration," submitted to *Software-Practice and Experience,* 1993.

[35]. A. Gibbons, *Algorithmic Graph Theory,* Cambridge, MA: Cambridge University Press, 1985.

[36]. L. R. Ford and D. R. Fulkerson, *Flows in Networks,* Princeton, NJ: Princeton University Press, 1962.

[37]. G. N. Brown, W. D. Grover, J. B. Slevinsky, and M. H. MacGregor, Mesh/arc networking: An architecture for efficient survivable self-healing networks," *Proc. IEEE ICC '94,* New Orleans, May 1994.

[38]. W. D. Grover, B. D. Venables, J. H. Sandham, and A. F. Milne, "Performance studies of a selfhealing network protocol in Telecom Canada long haul networks," *Proc. IEEE GLOBECOM'90,* San Diego, December, 1990.

[39]. W. D. Grover, B. D. Venables, M. H. MacGregor, and J. H. Sandham, "Development and performance verification of a distributed asynchronous protocol for real-time network restoration," *IEEE J-SAC,* vol. 9, no. 1, pp. 112–125, January 1991.

[40]. D. A. Dunn, W. D. Grover, and M. H MacGregor, "Comparison of k-shortest paths and maximum-flow routing for network facility restoration," *IEEE J-SAC Integrity of Public Telecommunications Networks,* vol. 12, no. 1, January 1994.

[41]. D. A. Dunn, W. D. Grover, and M. H. MacGregor, "Development and use of a random network synthesis tool with controlled connectivity statistics," *TRLabs WP-90-10,* August 1990.

[42]. H. Amirazizi, "Controlling synchronous networks with digital cross-connect systems," *Proc. IEEE GLOBECOM'88,* Hollywood, FL, pp. 1560–1563, 1988.

[43]. W. D. Grover and B. D. Venables, "Performance of the selfhealing network protocol with random individual link failures," *Proc. IEEE ICC'91,* pp. 660–666.

[44]. B. A. Coan, M. P. Vecchi, and L. T. Wu, "A distributed protocol to improve the survivability of trunk networks," *Proc. XIII International Switching Symposium,* pp. 173–179, May 1990.

[45]. C. Han Yang and S. Hasegawa, "FITNESS: A failure immunization technology for network service survivability," *Proc. IEEE GLOBECOM'88,* pp. 1549–1554, 1988.

[46]. H. Sakauchi, Y. Okanoue, and S. Hasegawa, "Spare-channel design schemes for self-healing networks," *IEICE Trans. Commun.,* vol. E75-B, no 7, July 1992.

[47]. T. Chujo. H. Komine, K. Miyazaki, T. Ogura, and T. Soejima, "The design and simulation of an intelligent transport network with

distributed control," *Network Operations and Management Symposium,* San Diego, February 1990.

[48]. B. D. Venables, R. S. Burnett, and W. D. Grover, "Node recovery via interactive scavenging with the selfhealing protocol," *Telecom Canada (Stentor) CR-90-16-03,* August 1990, reissued as *Committee T1 Contribution T1Q1.2/91-035,* November 30, 1991.

[49]. B. Pekarske, "1.5 second restoration using DS3 cross-connects," *Proc. of Trends in Network Restoration,* TRLabs, Edmonton, Canada, May 1990.

[50]. B. Pekarske, "1.5 second restoration using DS3 cross-connects," *Telephony,* September 10, 1990.

[51]. AT&T Communications, "C-bit parity—a proposed addition to applications in the DS3 standard," *Contrib. T1 Standards Proj.,* T1X1.4/88-003, February 1988.

[52]. W. D. Grover, "A frame-bit modulation technique for addition of transparent signalling capacity to the DS-3 signal," *IEEE Pacific Rim Conf. on Commun., Computers and Sig. Proc.,* Victoria, BC, pp. 319–322, June 1987.

[53]. Wideband and Broadband Digital Cross-Connect Systems Generic Requirements and Objectives, TR-TSY-000233, issue 2, Bellcore, September 1988.

[54]. T. H. Wu, H. Kobrinski, H. Ghosal, T. V. Lakshman, "A service restoration time study for distributed control SONET DCS Self-Healing Networks," *Proc. IEEE ICC '93,* Geneva, May 1993.

[55]. M. R. Garvey and D. S. Johnson, Computers and Intractability, New York: W. H. Freeman 1979, p. 216.

[56]. Digital Cross-Connect Systems in Transport Network Survivability, SR-NWT-002514, issue 1, Bellcore, section 7-3, January 1993.

[57]. A. Graves, P. Littlewood, and S. Carleton, "An experimental cross-connect system for metropolitan applications," *IEEE J. Sel. Areas in Comm,* vol. SAC-5, no. 1, pp. 6–17, January 1987.

[58]. Y. Rokugo, Y. Kato, H. Asano, K. Hayasi, et. al, "An asynchronous DS3 cross-connect system with add/drop capacity," *Proc. IEEE Global Conf. Commun.,* Hollywood, FL pp. 1555–1559, 1988.

[59]. Digital Cross-connect system (DCS) requirements and objectives, Bellcore, TR-TSY-000170, issue 01, November 1988.

[60]. W. D. Grover and T. E. Moore, "Precision time transfer in transport networks using digital cross-connect systems," *IEEE Trans. Com.,* vol. 38, no. 9, pp. 1325–1332, September 1990.

Index